# Past and Present Soil Erosion

## Archaeological and Geographical Perspectives

*edited by*

Martin Bell and John Boardman

Oxbow Monograph 22
1992

*Published by*
Oxbow Books
Park End Place, Oxford OX1 1HN



ISBN 0 946897 46 8

*Published and distributed by*
Oxbow Books, Park End Place, Oxford OX1 1HN
(Phone: 0865–241249; Fax: 0865–794449)

*Distributed in the United States of America by*
The David Brown Book Company, PO Box 5605, Bloomington, IN 47407
(Phone: 812–331–0266; Fax: 812–331–0277)



*Printed in Great Britain by*
The Short Run Press, Exeter

**CONTENTS**

## Contributors

M. J. Allen, Wessex Archaeology, Portway House, South Portway Estate, Old Sarum, Salisbury, Wilts. SP4 6EB.

M. Bell, Department of Archaeology, Saint David's University College, Lampeter, Dyfed, SA48 7ED.

J. Bintliff, Department of Archaeology, University of Durham, 46 Saddler Street, Durham, DH1 3NU.

J. Boardman, School of Geography, University of Oxford, Mansfield Road, Oxford, OX1 3TB.

A. G. Brown, Department of Geography and School of Archaeological Studies, University of Leicester, Leicester, LE1 7RH.

T. Burt, School of Geography, University of Oxford, Mansfield Road, Oxford, OX1 3TB.

J. A. Catt, Soil Science Department, AFRC Institute of Arable Crops Research, Rothamsted Experimental Station, Harpenden, Herts. AL5 2JQ.

D. A. Davidson, Department of Environmental Science, University of Stirling, Stirling, FK9 4LA.

R. E. Donahue, Department of Archaeology and Prehistory, University of Sheffield, Sheffield, S10 2TN.

R. Evans, Department of Geography, University of Cambridge, Downing Street, Cambridge, CB2 3EN.

P. J. Farres, Department of Geography, Portsmouth Polytechnic, Buckingham Building, Lion Terrace, Portsmouth, Hants. PO1 3HE.

D. D. Gilbertson, Department of Archaeology and Prehistory, University of Sheffield, Sheffield, S10 2TN.

L. Heathwaite, Department of Geography, University of Sheffield, Sheffield, S10 2TN.

M. Helluin, Centre de Géomorphologie du CNRS, Rue des Tilleuls, 14000 Caen, France.

C. D. Hunt, Department of Geographical and Environmental Sciences, The Polytechnic of Huddersfield, Queensgate, Huddersfield, HD1 3DH.

C. Kuzucuoglu, Laboratoire de Géographie Physique URA 141 CNRS, 1 Place Aristide Briand, 92195 Meudon Cédex, France.

L. Lespez, Laboratoire de Géographie Physique URA 141 CNRS, 1 Place Aristide Briand, 92195 Meudon Cédex, France.

R. I. Macphail, Institute of Archaeology, University College London, 31–34 Gordon Square, London, WC1H OPY.

J.-F. Pastre, Laboratoire de Géographie Physique URA 141 CNRS, 1 Place Aristide Briand, 92195 Meudon Cédex, France.

J. Pellerin, Centre de Géomorphologie du CNRS, Rue des Tilleuls, 14000 Caen, France.

R. C. Preece, Department of Zoology, University of Cambridge, Downing Street, Cambridge, CB2 3EJ.

T. A. Quine, Department of Geography, University of Exeter, Amory Building, Rennes Drive, Exeter, Devon, EX4 4RJ.

S. Seeliger, Department of Geography, Portsmouth Polytechnic, Buckingham Building, Lion Terrace, Portsmouth, Hants. PO1 3HE.

S. P. Theocharopoulos, Soil Science Institute, 15 Venizelou Str, 14123 Lykourissi, Greece.

B. Valadas, Université de Clermont-Ferrand, Institut de Géographie, UA 141 du CNRS, Clermont-Ferrand, France.

B. Van Vliet-Lanoë, Centre de Géomorphologie du CNRS, Rue des Tilleuls, 14000 Caen, France.

M. Wagstaff, Department of Geography, University of Southampton, Southampton, Hants. SO9 5NH.

J. Wainwright, Department of Geography, University of Keele, Keele, Staffordshire, ST5 5BG.

D. E. Walling, Department of Geography, University of Exeter, Amory Building, Rennes Drive, Exeter, Devon, EX4 4RJ.

S. J. Wood, Department of Geography, Portsmouth Polytechnic, Buckingham Building, Lion Terrace, Portsmouth, Hants. PO1 3HE.

E. Zangger, Fliederstrasse 21, CH-8006, Zürich, Switzerland.

## ACKNOWLEDGEMENTS

Most of the papers included in this volume were originally given at a meeting with the same title held at the Institute of Archaeology, University College, London on 15.5.91. The meeting was attended by 110 people and was organised under the auspices of the Quaternary Research Association and the Association for Environmental Archaeology. We are grateful to the Institute's Director, Professor David Harris, for permission to hold the meeting, and to Dr. Tony Barham and Ms Judy Medrington who helped to make the arrangements. We are grateful to the following who helped with registration and other arrangements on the day: Ronan Foley; Martin Harvey; Jenni Heathcote; Simeon Mellaliu; Chris Pine; Keith Wilkinson. The meeting chairpersons Professor John Lewin, Professor J. A. Catt and Dr. Susan Limbrey ensured the smooth running of a rather packed programme.

Each contribution to this publication has been reviewed by an anonymous referee and we are grateful to those people for their very prompt advice. We are most grateful to Dr. Brigitte Van Vliet-Lanoe who drew the cover illustration. Editorial help has been provided by Dr. Jennifer Foster, who also prepared the index. David Brown of Oxbow Books provided advice at all stages. We are grateful to the Print Unit of Saint David's University College, Lampeter, which provided reductions of the drawings. The whole volume has been word processed by Mrs. Maureen Hunwicks of the Archaeology Department, Lampeter, to whom we are especially grateful for the long hours she has worked and the trouble she has taken.

Martin Bell and John Boardman

# CHAPTER 1

# PAST AND PRESENT SOIL EROSION: LINKING ARCHAEOLOGY AND GEOMORPHOLOGY

JOHN BOARDMAN AND MARTIN BELL

*This introductory chapter outlines the aims and background to the publication. Geographically the book is largely concerned with lowland Britain, France and the Mediterranean. The timescales considered range from ten years to one hundred thousand years. Common themes and areas of debate which emerge from several contributions are identified. These include identification of causes and processes, the use of the term colluvium, integration of geographical and archaeological perspectives, and the relationship of erosion studies to the formation and conservation of the archaeological record.*

## Introduction

Soil erosion is a subject of increasing public and academic concern. This volume aims to place these contemporary concerns in the long time perspective provided by archaeology and Quaternary science. There are clear benefits in the development of a greater dialogue between those working on present-day processes and those concerned with the past dimension. The interpretation of past deposits is inevitably heavily dependent on modern analogues. Studies of past erosion complement contemporary studies by providing information on longer term rates than are available from the usually very brief history of instrumental and scientific records. It may, for instance, provide some indication of long-term erosion rates before human impact. One of the main areas of debate concerns the relative importance of people and climate as triggers for erosion. Archaeology helps us to broaden our own empirical experience of the relationship between people and nature. It enables us to look beyond our own contemporary world at societies of many different types and dates, including those which existed under climatic regimes which may have no precise modern analogue. Interpretation of the archaeological record nonetheless presents formidable theoretical difficulties. Modern analogues do not provide us with precise models for the interpretation of the very different societies of the past. Furthermore, the destroying hand of time, in which soil erosion is an important process, has wiped large areas of the slate clean.

The background to this book is in part explained by the research backgrounds of the two editors. John Boardman originally conducted research on Quaternary palaeosols but latterly has worked largely on present-day soil erosion. Martin Bell is an environmental archaeologist who has worked on evidence for prehistoric erosion. By coincidence both editors had worked on the same study area on the South Downs and it was the experience of comparing notes and deposits that emphasised the need for greater dialogue which forms the basis of this book. It grew out of a one day meeting held at the Institute of Archaeology, University College, London on 15 May 1991. The meeting was organised under the joint auspices of the Quaternary Research Association and the Association for Environmental Archaeology. The discussion papers given at that meeting form the basis of this volume but some additional papers have also been included. One objective was to give particular emphasis to processes operating on the scale of the individual field. Thus we are not so much concerned with wider catchment scale studies and the role of rivers in sorting eroded soil. One reason for this was that the archaeology of river valleys was the subject of a separate conference in January 1991. The proceedings of that conference (Needham and Macklin 1992) and this should complement one another. They both emphasise the need for integrated archaeological and earth science approaches but they do so on different spatial scales and in the context of different types of deposit.

## Geographical scope

In selecting contributions we tried to draw on areas where there is work on both past and present dimensions. Chapters 2–8 are concerned with Britain. These examples are mainly drawn from the lowland zone. However, Evans (1992) includes reference to work in the highland zone, particularly the role of people in upland peat erosion. One reason that the highland zone has received less emphasis in erosion studies is the current limited extent to which arable agriculture is practised. This has not always obtained; for instance, in parts of mid-Wales such as northern Dyfed, where there is virtually no arable today, arable agriculture was common until the increasing mobility of produce which began with the coming of the railways. It is clear from the archaeological record that parts of Wales which are today entirely pastoral had Iron Age and Romano-British economies which included a significant arable aspect (Caseldine 1990). Despite limited references to the highland zone in this volume it should be noted that there have been a number of important studies which calculate erosion rates on the basis of sediment sequences in lake basins (e.g. Edwards and Rowntree 1980; Dearing *et al*. 1981). In addition to the papers on lowland England two contributors concentrate on current research in France and five are about the Mediterranean where there has been a long history of work on past erosion but much debate as to its causes. One paper in this section also considers the contribution of Geographical Information Systems to studies of soil erosion (Davidson and Theocharopoulos 1992). The remainder of the volume is concerned with specific analytical techniques and modelling.

### Timescales

Deliberately we have included work with contrasting levels of spatial coverage, analytical detail and timescale. Observations of recent erosion incidence and rate on the South Downs are mostly limited to the last 10 years (Boardman 1992a). The caesium-137 technique relates to a timescale of the last *c*.35 years (Quine and Walling 1992). Calculation of erosion in the Slapton catchments is based on about the last 100 years (Heathwaite and Burt 1992) and erosion at Woburn to historical records mostly covering the last 300 years (Catt 1992). Work on prehistoric erosion from arable land in Britain is mostly concerned with the last four to five thousand years (Bell 1992; Allen 1992). The longest timescales considered in this volume are Preece's (1992) contribution which draws on evidence from the last twelve thousand years and Van Vliet-Lanoe *et al*. (1992) who consider the last glacial/ interglacial cycle, i.e. around one hundred thousand years. It needs to be emphasised that any measurement of erosion is relative to the timescale on which it is based. Calculation of long-term rates would ideally involve comparison of a range of sources of evidence, each of which is likely to have different levels of chronological precision. This is the case with dating techniques. Pottery from eroded sediment will only exceptionally provide a date more precise than a couple of centuries. Radiocarbon under ideal circumstances, seldom met in eroded sediments, might provide a date to within a few decades. Dendrochronology has the capacity for annual precision, but on present evidence at least, contexts with suitably preserved wood are rare.

### The causes of erosion

Land-use change increasingly appears to be a major control on erosion. In the 1960s the alleged synchronous pattern of erosion in the Mediterranean appeared to be a clinching argument in favour of the climatic hypothesis (Vita Finzi 1969). We are now more aware of the difficulties of precise dating and of the importance of thresholds in geomorphological systems (Schumm 1977). Complex responses may follow from an initial destabilizing impact, for example the introduction of cattle into a Colorado basin and overgrazing leading to changed hydrology and severe stream downcutting rapidly produced a series of terraces (Womack and Schumm 1977). In this, as in many other present day complex responses, one wonders what future geomorphologists and archaeologists will make of the evidence. Similarly, the major phase of erosion which has affected British arable landscapes from the late 1970s through the 1980s may seem to be a response to changed climate. In fact, there is no such evidence. The increase in erosion has occurred in what Newson (1989) has called 'a decade of drought'. It is a response to intensification of farming and a shift to winter cereals – driven by economic forces. Sadly, one effect of this intensification has been the destruction of archaeological remains by ploughing and erosion, so the past has become a little less easy to decipher. The intensification of farming in the wetter, less arable, west of Britain has also changed patterns of erosion on slopes and the nature of chemical and particulate inputs into channels and storage sites such as lakes (Heathwaite and Burt 1992).

The challenge now, and in the future, is to disentangle the erosional effects of climate, land-use and changing agricultural and industrial technologies, recognising that they often interact. Evans' (1992) discussion of the cause of eighteenth century gullying on the Pennine peat moorlands illustrates this problem. Gullying seems to have been caused by a combination of an ecological change to moss-dominated peat due to Little Ice Age damp conditions, susceptibility of mosses to sulphate pollution from industry, and overgrazing by sheep leading to erosion. The initial destabilisation has given rise to two centuries of gully development with impacts on ecology and water quality that have continued to the present (Ladadz *et al*. 1991).

### Methodology

This book raises the issue of what approaches and methodologies are suitable in attempting to understand the past. Wagstaff (1992) makes clear that the availability of modern analytical techniques does not necessarily solve the problem of the age of terrace sequences. As with other features of the landscape their form may have evolved through time influenced by erosion and deposition and periods of construction and cultivation. Both Wagstaff and Bell (1992) touch on the problem of the purpose of prehistoric terraces and field systems. Understanding *why* they were constructed is difficult. Perhaps we can learn something from studies of traditional agricultural systems and the range of ways in which are perceived hazards such as flooding and erosion (Bell 1992). Also, the fact that terrace systems continue to be an important response to the farming of steep slopes means that the present may be used (with care) as a key to the past. A recent edition of the journal *Méditerranée* (Provansal 1990) is devoted to modern terrace systems in the Mediterranean and Richter (1991) considers techniques of combating erosion on terraced vineyard slopes in the Mosel region. In dealing with the prehistoric field systems of Britain few obvious modern analogues survive and those that do are now farmed in very different ways (Boardman 1992a).

The continuity of some aspects of agricultural technology from prehistoric through historic times is striking in the Mediterranean and in the dry lands of the Middle East and North Africa. This applies not just to terrace systems but also to methods of water harvesting, some of which go back at least 4000 years (Rapp and Hasteen-Dahlin 1990). Such time-honoured methods are appropriate to modern Third World needs and therefore archaeological-geomorphological studies may have practical implications. It is also important to know why apparently successful agricultural systems failed (Zangger 1992).

**Processes**

One of the great revelations of landscape archaeology from Hoskins (1955) onwards (e.g. Aston 1985) has been the degree of continuity and time-depth in many landscapes. In a sense that emphasis on continuity has been reinforced by the recognition that present geomorphological processes form an essential part of understanding many archaeological deposits. This emphasis on continuity may have made archaeologists rather slow to take on board the implication of much modern geomorphological research that the history of landscape is essentially that of episodically changing processes, at its simplest, the contrast between stability (soil development) and instability (erosion or deposition). A number of contributors to this volume emphasise the apparently very episodic nature of erosion in their study areas (e.g. Catt 1992 and Zangger 1992) and the same point is emerging from recent work on river valley sediments (Macklin and Lewin 1992). The difficulty of reconstructing erosional episodes, because by definition they involve loss of evidence, means that emphasis has to be placed on soils and sediments.

The development of analytical techniques which enable soils and sediments to be investigated in terms of the process of formation and evolution is to be welcomed. Examples of modern approaches to soil and sediment analysis include micromorphology (Macphail 1992), lake sediment analysis (Heathwaite and Burt 1992), magnetic susceptibility (Brown 1992) and caesium-137 (Quine and Walling 1992). Of themselves, these techniques do not provide the answers. They provide data that have then to be interpreted. For example, the successful use of the caesium-137 technique for estimating rates of soil loss is based on several assumptions regarding the behaviour of caesium in the soil and on a calibration procedure.

The importance of process is exemplified by the use of the term 'colluvium' which emerged during discussion at the meeting. Archaeologists have tended to categorise non-alluvial deposits as colluvium and explicitly or implicitly assign them to mass movement (gravity dominated), rainsplash or cultivation-related processes. To the geomorphologist the emphasis on mass movement is surprising since the dominant processes seen on present day agricultural land are associated with the flow of water (rilling, gullying, the interrill processes of wash and splash) and wind where soils are dry and non-cohesive.

On soils with low aggregate stability rainsplash may move large quantities of material short distances. Soil loss, as a result of splash, measured at the base of a slope or plot, is very small compared to that due to wash. Bollinne, working on loamy loessic soils in Hesbaye, Belgium, states:

> 'From the comparison of the results of measurements of wash loss and the results of calculated splash loss it appears that splash loss may be generally neglected except in the most stable soils'. (Bollinne 1978, 83).

Splash may be important on interrill areas where soil aggregates are broken down by raindrop impact and particles are then transported to rills and gullies. However, most erosion seems to be due to rilling: Govers and Poesen (1988) report 78% of total interrill and rill erosion as due to rilling and similar high values from other studies that they review.

Sheetflow is also frequently described as an erosional process. On non-cohesive sandy soils sheetflow seems to be of some importance (De Ploey 1971; Morgan *et al.* 1987). On cohesive, silty soils rills form very readily. Experimental work has shown that thin liminar flows of water (without splash) may produce laminated silt loams (Mücher and De Ploey 1984), but this occurs under experimental conditions with very short, gentle slopes that are uncommon in present-day agricultural landscapes. Indeed, De Ploey (1984) points out that rills are frequently reported from slopes as gentle as 2-3 degrees.

On most present-day erodible landscapes rills are the most important erosional form and there is no reason to think that this was not also true in the past. Cultural features such as terrace systems may have reduced rilling to a minimum while they were well maintained. Stony horizons in fine grained deposits indicate rill or gully erosion either in the form of fans or channel deposits. The ability of rill flow to move rock fragments, as measured by the size of the largest stone it can move, exceeds by a factor of 10 the competence of interrill flow (De Ploey and Poesen 1989).

The strong suspicion is that the non-sorted character of colluvium (see for example, Brown 1992), is usually a result of cultivation and mixing of originally sorted sediments. Debris flows may occur on valley sides and are a well known component of alluvial fan sequences but they are an extremely uncommon occurrence on arable land unless the slopes are steep and/or terraced, e.g. in Tanzania and New Zealand (Morgan 1986, 206). The term 'colluvium' may still have its uses, not least because of its long currency and lack of a suitable alternative, but it should be accompanied by a process-based explanation or admission of ignorance. The terms 'hillwash' and 'ploughwash' raise similar questions for the geomorphologist: what are the processes responsible for their formation? Farres (1992) shows how depth of ploughing and depth of sediment combine to control the survival of original features of the sediment. This model, though theoretically convincing, will be difficult to apply in practice because of lack of information about the depth of deposition and depth of ploughing in the past. Van Vliet-Lanoe (1992) suggests figures for increasing depth of ploughing through time which will increase the effectiveness of the mixing process.

The rate of operation of processes is a theme central to modern geomorphology. However, the establishment of reliable estimates of present rates of erosion (let along those of the past), has proved difficult. During discussion at the London meeting it emerged that contributors had contrasting perspectives on the comparative extents of past and present erosion. The implication of many of Boardman's (1992a) conclusions is that, other factors being equal, present-day agricultural practices make landscapes more liable to extensive erosion than in the past. Evans (1992), by contrast, suggests that surviving eroded sediments indicate that on average erosion in the past must have been higher than at present.

Our knowledge of modern rates is based on a very short, and quite possibly atypical, period of observation. A further difficulty is that estimates of past rates based on volumes of eroded material preserved as alluvium and colluvium have been recognised as crude due to the problems of reliably dating the beginning and end of deposition, accounting for losses, and identifying the source area. Quine and Walling (1992) suggest that previous estimates seriously underestimate losses by a factor of between 4 and 7. The implication of this is that erosion was greater in the past than often appreciated. If nothing else these points do at least highlight the inadequacy of our existing knowledge and the need for more precise quantification of erosion.

One of the most important geomorphological contributions in recent years has been the recognition of lags within erosional systems. Erosion on valley side slopes or in headwater areas may have a delayed impact on channels and downvalley areas. In the USA this was clearly demonstrated by Trimble (1983) in Coon Creek and Meade (1982) in the Atlantic coastal zone. This makes attempts to link erosion, alluviation and land-use change very difficult (cf Hunt *et al.* 1992).

Accurate dating of erosion episodes is clearly necessary. The detail and complexity of erosional history may be partly revealed by investigation of closely dated sequences (cf Preece 1992). Only 'partly', because for any particular site the gaps in the sequence may be more significant than that which remains. However, the increasing number of dated sites tends to emphasise that the onset of colluviation (primary and secondary) is controlled by clearance or subsequent changes in agricultural operations which did not occur contemporaneously over wide areas (Bell 1982). Evans (1992) makes this point with respect to sites around Yeovil, Somerset.

The link between dated sequences from the past and modern geomorphological observations of erosion is interesting. On the eastern South Downs in the period 1982–1992 a very large proportion of the erosion is attributable to two rather short wet periods in 1982 and 1987 (Boardman, unpublished). Would such a pattern be detectable in a buried sequence of deposits, particularly one disturbed by subsequent cultivation? Clearly radiocarbon dating would be of little help and without the chance inclusion of car licence plates within the deposits (cf Costa 1975), it is difficult to see how the temporal pattern of sedimentation could be adequately reconstructed.

Bintliff (1992) notes the rarity of preserved major erosional and depositional sequences in the archaeological record. This is a continuing problem in south-east England with regard to the lack of information about the erosion of original woodland soils of the early Holocene (Bell 1992; Allen 1992). Remnants of the soil exist in subsoil hollows; elsewhere the episode is marked by a major unconformity. There is also the question as to the conditions that facilitated the large-scale removal of sediments from valleys: in this case the modern erosion seems an unsatisfactory key to past events (Boardman 1992a).

Bintliff's comment about lack of evidence for erosion in the archaeological record raises the question as to what sort of evidence we would expect. To the geomorphologist, using the present landscape as an analogy, the question is straightforward. For example, major erosion on the South Downs in 1987 resulted in the formation of gravel fans containing flint and chalk stones in footslope and valley bottom positions (Boardman 1988). The location of the latter was largely controlled by field boundaries. Deposition of fine grained sediments was often far down valley, in some cases over a kilometre from their source but again with locations influenced by cultural features which produced ponding of water. Rills and gullies on slopes, although impressive at the time, were filled in and mixed with soil material within months. In valley bottoms each erosional event leads to reworking of previously deposited sediments frequently removing fine grained material leaving lag gravels and gullies. The record of major events is therefore one of erosion and deposition in predictable locations but survival of the evidence would be extremely localised.

Comparisons between the geomorphology of present eroding landscapes and the archaeological record of past events are difficult for several reasons. One is that direct analogies are not possible. Field system morphology, farming techniques and soils have changed. This is emphasised by comparison of some of the hypothetical models advanced in this volume. Compare for instance the erosion process which would have operated within the past landscape in the cover illustration by Brigitte Van Vliet-Lanoe, in which the slopes are divided into small terraced fields, the valley floor is pasture and settlement is concentrated on the hilltops, with the modern situation on the South Downs, illustrated by Boardman 1992, Figure 2.1. In the latter some slopes are cultivated as single fields and there is more intensive land-use and some settlement in the valley and lower slopes. It is clear from artifact scatters in field surveys and dry valleys that the distribution and character of settlement has changed substantially.

Until recently, few geomorphologists or archaeologists have attempted to make the connection between processes in present day landscapes and prehistoric deposits; one notable exception is Oakley (1945) in his study of the erosional effects of a summer rainstorm in the Chilterns. In this context it is also worthy of note that the effects of a large rainstorm on a British arable landscape are unknown. There is no record of the impact of events of the order of 100 or 1000 year return period (Boardman 1992b). There is the possibility that erosion in the past was caused by events of a greater magnitude than any that we have investigated in modern times. Those that have occurred have affected arable landscapes at times in the year when there has been little bare ground (Arkell 1955; Potts 1982) and also, until recently observers have been little interested in the impact of storms on agricultural land in comparison to the effect on channels and property (Barnes and Potter 1958; Cross 1967; Hanwell and Newson 1970).

**Erosion and the archaeological record**

The way in which the archaeological record has formed is a subject of growing academic interest (Schiffer 1987). Erosion and deposition are a key factors in this regard and have very different effects both on contrasting facets of the landscape and at particular points of time. Valley bottom and footslope sites are frequently well preserved, individual phases of activity often being separated by sediment. Conversely on hill tops, convex slopes and the upslope sides of fields, erosion can be severe. Certain types of archaeological features may be totally eroded away and only the truncated bases of the deeper features survive. Eroded sediment contains artifacts which on some bedrock types can be very numerous. This is particularly the case on the chalk where recorded concentrations range from 64 to 256 per $m^3$ (Bell 1986). In other areas soil eroded from agricultural land contains few artifacts, which in some instances, such as in much of Wales, may relate to the paucity of ceramics until the Medieval period and also to the acidic nature of the soils which means that poorly fired pottery is less likely to be preserved. As regards the origins of artifacts in eroded soil, some clearly derive from the erosion of upslope occupation sites but there is also a good deal of evidence to suggest that artifacts were spread on the fields with domestic and farmyard manure. Experience shows that the vast majority of the artifacts from eroded soil are pottery, flints and stone but occasionally more unusual small finds are present, as illustrated by sites on the South Downs (Bell 1981). At Chalton, Hampshire prehistoric layers produced a copper alloy ring and fragments of shale bracelet blanks. At Itford Bottom, East Sussex, Romano-British layers produced fragments of gold and of bronze. Indeed the concentration of artifacts in eroded soil may sometimes be as great as that on occupation sites themselves. That clearly has implications, firstly for how we recognise an archaeological site from fieldwalking, and secondly for our definition of sites. In relation to scheduled ancient monuments or areas protected by planning restraints there must in some cases be a strong case for including the surrounding fields and valley sediments, in view of the number of artifacts and the evidence of land-use history which they contain. Consideration of the effects of erosion is essential to the interpretation of artifact scatters, as emphasised in a recent volume edited by Schofield (1991). The same point emerged from discussion at a seminar on 'Fieldwalking survey in Ireland' held at University College, Dublin on 28.3.92 and organised by Dr. Gabriel Cooney. At this meeting it was shown how fieldwalking has greatly extended the known distribution of lithic period sites and enabled the identification of areas of contrasting land-use. Successive speakers emphasised the importance of considering geomorphological factors which influence the extent of erosion in various sectors of a landscape and the difficulty of interpreting what are usually mixed assemblages from the ploughzone. There are obvious advantages here in combining fieldwalking with investigation of sequences from neighbouring environments of deposition in which individual phases of activity may be stratigraphically separated and are related to palaeoenvironmental evidence. This has been attempted in the Bullock Down Survey (Drewett 1982) and at Chalton, Hampshire (Cunliffe 1973; Bell 1981).

In the present volume the effects of erosion and formation processes are further considered by Wainwright (1992) in relation to the forces responsible for the entrainment of individual artifacts. Quine and Walling (1992) also note the potential of caesium-137 derived maps which show spatial variation in the distribution of recent erosion. This provides a means of quantifying the often insidious effects of modern agricultural practice on the archaeological record and defining those areas which are in need of particular protection. In so far as we can use the topographic distribution of the last *c*.35 years' erosion as a reliable guide to earlier periods it also has some predictive value in indicating those topographic areas of lesser erosion where the archaeological record may be better preserved.

Archaeologists are being put under increasing pressure to define those areas of the landscape where the archaeology is well preserved. This arises on the one hand from the need to define and conserve what remains of a diminishing resource (Darvill 1987) and on the other hand from the growing practice of Environmental Impact Assessment. The idea of this is that during the planning stage of building or engineering works there should be an assessment of the effects on various aspects of the environment including the archaeology, so that appropriate ameliorating action can be built into the proposals. This type of approach was pioneered in the United States from the 1960s and has recently been introduced in Europe in response to a European Economic Community directive. It is becoming increasingly necessary to identify and rapidly assess areas of buried archaeology. This is especially difficult where sites of several periods are contained in deep sedimentary sequences. Examples are the recently discovered sites in the Severn Estuary which are affected by a series of major engineering projects, each of which will affect large areas (SELRC 1991; 1992). Archaeology buried in eroded soil is an important part of this assessment problem. A partial solution to this problem is predictive modelling of site locations. Gardner and Donahue (1985) have constructed such models on the basis of landform and sediment dates and in this case they achieved impressive results particularly with excavations in 'colluvial foot slopes' where more-or-less continuous sedimentation had occurred (cf Brown 1992).

Predictive modelling does, however, need to begin from a satisfactory database. This we clearly do not have in relation to many aspects of deeply buried archaeology which have hitherto been little investigated. It will often be necessary to carry out auger and soil pit surveys in order to establish the extent of buried archaeology (Gingell and Lawson 1985; Bell 1990). The present volume includes interim reports on two of the first generation of these assessment exercises. Kuzucuoglu *et al.* (1992) discuss the sites revealed on the T.G.V. (French High Speed Train) route between Paris and Lille, which ultimately connects with the Channel Tunnel rail link between England and France. On the English side Preece (1992) describes a large scale investigation of the Quaternary sediment sequence at the entrance to the tunnel. Both of these assessments, and many others which concern deeply stratified sequences, emphasise the impossibility of separating out an archaeological component from other aspects of the environment affected, such as geology. Preece's project was sponsored by the then Nature Conservancy Council (now English Nature) because the site was a Site of Special Scientific Interest for Quaternary geological reasons. Nonetheless the archaeological implications of his work are substantial. This highlights the importance of trying to work towards an integrated approach, not only to the various aspects of assessment, but more importantly to the conservation of the whole cultural resource, natural as well as that altered by humanity (Bell and Walker 1992).

We hope to have demonstrated that the study of past and present soil erosion is not a purely esoteric academic exercise. It is necessarily politically engaged in various ways. In addition to issues of planning, assessment and conservation there is a potential contribution to debates about forms of land-use which may be more sustainable for the future. Several papers emphasise the role of human agency as a prime factor in accelerated erosion and the need for changed land-use.

**Acknowledgements**

We thank Dr. Tim Burt for useful comments on a draft of this chapter.

**References**

Allen, M. J. 1992 Products of erosion and the prehistoric land-use of the Wessex chalk. In: Bell, M. and Boardman, J. (eds.) *Past and Present Soil Erosion*,. Oxbow Books, Oxford, 37–52.

Arkell, W. J. 1955 Geological results of the cloudburst in the Weymouth district, 18th July, 1955. *Proceedings of the Dorset Natural History and Archaeological Society* 77, 90–95.

Aston, M. 1985 *Interpreting the Landscape: Landscape Archaeology in Local Studies*. Batsford, London.

Barnes, F. A. and Potter, H. R. 1958 A flash flood in western Derbyshire. *East Midland Geographer* 10, 2–15.

Bell, M. 1981 *Valley sediments as evidence of prehistoric land-use: a study based on dry valleys in south east England*. London University Ph.D. thesis.

Bell, M. 1982 The effects of land-use and climate on valley sedimentation. In: Harding, A. F. (ed.) *Climatic Change in Later Prehistory*. Edinburgh University Press, 127–142.

Bell, M. G. 1986 Archaeological evidence for the date, cause and extent of soil erosion on the chalk. *SEESOIL*, 3, 72-83.

Bell, M. 1990 *Brean Down Excavations 1983-1987*. English Heritage Archaeological Report 15, London.

Bell, M. 1992 Prehistory of erosion. In: Bell, M. and Boardman, J. (eds.) *Past and Present Soil Erosion*,. Oxbow Books, Oxford, 21–35.

Bell, M. and Walker, M. J. C. 1992 *Late Quaternary Environmental Change: Physical and Human Perspectives*. Longman, London.

Bintliff, J. 1992 Erosion in the Mediterranean lands: a reconsideration of pattern, process and methodology. In: Bell, M. and Boardman, J. (eds.) *Past and Present Soil Erosion*. Oxbow Books, Oxford, 125–131.

Boardman, J. 1988 Severe erosion on agricultural land in East Sussex, UK, October 1987. *Soil Technology* 1, 333–348.

Boardman, J. 1992a Current erosion on the South Downs: implications for the past. In: Bell, M. and Boardman, J. (eds.) *Past and Present Soil Erosion,*. Oxbow Books, Oxford, 9–19.

Boardman, J. 1992b The sensitivity of Downland arable land to erosion by water. In: Thomas, D. S. G. and Allison, R. J. (eds.) *Environmental Sensitivity*. Wiley, Chichester, in press.

Bollinne, A. 1978 Study of the importance of splash and wash on cultivated loamy soils of Hesbaye (Belgium). *Earth Surface Processes* 3, 71–84.

Brown, A. G. 1992 Slope erosion and colluviation at the floodplain edge. In: Bell, M. and Boardman, J. (eds) *Past and Present Soil Erosion*. Oxbow Books, Oxford, 77–87.

Caseldine, A. 1990 *Environmental Archaeology in Wales*. St David's University College, Lampeter.

Catt, J. A. 1992 Soil erosion on the Lower Greensand at Woburn Experimental Farm, Bedfordshire - evidence, history and causes. In: Bell, M. and Boardman, J. (eds.) *Past and Present Soil Erosion*. Oxbow Books, Oxford, 67–76.

Costa, J. E. 1975 Effects of agriculture on erosion and sedimentation in the Piedmont Province, Maryland. *Geological Society of America Bulletin* 86, 1281–1286.

Cross, D. A. E. 1967 The Great Till Flood of 1841. *Weather* 22, 430–431.

Cunliffe, B. W. 1973 Chalton, Hampshire: the evolution of a landscape. *Antiquaries Journal* 53, 173–190.

Darvill, T. 1987 *Ancient Monuments in the Countryside: an Archaeological Management Review*. English Heritage, London.

Davidson, D. A. and Theocharopoulos, S. P. 1992 A survey of soil erosion in Viotia, Greece. In: Bell, M. and Boardman, J. (eds.) *Past and Present Soil Erosion*. Oxbow Books, Oxford, 149–154.

Dearing, J. A., Elner, J. K. and Happey-Wood, C. M. 1981 Recent sediment flux and erosional processes in a Welsh upland lake-catchment based on magnetic susceptibility measurements. *Quaternary Research* 16, 356–372.

De Ploey, J. 1971 Liquefaction and rainwash erosion. *Zeitschrift für Geomorphologie* 15, 491–496.

De Ploey, J. 1984 Hydraulics of runoff and loess loam deposition. *Earth Surface Processes and Landforms* 9, 533–539.

De Ploey, J. and Poesen, J. 1989 Experimental geomorphology and the interpretation of stone-lines. *Geo-Eco-Trop* 11, (1987 1–4), 75–82, publ. 1989.

Drewett, P. L. 1982 *The Archaeology of Bullock Down, Eastbourne, East Sussex: the development of a landscape*, Sussex Archaeological Society Monograph 1, Lewes.

Edwards, K. and Rowntree, K.M. 1980 Radiocarbon and palaeoenvironmental evidence for changing rates of erosion at a Flandrian stage site in Scotland. In: Cullingford, R. A., Davidson, D. A. and Lewin, J. (eds) *Timescales in Geomorphology*. Wiley, Chichester, 207–223.

Evans, R. 1992 Erosion in England and Wales – the present the key to the past. In: Bell, M. and Boardman, J. (eds.) *Past and Present Soil Erosion*. Oxbow Books, Oxford, 53–66.

Farres, P. J. , Wood, S. J. and Seeliger, S. 1992 A conceptual model of soil deposition and its implications for environmental reconstruction. In: Bell, M. and Boardman, J. (eds) *Past and Present Soil Erosion*. Oxbow Books, Oxford, 217–226.

Gardner, G. D. and Donahue, J. 1985 The Little Platte drainage, Missouri: a model for locating temporal surfaces in a fluvial environment. In: Stein, J. and Farrand, W. R. (eds.) *Archaeological Sediments in Context*. Peopling the Americas Edited Volume Series Vol. 1. Centre for the Study of Early Man, Institute for Quaternary Studies, University of Maine at Orono, 69–89.

Gingell, C. and Lawson, A. J. 1985 Excavations at Potterne 1984. *Wiltshire Archaeological and Natural History Magazine*, 79, 101–8.

Govers, G. and Poesen, J. 1988 Assessment of the interrill and rill contributions to total soil loss from an upland field plot. *Geomorphology* 1, 343–354.

Hanwell, J. D. and Newson, M. D. 1970 *The Great Storms and Floods of July 1968 on Mendip*. Wessex Cave Club Occasional Publication Number 2.

Heathwaite, L. and Burt, T. 1992 The evidence for past and present erosion in the Slapton catchment, South West Devon. In: Bell, M. and Boardman, J. (eds.) *Past and Present Soil Erosion*. Oxbow Books, Oxford, 89–100.

Hunt, C. O., Gilbertson, D. D. and Donahue, R. E. 1992. Palaeoenvironmental evidence for agricultural soil erosion from late Holocene deposits in the Montagnola Senese, Italy. In: Bell, M. and Boardman, J. (eds.) *Past and Present Soil Erosion*. Oxbow Books, Oxford, 163–174.

Kuzucuoglu, C. Lespez, L. and Pastre, J.-F. 1992 Holocene colluvial deposits on the slopes of the Paris Basin. In: Bell, M. and Boardman, J. (eds.) *Past and Present Soil Erosion/* Oxbow Books, Oxford, 115–123.

Labadz, J. C., Burt, T. P. and Potter, A. W. R. 1991 Sediment yield and delivery in the blanket peat moorlands of the southern Pennines. *Earth Surface Processes and Landforms* 16, 255–271.

Macphail, R. I. 1992 Soil micromorphological evidence of ancient soil erosion. In: Bell, M. and Boardman, J. (eds.) *Past and Present Soil Erosion.* Oxbow Books, Oxford, 197–215.

Meade, R. H. 1982 Sources, sinks, and storage of river sediment in the Atlantic drainage of the United States. *Journal of Geology* 90, 235–252.

Morgan, R. P. C. 1986 *Soil Erosion and Conservation.* Longman.

Morgan, R. P. C., Martin, L. and Noble, C. A. 1987 Soil erosion in the United Kingdom: a case study from mid-Bedfordshire. *Occasional Paper* No. 14, Silsoe College, Cranfield Institute of Technology.

Mucher, H. J. and De Ploey, J. 1984 Formation of afterflow silt loam deposits and structural modification due to drying under warm conditions: an experimental and micro-morphological approach. *Earth Surface Processes and Landforms* 9, 523–531.

Needham, S. and Macklin, M. 1992 *Archaeology under Alluvium.* Oxbow Books, Oxford.

Newson, M. D. 1989 Flood effectiveness in river basins: progress in Britain in a decade of drought. In: Beven, K. and Carling, P. (eds.) *Floods: Hydrological, Sedimentological and Geomorphological Implications.* John Wiley, Chichester, 151–169.

Oakley, K. P. 1945 Some geological effects of a 'cloudburst' in the Chilterns. *Records of Buckinghamshire* 14, 265–280.

Potts, A. S. 1982 A preliminary study of some recent heavy rainfalls in the Worthing area of Sussex. *Weather* 37(8), 220–227.

Preece, R. C. 1992 Episodes of erosion and stability since the Late-glacial: the evidence from dry valleys in Kent. In: Bell, M. and Boardman, J. (eds.) *Past and Present Soil Erosion.* Oxbow Books, Oxford, 175–183.

Provansal, M. (ed.) 1990 L'agriculture en terrasses sur les versants Méditerranéens; histoire, conséquences sur l'évolution du milieu. Séminaire-Aix-en-Provence, 3 Février 1990. *Méditerranée* 71.

Quine, T. A. and Walling, D. E. 1992 Patterns and rates of contemporary soil erosion derived using caesium-137; measurement, analysis and archaeological significance. In: Bell, M. and Boardman, J. (eds.) 1992 *Past and Present Soil Erosion.* Oxbow Books, Oxford, 185–196.

Rapp, A. and Hasteen-Dahlin, A. 1990 Improved management of drylands by water harvesting in Third World countries. In: Boardman, J., Foster, I. D. L. and Dearing, J. A. (eds.) *Soil Erosion on Agricultural Land.* John Wiley, Chichester, 495–511.

Richter, G. (ed.) 1991 *Combating Soil Erosion in Vineyards of the Mosel Region.* University of Trier, Germany.

Schiffer, M. B. 1987 *Formation Processes of the Archaeological Record.* University of New Mexico Press, Albuquerque.

Schofield, J. (ed) 1991 *Interpreting Artefact Scatters: Contributions to Plough Zone Archaeology.* Oxbow Books, Oxford.

Schumm, S. A. 1977 *The Fluvial System.* Wiley-Interscience, New York.

SELRC 1991 *Severn Estuary Levels Research Committee Annual Report 1991.* Saint David's University College, Lampeter.

SELRC 1992 *Severn Estuary Levels Research Committee Annual Report 1992.* Saint David's University College, Lampeter.

Trimble, S. W. 1983 A sediment budget for Coon Creek basin in the Driftless Area, Wisconsin, 1853–1977. *American Journal of Science* 283, 454–474.

Van Vliet-Lanoe, B., Helluin, M., Pellerin, J. and Valadas, B. 1992 Soil erosion in western Europe: from the last interglacial to the present. In: Bell, M. and Boardman, J. (eds.) *Past and Present Soil Erosion.* Oxbow Books, Oxford, 101–114.

Vita-Finzi, C. 1969 *The Mediterranean Valleys: Geological Changes in Historical Times.* Cambridge University Press.

Wagstaff, M. 1992 Agricultural terraces: the Vasilikus Valley, Cyprus. In: Bell, M. and Boardman, J. (eds.) *Past and Present Soil Erosion.* Oxbow Books, Oxford, 155–161.

Wainwright, J. 1992 Assessing the impact of erosion on semi-arid archaeological sites. In: Bell, M. and Boardman, J. (eds.) *Past and Present Soil Erosion.* Oxbow Books, Oxford, 227–241.

Womack, W. R. and Schumm, S. A. 1977 Terraces of Douglas Creek, northwestern Colorado: an example of episodic erosion. *Geology* 5, 72–76.

Zangger, E. 1992 Neolithic to present soil erosion in Greece. In: Bell, M. and Boardman, J. (eds.) *Past and Present Soil Erosion.* Oxbow Books, Oxford, 133–147.

# CHAPTER 2

# CURRENT EROSION ON THE SOUTH DOWNS: IMPLICATIONS FOR THE PAST

JOHN BOARDMAN

*The 1980s on the South Downs were characterized by an increase in the rate and frequency of soil erosion on arable land. This was due to the widespread adoption of winter cereals and the continued intensification of agriculture. Rainfall necessary to initiate rilling occurs every autumn and major erosion events may result from as little as 30 mm in a day falling on wet bare soils, which have been prepared for, or drilled with, cereals. Severe erosion, as in October 1987, results from rainfall with a return period of 20–30 years and rates of soil loss reach 200 $m^3$ $ha^{-1}$. Erosion takes the form of rills and gullies; wind, sheetflow and mass movement are of negligible importance. Both valley-side slopes and valley bottoms are susceptible to erosion. The concentrated flow of water results in the sorting of sediments, typically coarse fans in footslope and valley-bottom positions with finer sediment, including organic matter, carried greater distances.*

*Comparisons between present and past erosion are difficult. Present day soils are stonier and this inhibits erosion. However, many characteristics of modern mechanized farming act to encourage erosion. Modern studies emphasise the critical importance of land use, timing and location of crops, and farming practices. In the past, systems of terracing or small fields would have reduced erosion to low values, although the role of fallow land is still unclear. The sorted character of most modern sediment distinguishes it from the typical 'colluvium' of the past, although this may simply be a function of subsequent mixing by cultivation.*

## The South Downs

The South Downs are a range of hills in south east England rising to about 200 m. They are composed of chalk, a soft Cretaceous limestone, with patches of Clay-with-flints and sandy Tertiary beds on some interfluves and dip slopes. The Downs are deeply dissected by dry valley networks, which were active under periglacial conditions; these give rise to valley-side relief of up to 150 m. A few valleys contain seasonal streams. The principal soils are classified as Andover 1 Association (Jarvis *et al*. 1984). These are mainly stony rendzinas, with a high proportion of silt of loessial origin; the stones are both chalk and flint. Valley-side soils are silty clay loams and silt loams, rarely thicker than 20 cm and frequently with A horizons of only 15 cm over chalk bedrock. The present thin soils are the remnants of a formerly thicker loess cover which has been removed by erosion during the Flandrian (Catt 1978; Evans 1990a). In valley bottoms, soils and buried soils are developed in sediments that were eroded from the slopes (e.g. Bell 1983). The mean annual rainfall on the South Downs is between 750 and 1000 mm, with a peak in October and November (Potts and Browne 1983).

Since the Second World War the dominant land use on the Downs has been arable farming, mainly the growing of wheat and barley. In the 1970s a major change of practice occurred with the adoption of autumn-sown cereals ('winter cereals') in preference to lower yielding spring-sown varieties. In some areas 60% of the landscape is now under winter cereals. Also, there has been an intensification of farming: fields have increased in size with the removal of hedges and grass banks, and larger and more powerful farm vehicles have enabled steeper slopes to be cultivated. Slopes of up to 25° are now under cereals.

## Current erosion

Wind erosion is rare because when soils are bare they tend to be damp; one case has been observed in the last decade where blowing was associated with freeze-drying of a silty valley-bottom soil. An increase in spring-sown crops could well lead to more wind erosion.

In the last ten years there has been an increase in the number of water erosion events on the Downs (Boardman 1990). Most of these occur between September and December on land prepared for, or drilled with, winter cereals. The coincidence of large areas of smooth ground, bare or with little vegetation cover, and the wettest months explains the increase in erosion. The most serious erosion tends to occur on large fields and often results in down valley flooding of property because of the lack of barriers to water movement in valley bottoms and the proximity of urban areas to cereal fields (Stammers and Boardman 1984).

Data for over 400 erosion events in the monitored area between Lewes and Brighton in the last decade have been collected; this includes site morphometry, soil, crop and management information and erosion amount. Six gauges within or close to the area allow estimates to be made of the rainfall.

Soil loss by sheet flow on the typical cohesive silty soils of the Downs appears to be minimal

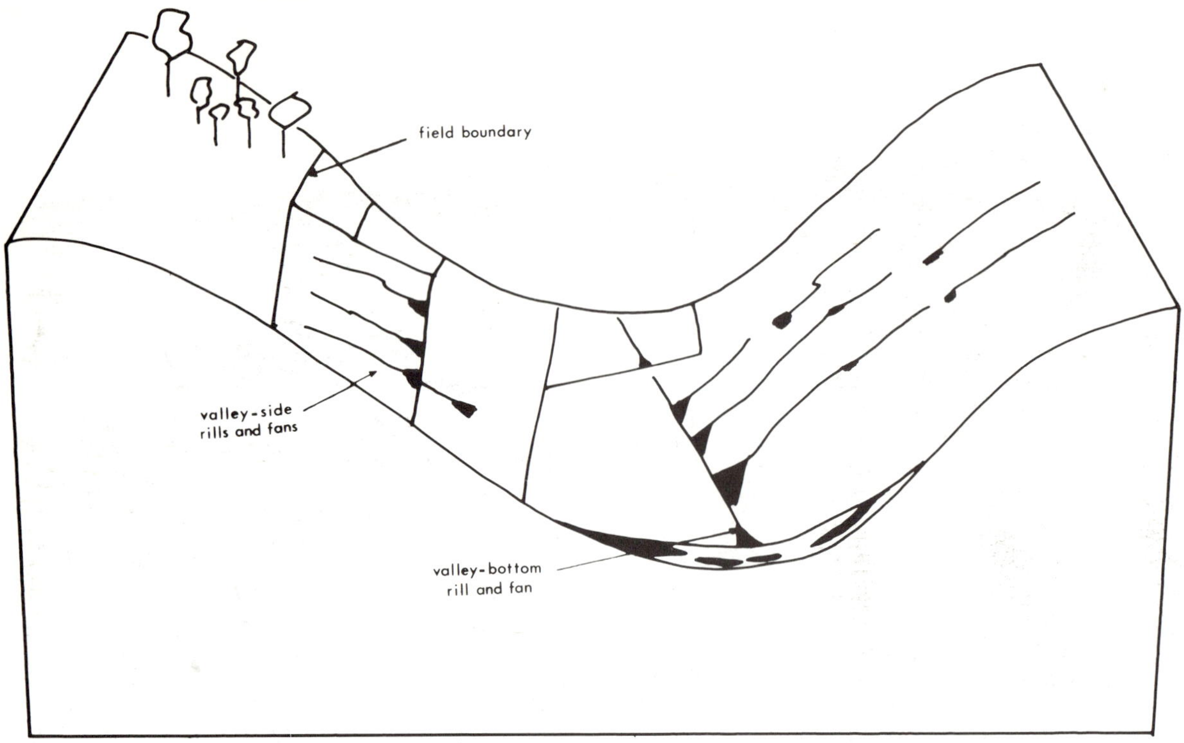


Figure 2.1 Typical location and form of erosion on agricultural land on the South Downs

although runoff can be extensive particularly on bare fields that have been rolled. Erosive sheetwash is more important on sandy soils developed in Tertiary deposits and loamy soils of the Weald, e.g. Wickham 1 and Curtisden Association. On the Downs, inter-rill erosion by thin sheet flow is presumed to occur once rill systems are established, but its contribution to total soil loss appears to be small. Rates of soil movement by creep, rain splash and sheet flow are insignificant compared to rilling. Similar conclusions have been reached for silty soils by Bollinne (1978) and Evans (1990b).

On bare smooth fields which have been prepared for, or drilled with, cereals rilling is initiated by about 30 mm of rain in two days. In the months when fields are at risk of erosion due to lack of vegetation (September to March), this amount of rainfall is to be expected on more than one occasion each year (Boardman 1992). Greater amounts of rainfall lead to more serious erosion e.g. 63 mm on 7 October 1987 (Boardman 1988). Rainfall amounts necessary to give rise to serious erosion on some fields have occurred in seven of the last nine years. Late winter rainfall peaks, which are unusual, may lead to erosion on land which has remained ploughed in preparation for spring cereals. This is because reduction in surface roughness by rainfall impact has to occur before rills develop (Boardman 1991).

Land use and farming practice change little from year to year and therefore the total amount of erosion is closely related to rainfall (Boardman 1990). However, over a time span of decades, the increase in winter cereals in some catchments has meant that serious flooding and erosion has resulted from progressively lower amounts of rainfall (Boardman and Evans unpublished). It is important to emphasise that erosion can be initiated by rainfall events with return periods of less than one year, and that the most serious erosion of modern times was a result of a daily rainfall event with a return period of perhaps 25 years. The impact of high magnitude rainfall events e.g. 1000 year return period, will depend on their timing with respect to the location and amount of bare ground.

On valley-side slopes serious erosion (>10 $m^3$ $ha^{-1}$ $yr^{-1}$) occurs on winter cereal fields where slope length is >200 m, maximum slope angle is >10° and relief within the field is >30 m. Evans (1980) points out that erosion occurs on slope convexities, and on the Downs sites of major erosion are frequently where fields link slope crests, convexities and valley bottoms (e.g. Boardman and Robinson 1985). Knowledge of the morphological, soil, rainfall and farming practice characteristics associated with erosion allow predictions to be made of erosion amounts using an Expert System approach (Harris and Boardman 1990).

The loss of soil is typically the result of the development of rill systems on valley-side fields and broad shallow rills in valley bottoms; less frequently gullies incise up to 2 m into valley sides and valley bottoms (Figure 2.1). Very high drainage density values have been recorded for some rill systems e.g. at Bevendean in 1982 (Boardman and Robinson 1985), and at Breaky Bottom in 1987, where on one 2 ha slope the drainage density was 650 km $km^{-2}$ (Boardman 1988). On valley sides rills frequently reach bedrock at about 20 cm and then widen; in

Figure 2.2 Rills on winter cereal field, Housedean, near Lewes (photograph: John Holloway). Many rills are developed along wheelings resulting from drilling. Later wheelings (spraying) are unrilled. Fans are mainly composed of gravel size chalk and flint; fines have been carried down valley to right.

Figure 2.3 Broad shallow rill development in valley bottom at Rottingdean, East Sussex, October 1987. Runoff occurring as a result of minor rainfall event. Temporary dams erected by Brighton Borough Council to prevent further flooding of houses can be seen to have been breached.

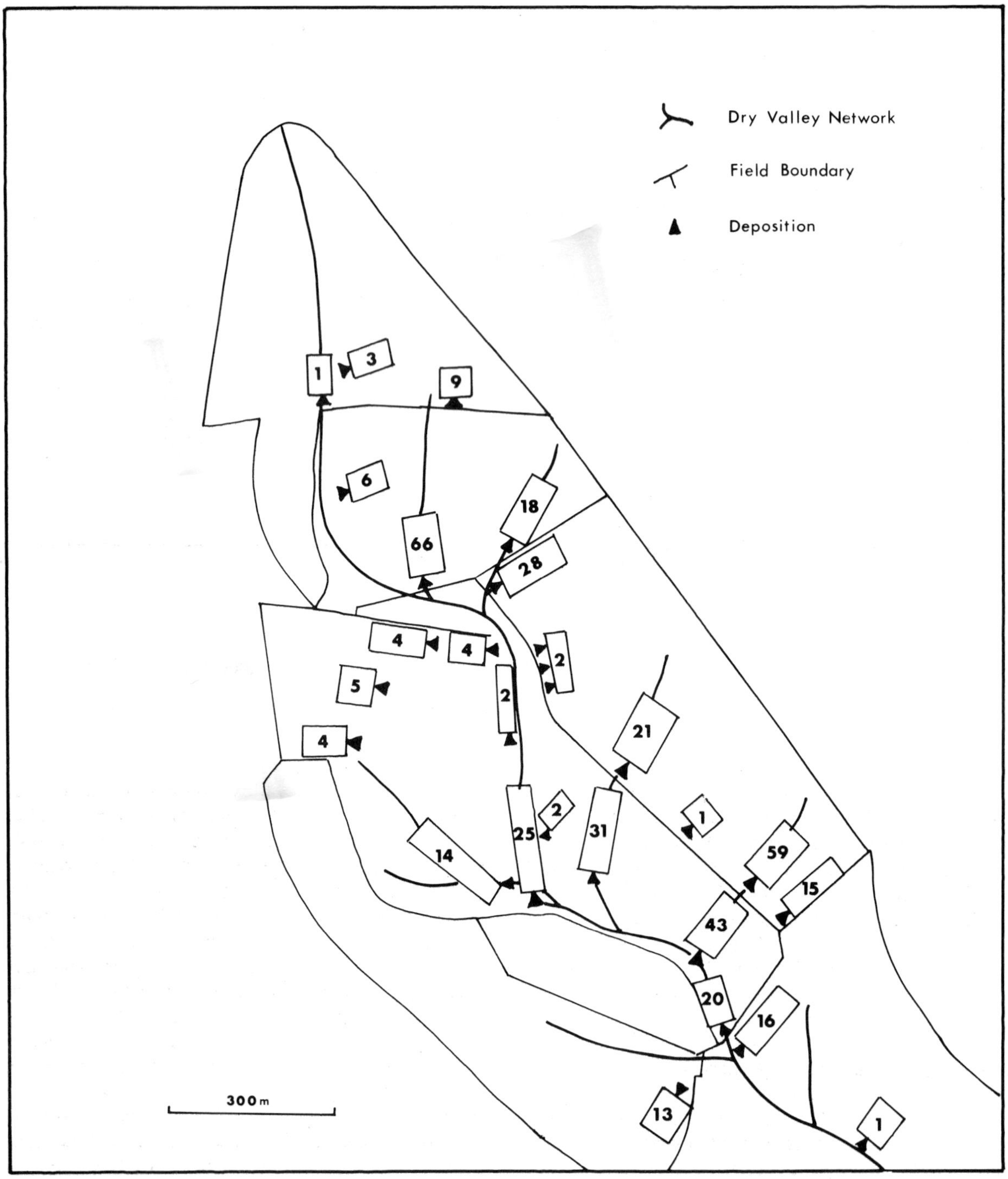


Figure 2.4 Location of erosion on fields in the Houndean dry valley, near Lewes. The figures are cumulative amounts of soil moved ($m^3$) 1982-90. Most erosion occurred on winter cereals with a little on oil seed rape, grass leys and spring cereals.

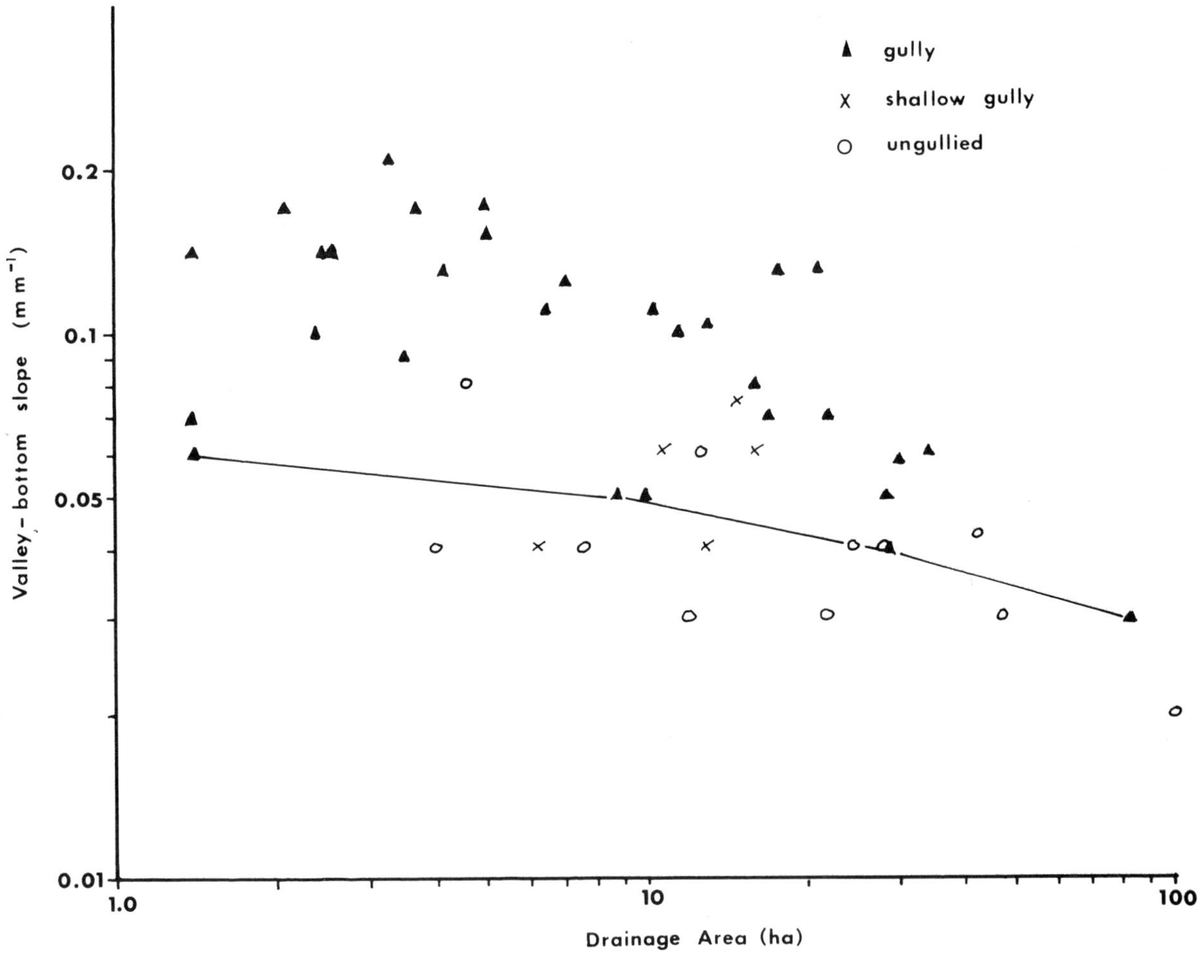


Figure 2.5 The susceptibility of valley bottoms and valley-side depressions to rilling (ephemeral gullies), eastern South Downs, 1982–90.

valley bottoms, the Bt horizon or plough pan in deeper silty Charity series soils also acts as an inhibitor on depth increase. The location of rills is often influenced by the lines of compaction along the wheel tracks of vehicles ('wheelings') (Figure 2.2).

In valley-bottom and depression locations, concentrated flow of water frequently leads to erosion (Figure 2.3). Valley-bottom rills of this type are referred to by Foster (1990) as 'ephemeral gullies' and have been recognized elsewhere as major forms of erosion (Auzet *et al.* 1990; De Ploey 1990). In areas where slopes are relatively short and gradients less than 10°, most erosion occurs in valley bottoms (Figure 2.4). De Ploey (1990) suggests that this is especially likely where sediment transport on slopes is below the transporting capacity of the runoff. However, cases of serious down-valley flooding on the South Downs are generally associated with both valley bottom and valley-side rilling (e.g. Boardman and Robinson 1985; Boardman 1988).

The susceptibility of valley bottoms to gullying has been examined by Patton and Schumm (1976); similar analysis, on the eastern South Downs shows most depressions or valley bottoms are at risk of erosion (Figure 2.5). Gullying has occurred in valley bottoms with gradients as low as 0.04. The analysis relates to rainfall conditions of the last nine years at arable sites in the base of dry valleys generally growing winter cereals. During this period of time, the maximum rainfall event to affect the area had a return period of at most 25 years; higher magnitude events might be expected to give rise to gullying in low gradient, otherwise stable, valley bottoms.

The amount and type of material moved by water in rills depends on the soil texture, rill depth and the transporting capacity of the water. Soil texture varies, but typically 60–80% of the fine fraction (<2 mm) is silt. Many soils on the Downs are stony and the stone content (>2 mm) varies from 21 to 81%, with a median value of 39% ( n = 23). On valley-side slopes, where frost-shattered chalk underlies the A horizon of the soil, incision of rills results in transport of chalk stones (Figure 2.6). Also, on thin soils, stones are incorporated into the A horizon by ploughing and even shallow rilling gives rise to their movement.

Deposition may occur on valley-side slopes due to

local changes in gradient. In valley bottoms and valley-side depressions, rill incision frequently encounters stony infills of flint and chalk. At these sites there is a tendency for lag deposits to form, with the winnowing out of fine grained components. Gravel size material is often moved short distances and valley-side footslope zones may be mantled with such deposits (Figure 2.6). The location of depositional sites is frequently controlled by field boundaries although silt, clay and organic matter may travel long distances along valley axes until effective ponding of water occurs. The typical depositional form is the fan displaying relatively good sorting (Figure 2.7). Some fans fine in a down-valley direction, others are composed entirely of either coarse (sand and gravel) or fine components of the soil. In some cases it is clear that coarse textured fans receive later inputs of fine material during the same storm event or later in the season. Fans are generally 5–15 cm thick and mixing of sorted sediments is brought about by subsequent cultivation. Soils appear to be thinning on slopes and thickening in valley bottoms.

Average rates of erosion may be calculated in various ways. The annual median rate on all eroded fields in a monitored area of 36 $km^2$ on the eastern South Downs, ranges from 0.5 to 5.0 $m^3$ $ha^{-1}$ (Boardman 1990). On individual fields rates range from very small values to just over 200 $m^3$ $ha^{-1}$ $yr^{-1}$. These figures relate to a nine-year observation period (1982–90). Measured rates of erosion are based on volumes of rills and gullies: the soil may be redeposited within, or be lost from, the field.

**Present and past erosion: some comparisons**

If we can understand the operation of processes that control erosion in the present landscape we can use the knowledge to interpret the evidence for erosion in the past. The greatest problem is the paucity of information with regard to processes in the past. Some of this information is available from archaeological studies, but often it is not possible to link sediments, agricultural practice and rainfall, as in studies of present day erosion. However, we shall attempt some tentative comparisons (Table 2.1).

In terms of field size, gradient, land use and mechanization, it is likely that modern farming systems greatly increase the risk of erosion, as against those of the past. Palaeoeconomic and environmental evidence suggests that until recently, mixed farming was the norm on the Downs (Bell, pers. comm.). Not only would this minimize erosion on grazed areas but it would also lessen the probability of runoff passing from one arable field to another. On areas of the Downs which at present have very high proportions of land under winter cereals this factor is important: a great deal of valley-bottom erosion (and flooding) results from the linking of erodible elements in the system (e.g. Boardman 1988).

There is little information about the earliest phase of erosion except that remnants of Bt horizons of woodland soils survive in subsoil hollows at many valley-bottom sites e.g. Kiln Coombe, Itford and Chalton (Bell 1983) and Asham (Ellis 1986). This type of evidence for the truncation of early Flandrian soils and substantial valley-bottom erosion occurs at many sites and presumably is to be associated with primary woodland clearance. The relatively stoneless loess cover would suffer extensive erosion where clearance and cultivation was taking place on valley-side slopes. Runoff from the slopes would be an essential element in this erosional episode. Rates of erosion on loessic soils at the present time are high (e.g. Bollinne 1978).

The evidence of surviving prehistoric field systems on the South Downs suggests that field size was small. For example, on Balmer Down (TQ 368110) fields were 1–2 ha in size and occupy valley-side locations. Typical slope lengths on the fields are *c.* 130 m and gradients *c.* 4°. In the winter of 1990–91 erosion was measured in the area of the field system on winter cereals (Figure 2.8). On several fields, where bounding grassy banks remain intact, erosion was minimal. Small spreads of organic-rich silt at downslope boundaries gave rates of soil movement of around 0.02 $m^3$ $ha^{-1}$ and most erosion was along wheelings (Fields 5–9, Figure 2.8). On the southern fields there was more substantial erosion: on fields 17, 18 and 19 the field boundaries were degraded and cultivation is carried out across them, wheelings were eroded and the effective length of slope was over 300 m. Rates of soil movement were 15.7, 11.1 and 21.3 $m^3$ $ha^{-1}$ respectively. At the southern end of field 18 and also field 15 runoff had concentrated and broken through the field boundary. Valley-side erosion, uninfluenced by former field boundaries, had occurred at site 14, where rates of soil movement were around 3.8 $m^3$ $ha^{-1}$.

Observations in one year of modern farming on and around former field systems suggest that rates of erosion are very low on small fields with intact boundaries. An increase of several orders of magnitude occurs if fields are enlarged and boundaries break down. The creation of field systems may have been a response to soils that were becoming thin and stony due to erosion (Bell 1983). On both small and enlarged fields erosion is concentrated on wheelings, except at sites of boundary break-down. Vehicle wheel tracks are a modern phenomena and rates of erosion would be lower in their absence.

Later erosion, associated with cultivation, affected already thin soils and the deposition of flinty lenses which would result from their erosion are not uncommon e.g. at Kiln Coombe, Chalton (Bell 1983; 1992) and Ashcombe (Allen 1984). The flint gravel accumulations are analogous to those produced by rilling and gullying processes at the present time (Figures 2.2, 2.3, 2.6 and 2.7). Plot experiments also have shown that movement of gravels implies rill and gully transport rather than interrill processes (Poesen 1987). Rates of erosion during these phases of secondary erosion are unlikely to have been higher than those at present given the differences in field morphology discussed previously.

The long history of erosion on the Downs means that in general soils have become stonier as the loessic

Figure 2.6 Severe erosion on fields recently drilled with winter cereals, Rottingdean, East Sussex, October 1987. Rills and gullies frequently cut through soil to chalk bedrock. Rill direction is controlled by wheelings and direction of maximum slope. Footslope and valley bottom are mantled with gravel fans.

Figure 2.7 Rills and gullies on grass ley at Bevendean, Brighton, November 1982. Fans of chalk and flint in valley bottom.

| | Present | Past |
|---|---|---|
| Field size | 5–30 ha | Often small e.g. 1–2 ha |
| Slope angle | Up to 25° | On field systems <5° |
| Land use | Emphasis on winter cereals | Mixed: arable, grazing and fallow(?) |
| Mechanization | Compaction: wheelings and rolling | No |
| | Fine tilth: power harrow | No |
| | Weed free | Weeds |
| Soils | Silty and stony | Silty |

Table 2.1 Some comparisons between present and past

cover in which early Flandrian soils developed was stripped; they have therefore become less erodible. In the present landscape there seems to be a threshold, perhaps around 60% stones, at which soils become even less erodible. Few fields have yet reached this level.

The view advanced here, is that modern farming systems, as exemplified by those of the 1980s on the Downs, optimize the risk of erosion and tend to maximize rates of erosion for this particular geomorphological environment. However, others have suggested that rates of erosion in the past may have been high. Evans (1990a, 1992) points out that erosion on fallow land would have been considerable. This would depend on surface roughness and weed cover: little is known about these aspects. On land which is left bare and ploughed, erosion occurs if there is sufficient rainfall to reduce the roughness (Papy and Boiffin 1989; Boardman 1991), but rates are up to an order of magnitude less than on bare smooth cereal fields (Boardman 1988). Under the Medieval Open Field system, the fallow field was available for grazing of animals on stubble and weeds which would protect it from erosion but, it was also ploughed (Harvey 1980, p. 45) and according to Vogt (1953) this led to serious erosion. Animals would create poached and compacted areas which cause erosion. The picture is therefore unclear. In the past, lower drilling densities and more footpaths and tracks along field edges would encourage erosion (Professor R.P.C. Morgan, pers. comm.). In this context, rates of erosion in the past may simply appear to be high because we are comparing long periods of time which are likely to have contained low-frequency, high-magnitude rainfall events, with short periods of modern time during which monitoring has been carried out.

**Discussion**

Many of the foregoing remarks assume that observations of the interaction between agriculture and erosion in the present landscape can be used as 'a key to the past'. This is partially true, the main limitation being lack of knowledge about the past.

Land use and farm practices of the past were substantially different and make extrapolation from the present difficult: autumn sowing of cereals was widely practised in the Iron Age and Romano-British periods but prior to that erosion had occurred on spring cereals, fallow and perhaps other crops, which at the present time give rise to little erosion on the Downs. There is also an assumption that the present climate is sufficiently similar to that of past periods and any differences can be ignored. This is a dangerous assumption. Relatively small shifts of climate in the Little Ice Age have been shown to have affected geomorphological processes considerably e.g. increased incidence of storms and floods in Norway and Scotland (Grove and Battage 1983; Whittington 1985). Recent work on the South Downs shows that a late winter rainfall peak produces a different pattern of erosion to that of other years (Boardman 1991).

In Germany, Bork (1989) has suggested that a phase of spectacular gullying in the early fourteenth century may have been produced by 'one single disastrous rainfall event'. Similar gullying is reported by Vogt (1953) in France. Curiously, there appear to be no descriptions of similar large scale gullies associated with agriculture during the past millennium in Britain, although it may be that they have gone unrecognised or undescribed.

Artifact distributions on slopes and in valley bottoms suggest short-distance transport within downland systems, at least during periods of occupation and cultivation. However, unconformities within the sediment sequence represent phases of major erosion and loss of soil from the downland system.

At many valley-bottom archaeological sites there is evidence in the form of an unconformity of

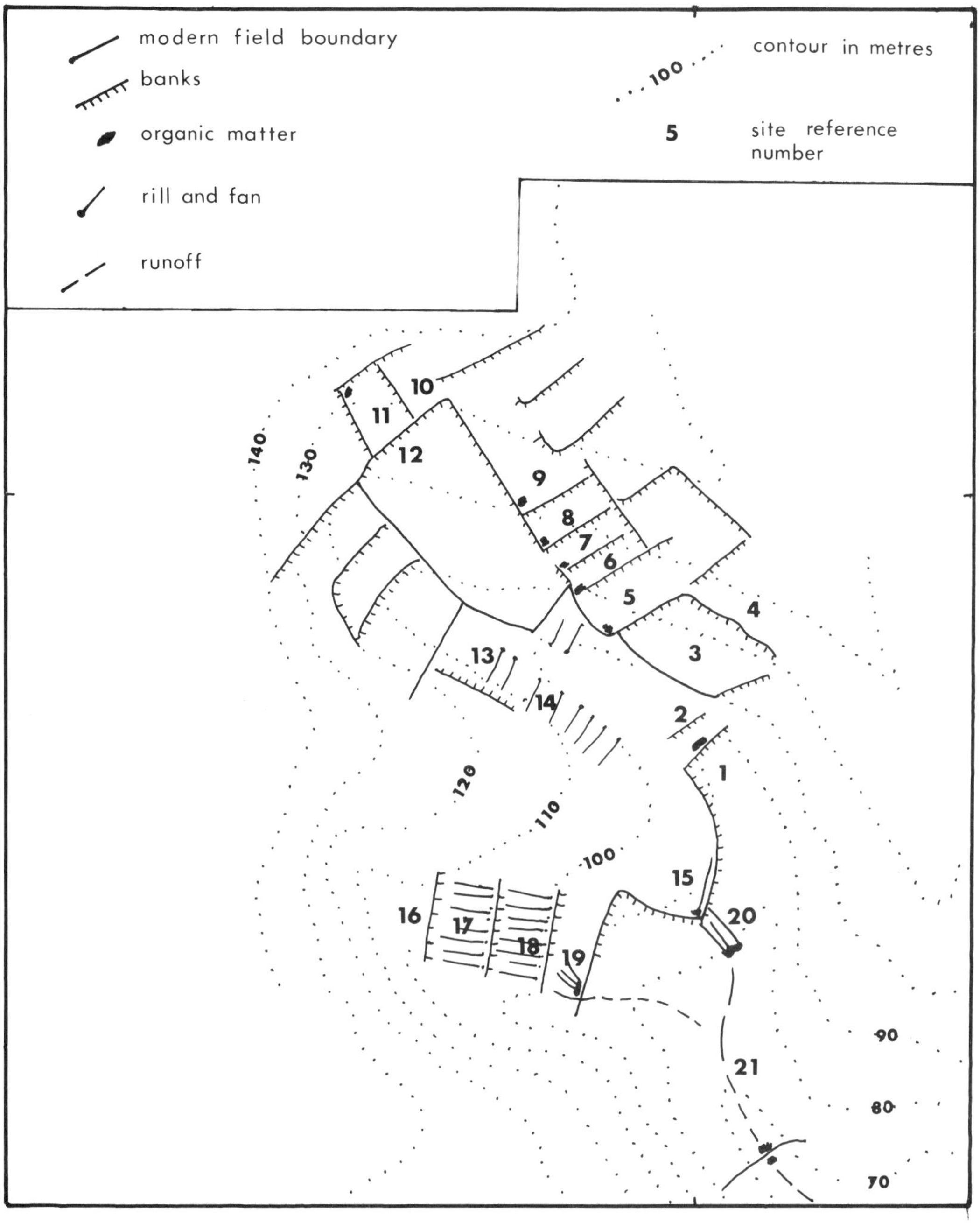


Figure 2.8 Erosion on and around a Celtic field system at Balmer, South Downs, January 1991.

substantial erosion which has removed early Flandrian sediments and soils e.g. sites in Bell (1992). There is also evidence for later erosion in one valley at Chalton, where there are no deposits from Neolithic to Medieval times. Bell suggests that these have been removed by a seasonal stream which is known to have existed at the site. At some sites such streams may have been active in the early Flandrian but not later.

The inadequacy of the modern landscape as an analogue for the past is demonstrated by the problem of valley-bottom gullying. Most valley bottoms with gradients of less than 0.04 m $m^{-1}$ are not at risk of gullying under present conditions (Figure 2.5), but there is evidence of gullying in the past at some such sites e.g. at Ashcombe with a gradient of 0.02 m $m^{-1}$ (Allen 1984). There are several possible explanations for this anomaly: rainfall in the 1980s may not have been great enough; water tables have been lowered and soils are more frequently dry; valley-bottom gradients have been smoothed by sedimentation. Soils have also become stonier at these sites.

It is difficult to envisage conditions under which sufficient runoff was generated in valley-bottom locations during primary woodland clearances to produce substantial erosion and truncation of the soil profile. Indeed, there is little that can be said about the relationship of primary woodland clearance to erosion on the South Downs since on many of the sites studied so far the evidence has been removed by later erosion (Bell 1992). Many of our suppositions are, in fact, based on other models, for example from the impact of European agriculture on the eastern USA (e.g. Wolman 1967).

Finally, reconstruction of past landscapes and geomorphological processes based on what remains is unsatisfactory. Field systems such as that at Balmer evolved through time and statements about their susceptibility to erosion have to be viewed in that context.

**Acknowledgement**

I thank Drs R. Evans, M. Bell and M. A. Fullen for many useful comments on a draft of this paper and John Holloway for Figure 2.2.

**References**

Allen, M. J. 1984 *Excavations at Ashcombe Bottom: first interim report August 1984.* Lewes Archaeological Group.

Auzet, A. V., Boiffin, J., Papy, F., Maucorps, J. and Ouvry, J. F. 1990 An approach to the assessment of erosion forms and erosion risk on agricultural land in the northern Paris Basin, France. In: Boardman, J., Foster, I. D. L. and Dearing, J. A. (eds.) *Soil Erosion on Agricultural Land.* John Wiley, Chichester, 383–400.

Bell, M. G. 1983 Valley sediments as evidence of prehistoric land-use on the South Downs. *Proceedings of the Prehistoric Society* 49, 118–150.

Bell, M. G. 1992 The prehistory of soil erosion. In: Bell, M. and Boardman, J. (eds.) *Past and Present Soil Erosion.* Oxbow, Oxford, 21–35.

Boardman, J. 1988 Severe erosion on agricultural land in East Sussex, UK, October 1987. *Soil Technology* 1, 333–348.

Boardman, J. 1990 Soil erosion on the South Downs: a review. In: Boardman, J., Foster, I. D. L. and Dearing, J. A. (eds.) *Soil Erosion on Agricultural Land.* John Wiley, Chichester, 87–105.

Boardman, J. 1991 Land use, rainfall and erosion risk on the South Downs. *Soil Use and Management* 7(1), 34–38.

Boardman, J. 1992 The sensitivity of downland arable land to erosion by water. In: Thomas, D. S. G. and Allison, R. J. (eds.) *Environmental Sensitivity.* John Wiley, Chichester, in press.

Boardman, J. and Robinson, D. A. 1985 Soil erosion, climatic vagary and agricultural change on the Downs around Lewes and Brighton, autumn 1982. *Applied Geography* 5, 243–258.

Bollinne, A. 1978 Study of the importance of splash and wash on cultivated loamy soils of Hesbaye (Belgium). *Earth Surface Processes* 3, 71–84.

Bork, H-R. 1989 Soil erosion during the past millennium in Central Europe and its significance within the geomorphodynamics of the Holocene. *Catena Supplement* 15, 121–131.

Catt, J. A. 1978 The contribution of loess to soils in lowland Britain. In: Limbrey, S. and Evans, J. G. (eds.) *The Effect of Man on the Landscape: the Lowland Zone. Council for British Archaeology Research Report* 21, 12–20.

De Ploey, J. 1990 Threshold conditions for thalweg gullying with special reference to loess areas. *Catena Supplement* 17, 147–151.

Ellis, C. 1986 The postglacial molluscan succession of the South Downs dry valleys. In: Sieveking, G. de G. and Hart, M. B. (eds.) *The Scientific Study of Flint and Chert.* Cambridge University Press, Cambridge, 175–184.

Evans, R. 1980 Characteristics of water-eroded fields in lowland England. In: De Boodt, M. and Gabriels, D. (eds.) *Assessment of Erosion.* John Wiley, Chichester, 77–87.

Evans, R. 1990a Soil erosion: its impact on the English and Welsh landscape since woodland clearance. In: Boardman, J., Foster, I. D. L. and Dearing, J. A. (eds.) *Soil Erosion on Agricultural Land.* John Wiley, Chichester, 231–254.

Evans, R. 1990b Water erosion in British farmers' fields - some causes, impacts, predictions. *Progress in Physical Geography* 14(2), 197–219.

Evans, R. 1992 The present: the key to the past. In: Bell, M. and Boardman, J. (eds.) *Past and Present Soil Erosion.* Oxbow, Oxford, 53–66.

Foster, G. R. 1990 Process-based modelling of soil erosion by water on agricultural land. In: Boardman, J., Foster, I. D. L. and Dearing, J. A. (eds.) *Soil Erosion on Agricultural Land.* John Wiley, Chichester, 420–445.

Grove, J. M. and Battagel, A. 1983 Tax records from western Norway, as an index of Little Ice Age environmental and economic deterioration.

*Climatic Change* 5, 265–282.
Harris, T. M. and Boardman, J. 1990 A rule-based Expert System approach to predicting waterborne soil erosion. In: Boardman, J., Foster, I. D. L. and Dearing, J. A. (eds.) *Soil Erosion on Agricultural Land.* John Wiley, Chichester, 401–412.
Harvey, N. 1980 *The Industrial Archaeology of Farming in England and Wales.* Batsford, London.
Jarvis, M. G., Allen, R. H., Fordham, S. J., Hazelden, J., Moffat, A. J. and Sturdy, R. G. 1984 Soils and their use in South East England. *Soil Survey of England and Wales Bulletin* 15, Harpenden.
Papy, F. and Boiffin, J. 1989 The use of farming systems for the control of runoff and erosion. *Soil Technology* 1, 29–38.
Patton, P. C. and Schumm, S. A. 1975 Gully erosion, northwestern Colorado: a threshold phenomena. *Geology* 3, 88–90.
Poesen, J. 1987 Transport of rock fragments by rill flow – a field study. *Catena Supplement* 8, 35–54.
Potts, A. S. and Browne, T. E. 1983 The climate of Sussex. In: Geographical Editorial Committee (ed.) *Sussex: Environment, Landscape and Society.* Alan Sutton, Gloucester, 88–108.
Stammers, R. L. and Boardman, J. 1984 Soil erosion and flooding on downland areas. *The Surveyor* 164, 8–11.
Vogt, J. 1953 Erosion des sols et techniques de culture en climat tempéré maritime de transition (France et Allemagne). *Revue Geomorphologie Dynamique* 4, 157–183.
Whittington, G. 1985 The Little Ice Age and Scotland's weather. *Scottish Geographical Magazine* 101(3), 174–178.
Wolman, M. G. 1967 A cycle of sedimentation and erosion in urban river channels. *Geografiska Annaler* 49A, 385–395.

# CHAPTER 3

# THE PREHISTORY OF SOIL EROSION

MARTIN BELL

*The relationship between soil erosion and prehistoric communities is considered initially from the point of view of perception and its relationship to timescales. These issues are then considered in the more specific context of research on lynchets and dry valley fills in chalk and limestone areas. Comparisons are made between individual prehistoric sediment increments and those laid down during recent erosion events. There are problems in such comparisons, in terms of the underlying uniformitarian assumptions, the difficulty of identifying the processes involved, and of evaluating the effects of erosion on people. An outline is given of evidence for the extent of erosion on the chalk. Quantification using surviving eroded soil must often represent a minimum figure in view of evidence from some valleys for the total removal of climax woodland soils. A case study on limestone at Brean Down, Somerset establishes the depth of soil eroded and its relationship to the history of human activity.*

## Introduction

Viewed against the background of human existence studies of present day soil erosion might be seen as chance snapshots of episodic processes at particular points in time. Contemporary observations and instrumental records can be complemented by the greater time depth provided by historical, archaeological and geological records. These studies also enable us to consider the relationship between soil erosion and a wider range of societies and forms of land-use than is observable in the contemporary world.

Soil erosion is the result of natural processes such as gravity, wind and water. These processes are particularly active when a protective vegetation cover is removed. In some parts of the world this comes about as the result of climatic change, which for example can be seen as a major influence on the history of erosion and river valley alluviation in parts of the American South West and prairies before the arrival of European agriculture (Butzer 1977; Knox 1984). Vegetation cover may also be affected by a wide range of other factors such as the activities of wild animals, storm events, disease, etc. In North West Europe, however, it is clear from the palaeoenvironmental record (Bell and Walker 1992) that human activity has been the most important factor in transforming environments by reducing natural vegetation cover and increasing erosion, at least over the last five millennia. These activities, particularly the creation of arable land, have increased sediment supply with the result that the effects of secular climatic changes and individual storm events are more likely to register in the soil erosion record. Over this time period natural ecosystems have been replaced by cultural landscapes. These are to some extent the cumulative, in a sense accidental, products of long histories of human activity, but also the result of more deliberate social transformation of landscape. It follows that our contemporary environment is the product, not only of natural factors, but the history of their interrelationship with human activity. Understanding of contemporary processes will be enhanced by a long time perspective and by integration of natural and cultural factors.

## Perception

Archaeologists have made significant contributions to the documentation of past erosion and soil history (eg Limbrey 1975) but there has been less consideration of the difficult issue of the human consequences of these changes. When this has been discussed, the stance has often been deterministic and has tended to implicate erosion in social upheaval and economic collapse, particularly in relation to Classical Mediterranean civilisation. The difficulties of thinking about past erosion perception and response are exacerbated by the remarkable lack of hard information about how farming communities, even today, perceive erosion (Boardman 1990b).

When considering the past we cannot simply reconstruct the environment and then infer from this the human consequences of change. People were not responding to the actual environment which we may be able to reconstruct with more-or-less accuracy from the palaeoenvironmental evidence. Their response was to the *perceived* environment, which reflected that community's lived world of experience. This has a temporal dimension, the period over which they have relevant environmental knowledge (see next section). It also has a spatial dimension, the area from which a particular group received environmental knowledge as a result of their own direct observations and information gained from other groups. That knowledge constitutes the group's 'behavioural environment', ie that part of their perceived environment which influences individual behaviour and decision making.

Environmental hazards, such as soil erosion, are

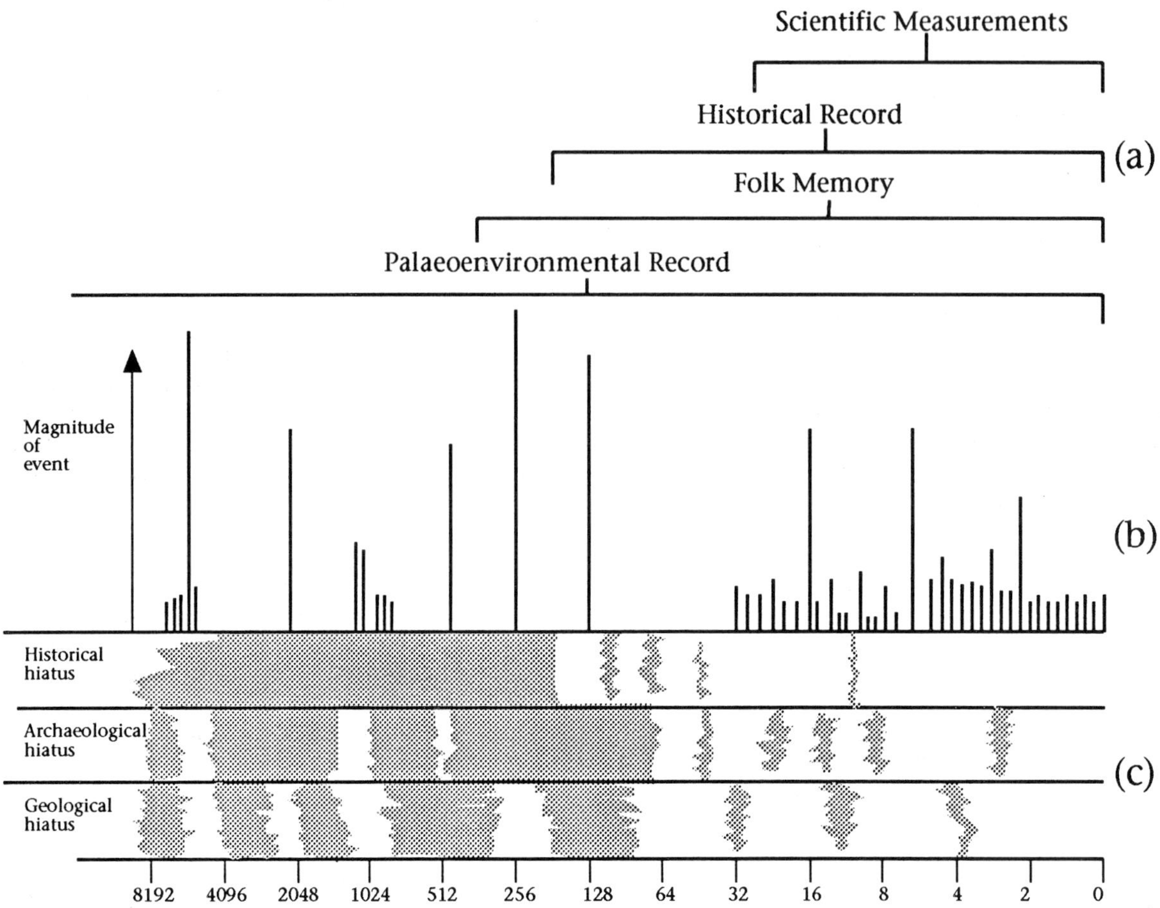


Figure 3.1 Hypothetical model of the record of environmental phenomena against time showing (a) the various sources of evidence available, (b) the hypothetical magnitude of events recorded, (c) hypothetical periods when there are gaps in the various data sets

necessarily defined in relation to a subject: that which the hazard affects. This raises the issue of how we investigate past hazard perception in the archaeological record. We can develop theoretical frameworks which will be informed by present-day hazards research (eg Burton *et al.* 1978; Hewitt 1983), by the ethnohistorical record and by social theory. Ultimately, however, the validity of our theories about past people/environment relationships can only be evaluated by rigorous scientific testing against the evidence for a group's behavioural response to particular environmental scenarios and against clues to the group's environmental perception which are embedded in their art, monuments, etc. Such an approach, which aims to integrate scientific study of past societies and environments with an awareness of social and perceptual issues, has recently been described by Renfrew and Bahn (1991) as 'cognitive-processual archaeology'.

It must be acknowledged that in many cases the archaeological record is not of sufficient quality or precision to enable issues of perception and response to be addressed in any detail. This emphasises the need to concentrate studies of people's relationship to environmental change in areas such as the wetlands (Coles and Coles 1989) and lakes, where evidence of appropriate quality may be preserved. In the case of soil erosion, particular interest attaches to the interfaces of these areas with arable land.

**Timescales**

Human perception of environmental change and decisions to adapt depend partly on issues of timescale, in particular whether the change is sufficiently rapid to be perceived and on the expectation of its recurrence. One difficulty is that for various periods of the past we have data of differing types and levels of precision and there is a need, therefore, to consider the problems of interpreting and integrating these varied data sets. Figure 3.1 is a hypothetical model of the record of environmental change against time. In soil erosion studies the period of detailed scientific record is mostly very short; for instance about 10-15 years on the South Downs. On its own this is too short a period for reliable estimates of the frequency of stochastic phenomena.

Before this we have some historical records of particular events, but inevitably these are likely to be the more extreme events and those for which an aware observer happened to be present and had the motivation to record what was seen. Historical records of flooding and erosion on the English chalk include the Till flood in Wiltshire in 1841 (Cross 1967), the Louth flood of 1920 (Latter 1932) and the Chiltern flood (Oakley 1945). One suspects that a wealth of other information is hidden in the pages of local papers and histories. They are worth discovering because of the insights they provide into past perception of environmental change. The Till flood left 3 dead, 200 people homeless and was followed by a remarkable charitable response. A local vicar, in a poem sold to aid the distressed, provides a revealing glimpse of early Victorian environmental perception by drawing a moral lesson and hinting at retribution for the 'wasted hours and squandered days' of those affected.

In the period before relevant historical accounts folk memory may preserve knowledge of environmental events. Studies of these oral traditions are most important in those parts of the world, such as the Americas and Australia, which have only recently adopted literacy. In these areas comparison of the palaeoenvironmental and ethnohistorical record demonstrates that traditions of major environmental events may be preserved for at least 300-400 years (Bell and Walker 1992). An example relevant to the soil erosion theme is the folk memory of the engulfing of a native American site at Ozette, Washington around AD1750 prior to European contact (Coles and Coles 1989; Gleeson and Grosso 1976). Folk memory, as with historical sources, may only record the most extreme events.

For the great bulk of human existence we have no instrumental scientific or historical records, or folk memory of environmental changes. For these longer term perspectives we have to rely on archaeological and geological information. This varies in its quality and precision both temporally and contextually. Some contexts such as lakes and peat bogs provide long and temporally precise records of environmental phenomena such as soil erosion. By contrast, the evidence for early erosion derived from field boundaries (eg lynchets and terraces) and valley fills often has hiatuses. In general the prehistoric record tends to be characterised by particular sites and phases for which there is good resolution, events of various magnitudes being recorded. During other phases only major events may leave a trace and during some no evidence will survive.

One way in which it may be possible to achieve a more integrated scientific and social perspective regarding long-term soil erosion is to concentrate on those parts of the archaeological record which are well preserved and also to make sure that we give adequate consideration to the relatively recent time periods for which we have overlapping (Figure 3.1) and partly independent sources of evidence.

**Erosion on calcareous strata**

Having noted some of the philosophical issues pertinent to the archaeological study of erosion and to the theme of this volume as a whole, the second part of this paper is concerned with the comparison of past and present evidence. This discussion concentrates, though by no means exclusively, on the chalk landscape of the South Downs.

Chalk and limestone country is characterised by a valley morphology which was largely established during the Pleistocene and, due to the permeable nature of the bedrock, many valleys do not carry streams today. The rarity of permanent watercourses means that, by comparison with other bedrocks, the products of erosion are more likely to be preserved in dry valleys and along the boundaries of fields as lynchets.

*Lynchets*

These linear features represent the boundaries of ancient fields (Figure 3.2). They may be negative, at the upper edge of a field where erosion has taken place, or positive at the lower edge where soil has accumulated. Individual boundaries separating two fields frequently have both positive and negative elements. On the chalk Downs the fields formed by lynchets are of two basic types. There are 'Celtic' fields, brickwork-like patterns of small rectangular fields, generally dating between the middle Bronze Age and the Romano-British period (see Boardman 1992, Figure 2.8). Later there are strip lynchets which define long linear fields in step-like terraces often on steep hillsides; these date to the Medieval period and are an adaptation of the open field system to steep slopes (Taylor 1966). Deposits forming lynchets often contain large numbers of artifacts, particularly pottery which can be dated by pottery fabric comparisons with assemblages on nearby stratified sites. The assumption is that many of the artifacts derive from the practice of mucking out domestic and farmyard middens onto the fields, although some must also derive from the erosion of settlement horizons.

Lynchet formation remains an imperfectly understood process. Unvegetated soil in arable fields is subject to a range of slope processes. Soil disturbance on a slope, whether wetting, drying, freezing, thawing, rainsplash, animal disturbance or tillage will generally give rise to a greater component of downslope than upslope movement because of gravity. Observation of present-day erosion (see below) indicates that small-scale rill erosion during high rainfall is perhaps the most important factor.

The extent to which cultivation implements and techniques themselves contributed to erosion remains unclear. In prehistory the main cultivation implement in Britain and Northern Europe was a light ard which scratched the soil, rather than inverting it. Even so, wear on ard tips and the asymmetry of ard marks shows that these implements could have been tilted so that more soil was pushed to one side than the other (Fowler 1983, p177). This raises the issues of whether gravity was habitually used to turn the furrow downhill, or conversely whether these light ploughs were capable of turning soil upslope in order to counteract the effects of erosion. The discovery on several sites, most notably Gwithian, Cornwall (Megaw *et al.* 1961), of ard marks running in two directions at right angles suggests that cross-ploughing was the norm in prehistory. One set of furrows would often have been roughly at right angles to the contours and could have channelled water, thus exacerbating erosion. Various of the issues raised in this paragraph could be further addressed using experimental archaeology.

Figure 3.2 Excavated lynchet at Bishopstone, East Sussex. Note the presence of a ditch and two postholes which may represent early boundaries along which lynchet accumulation began (Photo. Brenda Westley).

Lynchets have an obvious similarity to terraces which, in many parts of the world, are designed to reduce erosion. Archaeologists have not, however, addressed the problem of the extent to which they represent a response to a perceived problem. Observations at Balmer (Boardman 1992, p.14) show that lynchet systems of Romano-British date are still remarkably effective in reducing erosion by comparison with that on adjacent slopes without lynchets. Even so artifact distributions in excavated lynchets (e.g. Bell 1977; Fowler and Evans 1967) do show that they built up over long periods, with no evidence of hand digging to create terraces. There is, however, evidence that people created barriers which would have reduced soil movement downslope. These take the form of low boulder walls at Overton and Fyfield in Wiltshire (Fowler and Evans 1967), and fences at Bishopstone (Bell 1977) and Bullock Down, Sussex (Drewett 1982, Figure IX) It has also been noted that some exceptionally flinty lynchets could represent attempts to combine stone clearance with the need to trap eroding soil (Bell 1981). Until earlier this century stone picking was a traditional activity in downland agricultural communities. The problem is that some valley centre lynchets which are flinty, such as Itford Bottom (Bell 1983, Figure 10) might represent sediments from which fines has been removed. However, John Boardman (pers. comm.) points out that gravel transport and the removal of fines is unlikely to have taken place on the short and interrupted slopes represented by 'Celtic' fields. It may prove possible to resolve some of the problems associated with the origins of these flinty lynchets analytically (see future analyses below).

A number of excavated lynchets produced no sign of formation against a humanly created barrier: presumably those fields were simply separated by a grassy bank. In fact experimental cultivation at Butser Iron Age farm demonstrated lynchet formation on a 12 degree slope, and showed that an important factor in trapping soil was lusher vegetation growth along the lower field boundary (Fowler 1983, Plate VI). We should also consider the possibility that lynchet formation may have been encouraged by the deposition of weeds and compost on the baulks between fields. Attractive as this idea is, it is not supported by mollusc evidence which suggests rather dry grassy banks (Fowler and Evans 1967; Thomas 1977; Bell 1983, Figure 15).

The very substantial step-like nature of some Medieval strip lynchets, does suggest that the heavier ploughs of this period were deliberately employed to gradually erode soil from the upper part of a field in order to encourage the development of a flatter terrace for cultivation.

The period over which some lynchets formed suggests long-term arable activity. Had there been long pastoral hiatuses one would expect that these would have produced sufficient earthworm sorting to have left a visible trace; they should similarly be detectable in the molluscan sequences. Prehistoric farmers knew ways of maintaining soil structure and fertility. From the middle Bronze Age, when clearly defined field systems become widespread, the abundance of artifacts in lynchets and valley sediments indicates manuring. On coastal sites there is evidence that the fertilising properties of seaweed were exploited from as early as the Bronze Age at Gwithian in Cornwall and from the Iron Age at Bishopstone, Sussex (Bell 1981b). Liming is attested both in a literary reference by Pliny (Bowen 1961) and by the discovery of Romano-British marl pits dug into the boundaries of 'Celtic' fields on Bullock Down, Sussex (Drewett 1982). Liming would have become increasingly necessary in this area as overlying loess was eroded away, exposing Tertiary-derived superficial deposits on the ridge tops. It is reasonable to infer that these activities represent a human response to a perceived problem.

*Dry valleys*

Several dry valleys on the South Downs have been investigated archaeologically (Bell 1983; Allen 1988, 1991) and contain several metres of Holocene sediment. These are very similar in appearance to the sediments forming lynchets and that similarity extends to the molluscan assemblages and the abundance of artifacts which they contain. Mollusca, confirmed on some sites by micromorphology (Macphail 1992, p.197), show that these sediments accumulated after human clearance of climax woodland, generally in an open, predominantly arable agricultural environment. It is instructive to compare these sediments to those derived from modern erosion events in the same area and to consider some of the problems inherent in such comparisons.

*Comparison of past and present*

We know the past through our lived experience of the present. Uniformitarian reasoning lies at the heart of palaeoenvironmental studies. Today, however, it is a methodological uniformitarianism which acknowledges that some environmental changes are both sudden and episodic and may not therefore be observable on a particular day, or even during the lifetime of an individual. We must also acknowledge that past environmental contexts may not have a precise modern analogue. This may particularly be the case with cultural landscapes, in which biological communities and humanly created landforms may reflect extinct patterns of activity and land-use. Despite these problems observation of contemporary erosion must play a key role in the study of earlier deposits.

Boardman's (1992, p.10) investigations show that most erosion on the South Downs today occurs during specific rainfall events during which runoff forms rills. It seems logical to envisage that in prehistory, other things being equal, the individual erosion increments from rainfall of a given magnitude may have been significantly smaller than those of today. Fields were much smaller and separated by lynchets and organic matter content and weed flora is likely to have been greater. These and other contrasts between past and present arable are further discussed by Boardman (1992, p.14). The artifactual evidence certainly indicates that prehistoric increments were small because the deposits accumulated over long periods. Evidence of sorting, which would be expected if water action had played a key role in deposition (Boardman 1992, p.16), may have been obliterated by subsequent tillage, even with a scratch plough, if the individual erosion increments were mostly small.

Sometimes, however, individual erosion increments can be identified. Within the valley sediment at Kiln Combe there were two 5–10 cm thick bands containing small rounded chalk pieces (Figure 3.3). These were clearly recognisable because they ran from calcareous deposits on the valley side over partly decalcified soils in the valley centre. One band sealed a soil with Beaker artifacts and a radiocarbon date of 3630±90 BP (Har-5469). The other band was within a soil containing Iron Age pottery. Both bands are very similar to the fans of chalk pieces which are today being laid down at the base of slopes where rills and gullies discharge onto the valley floor ( Boardman 1992, Figures 2.2, 2.6 and 2.7). This modern analogue implies prehistoric rilling on the 10 degree southern slope of Kiln Combe greater than the depth of the present day soils (15–20 cm), indeed during the periods in question the soils were probably up to twice their present depth.

A comparable lens of chalk granules 1–2 cm thick (Figure 3.4) and dating to the Post-Medieval period in the Bascombe Valley, Chalton (Bell 1983, p144, Layer 8) can be similarly explained by modern analogues. In January 1988 John Budden photographed comparable deposits derived from rilling of an arable field within the same valley (Figure 3.5). It is instructive to note that an earlier explanation of Bascombe, Layer 8 (Bell 1981a, p301) suggested that it might derive from a pastoral environment. The basis of this argument was that, in the days before the myxomatosis outbreak of 1953, rabbits are recorded as covering some slopes with chalk scree (Tansley 1949, p51; Williamson 1978, p122). Rabbits were a Norman introduction, the population of which does not seem to have exploded until the Post-Medieval period (Sheail 1971). Although the specific example of rabbits is irrelevant in terms of prehistoric processes it makes a more widely applicable point by illustrating how one ecological factor, such as grazing pressure, even by a small wild animal, can render inappropriate analogues derived from periods when different conditions obtained.

Fans deposited by rilling now seem a much more plausible explanation for the chalky bands in Kiln Combe and Bascombe, but their survival implies that deposition was succeeded by a long stable, probably grassland, period during which earthworm activity (Darwin 1881) buried the bands to such a depth that they were not mixed by later

Figure 3.3 Dry valley sediments at Kiln Combe, East Sussex: (a) basal buried soil; (b and c) layers of small chalk pieces probably derived from rills; (d) flinty deposit of Romano-British date; (e) chalky Medieval colluvium (Photo. Brenda Westley).

tillage; this would probably have taken one to two decades (Atkinson 1957). This could have occurred if erosion took place from valley side arable onto grassland on the valley centre, alternatively cultivation of both slope and valley floor might have ceased following the erosion, perhaps even as a consequence of it. At Kiln Combe it did not, however, cease immediately since probable ard marks within the chalky band indicate one or more acts of tillage following its deposition.

In view of the inference of prehistoric gullying it is noteworthy that archaeologists have not identified erosion gullies in the large expanses of chalk uncovered during excavations. Admittedly the excavated settlements do tend to be on the gently sloping hilltops and spurs rather than on the concave lower slopes where gullying is today particularly concentrated. It may also be that archaeologists have not really considered the possibility of natural gullying when interpreting excavated features. With hindsight the writer is tempted to suggest that an irregular gully uncovered in his excavation at Bishopstone, Sussex and originally interpreted as humanly dug (Bell 1977, p143, gully 885) may in fact be an erosion feature.

Sections across dry valleys (Bell 1983) clearly show that flint size and density increases from the valley margins towards its centre. At Itford Bottom the extent of this flinty deposit is indicated by traces of former flint diggings down the valley axis (Bell 1983, Figure 9). A flinty valley centre is also seen in cliff section at Saltdean (Bell 1986). A particularly flinty deposit at Kiln Combe was up to 90 cm thick (Figure 3.3) and very flinty deposits were recorded south of Butser Hill, Hampshire during the widening of the A3 in 1968 (Bell 1981, Plate XXXVII). In this latter case a temporary stream, for which the local name is a lavant, occasionally broke out in this part of the valley until the turn of the century (J. Budden, pers. comm.). The likelihood is that the lavant has washed away the fine sediment leaving a clast-supported flinty layer. Lower down the same valley, in Bascombe, the lavant is believed to have been responsible for the flushing out of sediments resulting in an hiatus between the Neolithic/Bronze Age and the Medieval periods (Figure 3.4). Historical evidence and Anglo-Saxon charters indicate that seasonal streams were more widespread and rose more frequently until water-tables were lowered by the extensive pumping of the chalk aquifer which began in the nineteenth century (Bell 1981a, p58).

Despite a similarity to the deposits at Chalton it is not, however, thought that streams produced by seasonally higher water-tables can account for most of the very flinty valley fills previously noted. In some cases they are in the upper parts of minor valleys, well above likely water-tables in prehistory. For instance the Kiln Combe valley is 1 km inland from the precipitous cliffs of Beachy Head. Here we are fortunate in having a study of the local water-table before massive twentieth century pumping (Whitley 1900). This suggests a water-table around 50 m below the excavated trench and rising only gradually away from the cliffs. A prehistoric water-table similar to that of the late nineteenth century is also suggested by the probable Bronze Age or Iron Age well which was exposed in the cliff face in 1971 at Belle Tout one and three-quarter km to the west: this was 43 m deep and went below the present wave cut platform (Bradley 1974; Drewett 1982).

A more appropriate modern analogue than seasonal streams for the origins of flinty valley fills is suggested by observation of erosion events, such as at Bevendean (Boardman and Robinson 1985) and Rottingdean (Boardman 1989). Both rainfall events produced so much runoff that the floors of these normally dry valleys carried torrents of water which carried away the finer sediment leaving a flinty residue (Boardman 1992, Figures 2.2 and 2.3). Rainfalls of sufficient magnitude to produce such dramatic events are of infrequent recurrence interval. The Bevendean event, for instance, can be calculated to have a 25 year recurrence interval in terms of the recorded levels of autumn (October to December) rainfall (Boardman and Robinson 1985). More precise calculation is impossible due to the importance of rainfall intensity, for which information is only available for the very recent past. For intense rainfall to have led to dramatic erosion implies a coincidence with particular soil conditions. The comparable events studied by Boardman occurred when a large part of the valley was under arable cultivation. This is a plausible scenario in the case of the very flinty deposit at Kiln Combe, which dates to the Romano-British period, when much of the surrounding area was covered by field systems with several settlements nearby (Drewett 1982). The practice of autumn sowing has also been shown by Boardman (1990a) to be frequently associated with erosion events. There is growing evidence from weed assemblages that autumn sowing of spelt was widely practised in the Iron Age and Romano-British periods (Jones 1981; Reynolds 1981).

*Future analyses*

Comparisons between past and present erosion on the South Downs have so far been based principally on field observations and uniformitarian assumptions which between them suggest hypotheses which now need to be tested analytically. Analyses are in particular needed to quantify the degree of sorting within deposits. Studies of the orientation of coarse particles in the manner of till fabric analysis could be profitable (Lowe and Walker 1984). Clear evidence of preferred orientation will be more easily obtained where the clasts are tabular, as on some limestones, than from work on rounded and irregular flint nodules (J. Boardman pers. comm.). Some flints are also tabular and pottery sherds might also prove suitable for orientation studies. Such investigations might clarify the main direction of movement,

Figure 3.4 Valley sediments at Bascombe, Chalton, Hampshire: (a) clay and flints with Neolithic/Bronze Age artifacts; (b) stone-free horizon with Romano-British and Medieval artifacts; (c) stone horizon, Medieval; (d) stone-free horizon; (e) layer of partly rotted chalk granules; (f) present topsoil (Photo. Brenda Westley).

Figure 3.5 Erosion on Church Down, Chalton, Hampshire in early January 1988 (Photo. John Budden).

whether from the valley side, as at the end of a gully or rill, or as a result of water movement along the main valley axis. This would help to identify the origins of stony deposits and distinguish those associated with running water from those derived from the valley sides by gravity or dumped by human agency, for example as part of erosion control measures. Lynchet sediments, for so long neglected, particularly need detailed analytical comparison both with the products of current erosion events and with earlier dry valley sediments, in order to test the field-derived inference that the two are essentially similar. Such analytical work would help to clarify whether there is a degree of deliberate construction in some lynchets and identify the origins of the very stony examples. It should also be noted that many of these problems could profitably be addressed using micromorphology (Macphail 1992, p.197).

*The extent of erosion on the Chalk*

This subject is relevant, not only to our understanding of long-term erosion rates but also because of its implications for the soils which would have been available to prehistoric communities. The writer has previously attempted to quantify the amount of erosion in nine valleys on the basis of the surviving sediments (Bell 1986). This indicated the loss of between 2.5 and 24 cm of soil and suggested that in some areas the Bronze Age soils were about twice their present depth. The method used was imprecise, as noted by Quine and Walling (1992), in particular because it quantifies on the basis of cross sections rather than spatial mapping. More precise measurements of erosion from limestone at Brean Down are given below. Even so, the original crude estimates of erosion in the chalk valleys are of the same order of magnitude as the results of Evans' (1990) much more extensive soil-map based studies, which indicate that for most of the mapped areas of England surviving colluvium accounts for erosion of between 8 and 14 cm of soil. The problem with any calculations based on surviving eroded soil is that it is difficult to quantify losses to the study area, which in some cases may have been considerable.

A particular problem concerns the fate of the original woodland soils which we know from molluscan and pollen studies once covered the chalk (Evans 1972; Waton 1982). The basal fills of some valleys do contain woodland Mollusca, eg Brook, Kent (Burleigh and Kerney 1982), Asham and the Devil's Dyke, East Sussex (Ellis 1986), although it is not always clear to what extent these are *in situ* or transported. However, in the valleys investigated by the writer, woodland Mollusca only survive in subsoil hollows, which on the basis of modern analogues can confidently be interpreted as tree-throw pits. The extent to which these features survive also varies topographically, for instance very well preserved examples were found on the valley floor in Itford Bottom but on either side of the valley there were only shallow, probably much truncated, examples such as were also found on the valley floors at Kiln Combe and Chalton. Excavation of the valley sediments at Coombe Bottom, Hambledon Hill, Dorset (Bell and Allen forthcoming) revealed no trace of the former woodland soil or of subsoil hollows (Figure 3.6). Here late Devensian chalky meltwater deposits are overlain directly by soil eroded about the Iron Age. A similar lack of subsoil hollows and evidence for former woodland was found during the investigation of valley sediments at Loscombe Wood, Dorset (Bell *et al.* 1990). Truncated subsoil hollows are found during hilltop excavations at the base of shallow rendsina soils, for instance at Bishopstone (Bell 1977, Figure 3).

The implication is that in some parts of the chalk the whole of the woodland soil, perhaps 50 cm or more thick, and much of the sediment derived from its erosion, which would have contained a woodland mollusc assemblage, has been totally eroded and removed from the site. One caveat is that some valleys, such as Chalton and Kiln Combe, have basal fills which are partly decalcified and do not preserve molluscs. In those situations it is possible that the Mollusca have been lost by decalcification rather than erosion. Despite this consideration there is persuasive evidence in some areas for an early erosion phase of woodland soils, which is often not represented by surviving sediments. This has archaeological implications, because it implies hiatuses during which evidence of human activity may not be preserved. This may explain the absence of evidence for Neolithic activity in the valleys investigated by the writer and Allen (1988; 1991). On the other hand the extent of Neolithic arable agriculture in Britain is questionable as Thomas (1991) has recently shown, and there is evidence from various palaeoenvironmental sources (eg Thomas 1982; Thorley 1981) that much of the South Downs, unlike parts of the Wessex chalk, was wooded until the Bronze Age.

This raises the issue of what has happened to the missing soil. Some must have been deposited elsewhere in the same valleys. Fine sediment, such as the formerly more extensive loess, may have been removed by short-lived floods during rainfall events and by seasonal streams which were more extensive in the past, although little of the soil eroded on the South Downs today reaches rivers (Boardman 1990a). The contribution of erosion from arable land in providing sediment for reworking by streams is illustrated by excavations at Everley Water Meadows beside Hambledon Hill, Dorset (Figure 3.7). This excavation revealed a late Bronze Age burnt mound on the side of a palaeochannel of the River Iwerne (Bell and Allen forthcoming). On the west side of the palaeochannel a series of layers of sediment with small chalk pieces were interleaved with the palaeochannel gravels and overlying finer sediments. A unknown proportion of the soil eroded from this slope is likely to have entered the Iwerne stream and to have been reworked. The stratigraphy

| | Volume of sediment | Area from which derived | Depth eroded (Auger) | Depth eroded (Section) |
|---|---|---|---|---|
| Unit 4a (Iron Age/ Romano-British) | 1457 m/3 | 6000 m/2 | 24 cm | 14 cm |
| Unit 6 (Middle Bronze Age) | 445 m/3 | 5400 m/2 | 8 cm | 17 cm |

Table 3.1 Brean Down, Somerset, quantification of eroded soil

indicates that the earliest of the chalky sediments was deposited at about the time of the Bronze Age activity. Subsequent chalky sediments probably date to the Iron Age and Romano-British periods when a field system on Stepleton spur to the west was under cultivation as a lynchet excavation has shown.

In considering the loss of soil we need to consider qualitative as well as quantitative changes. Catt (1978) has shown that Devensian loess is widespread in southern Britain and there is much evidence that thicker loess deposits were present prior to erosion which followed Holocene human clearance. This is indicated by the presence of silt horizons, the mineralogy of which indicates they are of loessic origin, in the base of valley sediment sequences at Newhaven (Catt in Bell 1976), Kiln Combe and Chalton (Catt in Bell 1981). Many valley sediments show a tendency for the basal deposits to contain a high proportion of silt and for the proportion of, first, flints and then chalk granules to increase upwards through the sequence (Figure 3.3). This can be interpreted in terms of the thinning of soils leading first to erosion of loess then to erosion of flinty subsoil horizons and finally to erosion of the underlying chalk. This sequence is supported by recent archaeologically related soil surveys of the Dorset chalk. Fisher (1991) working in the Cranbourne Chase area has shown that argillic brown earths were formerly much more extensive. On the Dorset Ridgeway, Staines (1990) records that soils buried by Bronze Age barrows and those forming lynchets are non-calcareous and clay-rich whereas present day soils on the slopes are only very thin chalky rendzinas.

*Erosion at Brean Down*

This case study concerns a site on Carboniferous Limestone in Somerset where recent excavations have identified and dated discrete erosion episodes (Bell 1990a; 1992). The Holocene sequence at Brean Down sandcliff has a palaeosol at the base overlain by a series of blown sands (derived from the intertidal area), occupation horizons and slope sediments. The earliest phase of soil erosion was from the basal palaeosol (Unit 8a) which contained Neolithic and Beaker artifacts. This phase was only attested by Macphail's (1990) micromorphological analysis, an unknown thickness of soil was lost downslope. This was followed by a phase of blown sand deposition during the Beaker period and then by a second soil erosion phase (Unit 6) during the middle Bronze Age. This erosion was so severe that it resulted in the total removal of the basal palaeosol from the upper part of the slope where underlying Devensian limestone breccia was exposed. The eroded soil was deposited at the base of the slope within and behind a settlement of circular stone houses. Micromorphological examination (Macphail 1990) shows that the soil derived from arable land.

The depth of soil eroded can be quantified in three ways (Table 3.1). Firstly the detailed, but spatially restricted, information from the excavated area suggests that along this particular cross-sectional line, 17 cm of palaeosol had been eroded. Secondly the deposits were mapped by auger survey (Crabtree 1990) and soil pits which enables the volume of sediment and the area from which it was derived to be quantified. That data indicates the loss of 8 cm of palaeosol. Thirdly we have the palaeosol itself where it was still preserved below Beaker aeolean sand, this was 20 cm thick. This latter figure, and that from the excavated area is likely to be the more reliable and indicates that some eroded sediment lies south of the auger mapped area where the sedimentary sequence interleaves with estuarine sediments. Once again this suggests the possibility of reworking and loss.

An interesting aspect of this middle Bronze Age erosion episode is that occupation of the settlement continued whilst the eroded soil was accumulating because it was interleaved with thin lenses of ash and blown sand. It is also noteworthy that despite the total removal of the arable resource immediatly adjacent to this settlement it continued in occupation intermittently for another 700 years. This prehistoric erosion was clearly both dramatic and perceptible but there is no evidence that it forced the community to change settlement location or way of life, presumably because other resources or social ties to this spot were more important than their arable resource. Occupation was later interrupted

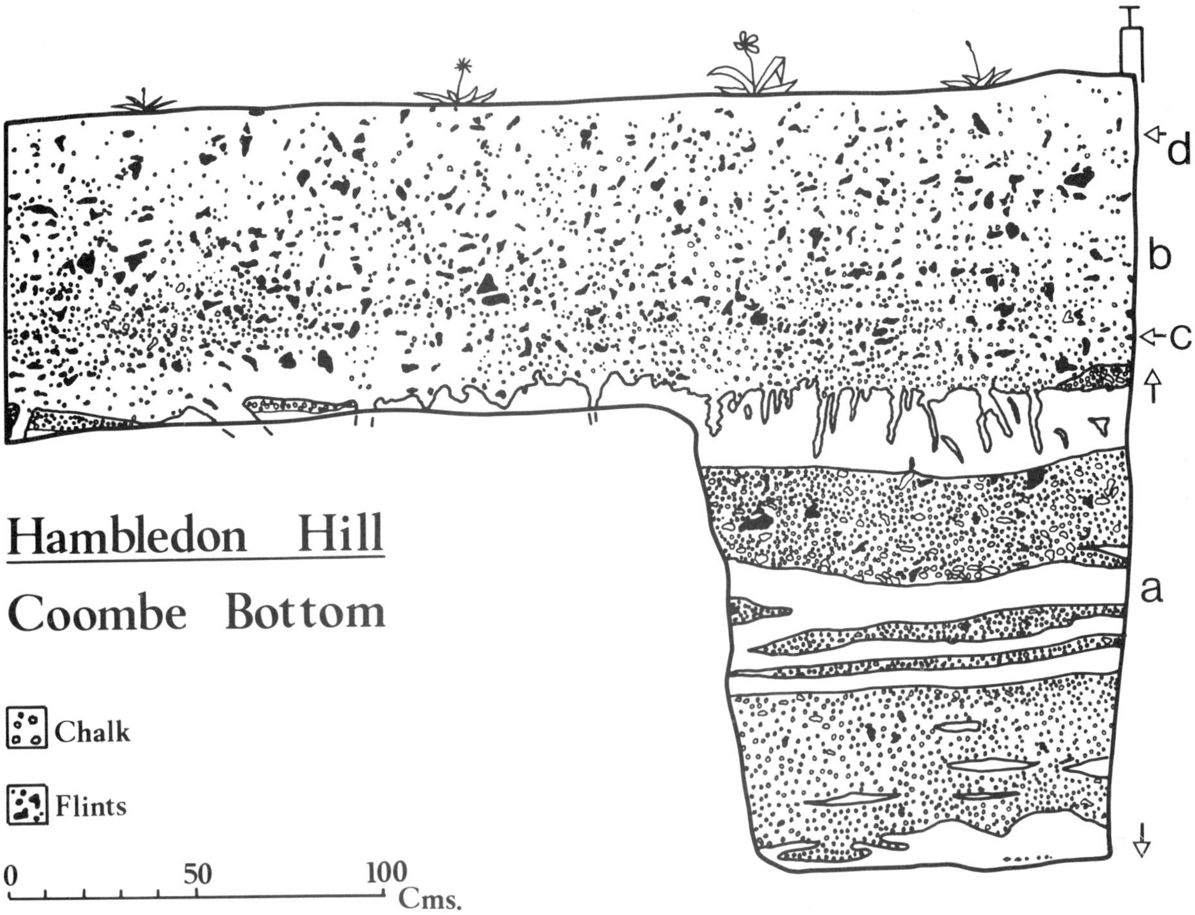


Figure 3.6 Valley sediments at Coombe Bottom, Hambledon Hill, Dorset: (a) Late Devensian sediments; (b) Holocene eroded sediment; (c) marks the depth to which Iron Age artifacts were found; (d) marks the depth to which Post-Medieval artifacts were found

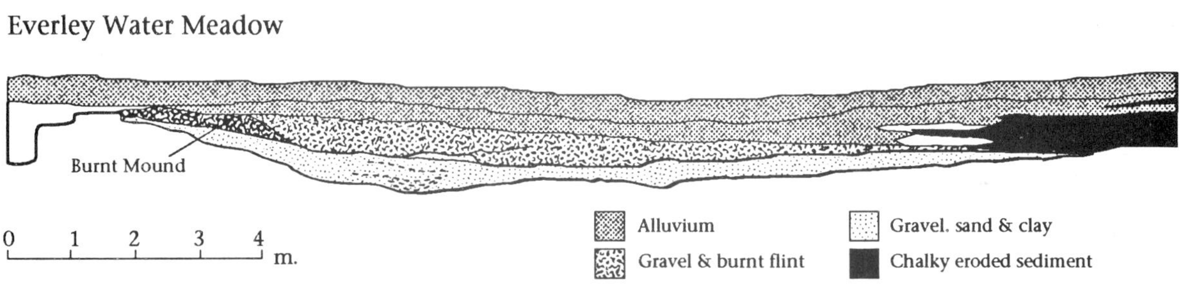


Figure 3.7 Everley Water meadow, east of Hambledon Hill, Dorset. Palaeochannels of the River Iwerne on the side of which is a late Bronze Age burnt mound. On the right (west) are lenses of chalky eroded sediment derived from Stepleton spur.

Figure 3.8 Brean Down, Somerset: (a) basal, truncated palaeosol; (b) middle Bronze Age blown sand; (c) late Bronze Age occupation; (d) Iron Age/Romano-British eroded soil on the surface of which are cultivation marks; (e) field wall (Photo. M. Bell).

by two phases of sand blow suggesting this community was more susceptible to the effects of coastal change than soil erosion.

Following late Bronze Age occupation the site was abandoned around 700 BC. This was followed by a little sand blow and then by a third erosion episode (Unit 4a) which appears to have taken place around the Iron Age, certainly between 700 BC and the fourth century AD. The topsoil from the slope had been mostly lost during the middle Bronze Age erosion episode and much of the sediment eroded during this phase was subsoil Pleistocene breccia largely derived from the slopes above the sandcliff. These slopes are far too steep and craggy to have been cultivated and the trigger for this erosion phase is uncertain; perhaps the most likely scenario is overgrazing of the slope. Another possibility is suggested by the construction of an Iron Age hillfort on the east end of the Down during the late Iron Age. Hillfort construction is one event in landscape history which seems inherently likely for defensive reasons to have been associated with extensive clearance of the surrounding slopes. The depth of soil eroded during this phase may be quantified in two ways (Table 3.1). Firstly quantification may be based on the excavated cross section which indicates the removal of 14 cm of subsoil/soil. Secondly quantification may be based on the spatial picture derived from auger survey which indicates that these deposits were more extensive, and in places thicker, inland of the excavated area. This method indicates the removal of 24 cm of subsoil/soil.

Sometime after the deposition of this eroded sediment, and presumably after it had undergone some pedogenesis, it created the opportunity for further arable activity on the sandcliff (Figure 3.8). This is indicated by the micromorphological evidence, the homogeneous nature of the layer without any trace of sorting, and by the presence of plough furrows running along the contours parallel to a stone field boundary wall. That field was engulfed in blown sand after which it was given over to the graves of a sub-Roman cemetery. From then to the present day there was no evidence of significant soil erosion.

The Brean sequence is significant as an example of erosion concentrated in particular brief episodes which would have been humanly perceptible. In the case of the middle Bronze Age episode this did not lead to a dramatic human response. In the case of the last episode, which in a sense reversed the earlier loss of soil from the slope, it was opportunistically exploited by the reestablishment of arable.

*Erosion at All Cannings Cross*

Excavations during the 1920s on the chalk at All Cannings Cross in Wiltshire (Cunnington 1923) revealed an Iron Age settlement located in a combe at the foot of the Downs. The settlement persisted from about the seventh to the second centuries BC during which there were periodic innundations of parts of the settlement by eroded soil, presumably derived from arable fields upslope. The sediments separated features of different phases and preserved clay domes over pits and ovens which elsewhere would normally have been totally lost by erosion. Here, as at Brean, the erosion must have been perceptible but the settlement continued regardless. Perhaps preventative measures were thought to be unnecessary because innundations were not sufficiently frequent to justify a response.

**Conclusions**

It has been the purpose of this paper to argue that archaeology can make a significant contribution to the long-term perspective in studies of soil erosion by helping to date past sediments and identify the land-use regimes under which they occurred. It is clear that the problems of soil erosion which are increasingly being documented in the contemporary world have a long history, especially on the chalk. It is difficult to escape the conclusion that to some extent we are relearning lessons about autumn sowing and perhaps long-term arable which were last learnt the hard way during the Iron Age and Romano-British periods. It has been argued that modern analogues are an essential element in our attempts to understand the earlier deposits, but problems with uniformitarian reasoning have been noted. There is a need for existing, mostly field-derived models, to be tested by analytical comparison. It has also been argued that archaeological studies of erosion need to focus on how these processes interacted with the lives of past communities. This perceptual aspect is one of considerable difficulty but the case studies of Brean Down and All Cannings Cross give us a glimpse of some alternative human reactions to erosion. Even in situations where erosion was perceptible it may, in the past as today, only have represented an insidious long-term threat to agriculture. That in a sense is the *raison d'etre* for this volume in demonstrating the contribution of a long-term perspective to proper understanding and future management.

**Acknowledgements**

I am grateful to the following with whom I worked on projects, the results from which have been included in this paper:- Roger Mercer (for permission to publish two drawings from the Hambledon Hill Project), the late Dr Bob Smith (who supervised work at Everley Water Meadow) and Mike Allen (who assisted with analytical aspects of Hambledon Hill valley sediments); Dr Keith Crabtree and Dr Richard Macphail (Brean Down); John Budden (Chalton). Stimulating discussion with Dr John Boardman is likewise acknowledged.

## References


Allen, M.J. 1988 Archaeological and environmental aspects of colluviation in south-east England. In: Groenman-van Waateringe, W. and Robinson, M. (eds.) *Man-made Soils*. British Archaeological Reports, International Series 410, Oxford, 67–92.

Allen, M.J. 1991 Analysing the landscape: a geographical approach to archaeological problems. In: Schofield, A.J. (ed.) *Interpreting Artifact Scatters: Contributions to Ploughzone Archaeology*. Oxbow Monograph 4, 39–57.

Atkinson, R.J.C. 1957 Worms and weathering. *Antiquity* 31, 219–233.

Bell, M. 1976 The excavation of an early Romano-British site and Pleistocene landforms at Newhaven, Sussex. *Sussex Archaeological Collections* 114, 218–305.

Bell, M. 1977 *Excavation at Bishopstone. Sussex Archaeological Collections* 115, Lewes.

Bell, M. 1981a *Valley Sediments as Evidence of Prehistoric Land-use: A Study Based on Dry Valleys in South East England*. London University Ph.D.

Bell, M. 1981b Seaweed as a prehistoric resource. In: Brothwell, D and Dimbleby, G. (eds.) *Environmental Aspects of Coasts and Islands*. British Archaeological Reports, International Series 94, 117–26.

Bell, M. 1983 Valley sediments as evidence of prehistoric land-use on the South Downs. *Proceedings of the Prehistoric Society* 49, 119–150.

Bell, M. 1986 Archaeological evidence for the date, cause and extent of soil erosion on the chalk. SEESOIL 3, 72–83.

Bell, M. 1990a *Brean Down Excavations 1983-1987*. English Heritage Archaeological Report 15, London.

Bell, M. 1992 Brean Down excavations 1989 and 1991. In: *Severn Estuary Levels Research Committee Annual Report 1991, Lampeter.*

Bell, M. , Watson, N. and Jones, J. 1990 The land molluscs from Winterbourne Steepleton. In: Woodward, P.J. *The South Dorset Ridgeway: survey and excavations 1977–84*. Dorset Natural History and Archaeological Society, Monograph 8, 114–117.

Bell, M. and Allen, M. forthcoming Valley sediments and molluscs from around Hambledon Hill, Dorset.

Bell, M. and Walker, M. J. C. 1992 *Late Quaternary Environmental Change : Physical and Human Perspectives*. Longman, London.

Boardman, J. 1989 Severe erosion on agricultural land in East Sussex, UK, October 1987. *Soil Technology* 1, 333–348.

Boardman, J. 1990a Soil erosion on the South Downs: a review. In: Boardman, J., Foster, I.D.L. and Dearing, J.A. (eds.) *Soil Erosion on Agricultural Land*. Wiley, Chichester, 87–105.

Boardman, J. 1990b *Soil Erosion in Britain: costs, attitudes and policies* . Education Network for Environment and Development, University of Sussex, Falmer.

Boardman, J. 1992 Current erosion on the South Downs: implications for the past. In: Bell, M. and Boardman, J. (eds.) *Past and Present Soil Erosion*. Oxbow Books, Oxford, 9–19.

Boardman, J. and Robinson, D.A. 1985 Soil erosion, climatic vagary and agricultural change on the Downs around Lewes and Brighton, autumn 1982. *Applied Geography* 5, 243–258.

Bowen, HC. 1961 *Ancient Fields*. British Association for the Advancement of Science, London.

Bradley, R. 1974 A chalk-cut shaft at Belle Tout. *Sussex Archaeological Collections* 112, 156.

Burleigh, R. and Kerney, M.P. 1982 Some chronological implications of a fossil molluscan assemblage from a Neolithic site at Brook, Kent, England. *Journal of Archaeological Science*, 9, 29–38.

Burton, I., Kates, R.W. and White, G.F. 1978 *The Environment as Hazard*. Oxford University Press, Oxford.

Butzer, K.W. 1977 *Geomorphology of the lower Illinois valley as a spatial-temporal context for the Koster site*. Illinois State Museum, Reports of Investigations 34.

Catt, J. 1978 The contribution of loess to soils in lowland Britain. In: Limbrey, S. and Evans, J.G. (eds.) *The Effect of Man on the Landscape: the Lowland Zone*. CBA Research Report 21, 12–20.

Coles, B. and Coles, J. 1989 *People of the Wetlands*. Thames and Hudson, London.

Crabtree, K. 1990 The auger survey. In: Bell, M. *Brean Down Excavations 1983–7*. English Heritage Archaeological Report 15, London, 90–94.

Cross, A.E. 1967 The Great Till flood of 1841. *Weather* 22, 430–433.

Cunnington M.E. 1923 *The Early Iron Age inhabited site at All Cannings Cross Farm, Wiltshire*. G. Simpson, Devizes.

Darwin, C. 1881 *The Formation of Vegetable Mould Through the Action of Worms with Observations on their Habits*. Faber and Faber, London.

Drewett, P.L. 1982 *The Archaeology of Bullock Down, Eastbourne, East Sussex; the Development of a Landscape*. Sussex Archaeological Society Monograph 1, Lewes.

Ellis, C. 1986 The postglacial molluscan succession of the South Downs dry valleys. In: Sieveking, G. de G. and Hart, M.B.(eds.) *The Scientific Study of Flint and Chert*. Cambridge University Press, Cambridge, 175–184.

Evans, J.G. 1972 *Land Snails in Archaeology*. Seminar Press, London.

Evans, R. 1990 Soil Erosion: its impact on the English and Welsh landscape since woodland clearance. In: Boardman, J., Foster, I.D.L. and Dearing, J.A. (eds.) *Soil Erosion on*

*Agricultural Land.* Wiley, Chichester, 231–254.
Fisher, P.F. 1991 The physical environment of Cranbourne Chase. In: Barrett, J., Bradley, R. and Hall, M. (eds.) *Papers on the Prehistoric Archaeology of Cranbourne Chase.* Oxbow Monograph 11, 11–19.
Fowler, P.J. 1983 *The Farming of Prehistoric Britain.* Cambridge University Press, Cambridge.
Fowler, P.J. and Evans, J.G. 1967 Plough-marks, lynchets and early fields. *Antiquity* 41, 289–301.
Gleeson, P. and Grosso, G. 1976 Ozette site. In: Croes, D. R. (ed.) *The Excavation of water saturated archaeological sites (wet sites) on the north-west coast of North America.* National Museum of Man, Mercury Series, Archaeological Survey of Canada, Paper 50, Ottawa.
Hewitt, K. (ed.) 1983 *Interpretations of Calamity .* Allen and Unwin, Boston.
Jones, M. 1981 The development of crop husbandry. In: Jones, M. and Dimbleby, G.W. (eds.) *The Environment of Man: the Iron Age to the Anglo-Saxon Periods.* British Archaeological Reports 87, Oxford, 95–127.
Knox, J.C. 1984. Responses of river systems to Holocene climates. In: H.E. Wright Jnr. (ed.) *Late Quaternary Environments of the United States, 2 The Holocene.* Longman, London, 26–41.
Latter, P.R. 1932 The Louth Flood. *Lincolnshire Magazine* 1932, 1, no 2, 73–77.
Limbrey, S. 1975 *Soil Science and Archaeology.* Academic Press, London.
Lowe, J.J. and Walker, M.J.C. 1984 *Reconstructing Quaternary Environments.* Longman, London.
Macphail, R.I. 1990 Soil history and micromorphology. In: Bell, M. *Brean Down Excavations 1983–1987.* English Heritage Archaeological Report 15, London, 187–196.
Macphail, R. 1992 Soil micromorphological evidence of ancient soil erosion. In: Bell, M. and Boardman, J. (eds.) *Past and Present Soil Erosion.* Oxbow Books, Oxford, 197–215.
Megaw, J.V.S., Thomas, A.C. and Wailes, B. 1961 The Bronze Age settlement at Gwithian, Cornwall. *Proceedings of the West Cornwall Field Club* 2, 200–215.
Oakley, K.P. 1945 Some geological effects of a cloudburst in the Chilterns. *Records of Buckinghamshire* 14, 265–280.
Quine, T.A. and Walling, D.E. 1992 Patterns and rates of contemporary soil erosion derived using Caesium-137: measurement, analysis and archaeological significance. In: Bell, M. and Boardman, J. (eds.) *Past and Present Soil Erosion.* Oxbow Books, Oxford, 185–196.
Renfrew, C. and Bahn, P. 1991 *Archaeology: Theory, Methods and Practice.* Thames and Hudson, London.
Reynolds, P.J. 1981 Deadstock and livestock. In: Mercer, R. (ed.) *Farming Practice in British Prehistory.* Edinburgh University Press, Edinburgh, 97–122.
Sheail, J. 1971 *Rabbits and their History,* David and Charles, Newton Abbot.
Staines, S.J. 1990 The soils of Winterbourne Steepleton, and the soil sections at Cowleaze and Loscombe. In: Woodward, P.J. *The South Dorset Ridgeway.* Dorset Natural History and Archaeological Society, Monograph 8,117–121.
Tansley, A.G. 1949 *Britain's Green Mantle.* Allen and Unwin, London.
Taylor, C. 1966 Strip Lynchets. *Antiquity* 40, 277.
Thomas, J. 1991 *Rethinking the Neolithic.* Cambridge University Press, Cambridge.
Thomas, K.D. 1977 A preliminary report on the Mollusca from the lynchet section. In: Bell, M. *Excavations at Bishopstone, Sussex Archaeological Collections* 115.
Thomas, K.D. 1982 Neolithic enclosures and woodland habitats on the South Downs in Sussex, England. In: Bell, M. and Limbrey, S. (eds.) *Archaeological Aspects of Woodland Ecology.* British Archaeological Reports, International Series 146, Oxford, 147–170.
Thorley, A. 1981 Pollen analytical evidence relating to the vegetation history of the chalk. *Journal of Biogeography* 8, 93–106.
Waton, P. 1982 Man's impact on the chalklands: some new pollen evidence. In: Bell, M. and Limbrey, S. (eds.) *Archaeological Aspects of Woodland Ecology.* British Archaeological Reports, International Series 146, 75–91.
Whitley, H.M. 1900 The underground water-levels of the South Downs between Eastbourne and the River Cuckmere. *Minutes of Proceedings of Institute of Civil Engineers* 142, 342–351.
Williamson, R. 1978 *The Great Yew Forest: the Natural history of Kingley Vale.* Macmillan, London.

# CHAPTER 4

# PRODUCTS OF EROSION AND THE PREHISTORIC LAND-USE OF THE WESSEX CHALK

MICHAEL J. ALLEN

*Archaeological investigations of colluvium aim to determine past human land-use. However, to do so, it is necessary to understand soil erosion using various analytical techniques. The data and interpretations are of use in reconstructing the past environment and land-use patterns. This paper discusses the nature and variety of colluvial deposits, the complexity of erosion and deposition regimes and their use as indicators of past human activity, as well as providing a chronological dimension to colluvial deposits. Several types of colluvial deposit are recognised in Wessex and a classification is presented. Human activities that have induced erosion are examined in relation to environmental systems and thresholds.*

## Introduction

Archaeological and palaeoenvironmental investigations of Post-glacial soil erosion have been conducted by the writer on a number of chalkland sites in southern England (Allen 1984; 1988; 1991). Recently (1987-1992) this work has been concentrated largely within the area of Wessex (Figure 4.1) as a part of post-graduate research (Allen in prep.) and archaeological investigations conducted by Wessex Archaeology. This research includes over fifteen chalkland sites in Wessex (Table 4.1) with thirteen detailed molluscan sequences; a further ten sites have been investigated elsewhere in southern England. This relatively large data-base of broadly dated Post-glacial colluvial sequences enables not only a general overview to be prepared but also the recognition of common patterns in colluvial sequences. The colluvial sequences considered are largely restricted to deposits which have accumulated in topographical locations such as dry valleys, lynchets, plateau edges and footslopes (i.e. base of slope), rather than those which have washed into archaeological features such as ditches. This is because colluvium from archaeological features may record restricted, and intensive, human activity, whereas deposits which are not directly associated with known archaeological sites are likely to be more representative of the region than a specific locality.

Although the initial aims of this work were primarily to reconstruct land-use patterns and the history of human activity in the landscape, such investigations lead inevitably to an attempt to understand past soil erosion events. An archaeological approach enables erosion events to be dated and related to known cultural horizons. Archaeological and palaeoenvironmental studies of past soil erosion can also be used to place modern soils and erosion in a temporal context. Thus, just as study and observations of modern erosion regimes and analogous deposits are essential to understand past soil erosion (Allen 1988; 1991), the recognition of past events may enable us to determine the reasons for, and mechanisms of, erosion which are applicable to modern day scenarios (e.g. Bell 1986; Evans 1992).

Most colluvial studies discussed in this paper have involved excavations of deposits which have been dated by artifact distributions (Bell 1983; Allen 1988), however we still know little about the *rates* of deposition. Palaeoenvironmental interpretation has primarily been via molluscan analysis (e.g. Kerney *et al.* 1964; Evans 1966; Bell 1983), but also magnetic susceptibility (Allen 1986; 1988; Allen and Macphail 1987), limited soil micromorphological studies (e.g. Fisher and Macphail 1985; Macphail 1992; Kwaad and Mücher 1979) and field observations (e.g. Bell 1982).

The term 'colluvium' is used in this paper in view of its currency over a number of years (Evans 1968; 1972; Bell 1981a; 1983; 1986; Allen 1984; 1988; 1991) and describes material that is moved downslope by gravity, usually lubricated by temporary flushes of water that is not in a permanent, well-defineed channel.

## Nature and diversity of deposits

Recent work continues to emphasise that we cannot predict the depth or type of colluvial deposits on the chalk (Allen 1988; 1991) although caesium-137 studies (Quine and Walling 1992) show that topographic conditions explain some of the variation in colluvial rates and quantities. Potential receptor sites are often so subtle as not to be recognised on even detailed contour surveys (e.g. 1:25,000 OS series), but can usually be located in the field. Even so, on the slopes of Strawberry Hill (Figure 4.1), a combe on the northern scarp of Salisbury Plain was so small that it could not even be recognised in the field, yet when trenched by a JCB over 3 m of deposits were revealed (Figure 4.2). Conversely, other larger well-defined valleys, notably Stonehenge Bottom, produced no colluvium (Richards 1990, 210–211) despite being located in areas of suitable topography and known

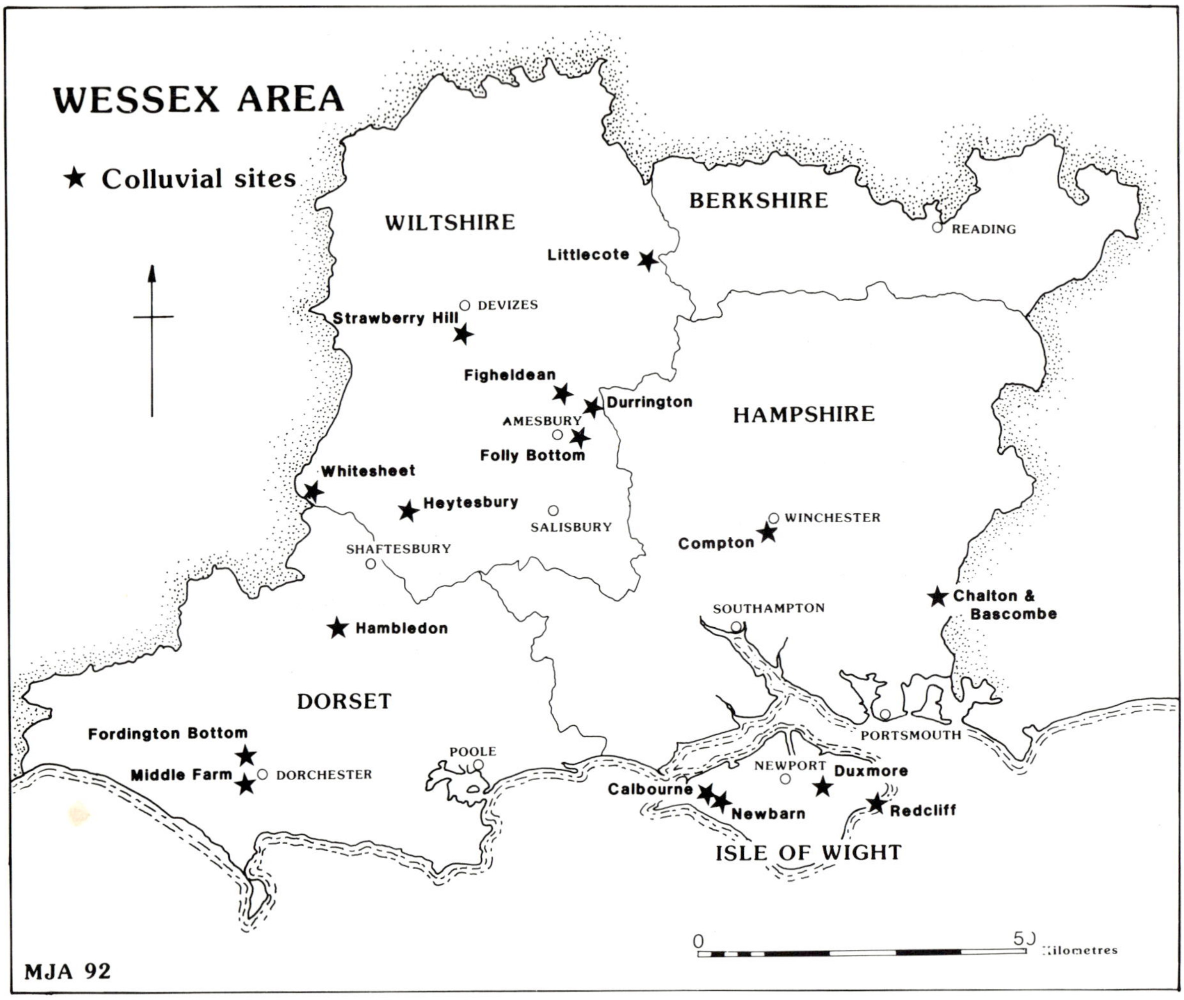


Figure 4.1 Map of Wessex showing colluvial sites investigated.

intensive prehistoric activity. Superficially it would appear that the depth and type of deposits bears little relationship to valley and slope morphology, and the intensity of past land-use. Whether deposits survive in any particular location appears to depend more on the nature of past erosive episodes rather than site topography. Locations that have the potential to receive deposits may be subsequently the source of material for later erosive episodes. There is the potential for net removal of material; large single events may remove decades, possibly centuries, of sediment accumulation.

The nature of the deposits preserved is a combination of a large number of factors of which slope and valley morphology play a part. The local soils, i.e. the source material for the deposits, provide one set of variables. Particle size, structure, organic matter content and stoniness contribute to their erodibility and the composition of the final deposit. Soils, however, cannot be seen as a constant through time or space. Human activities such as forest clearance, soil tillage and even intensive grazing will change nutrient and organic matter levels, and soils may be continually thinned by erosive processes. Many valleys contain a number of different soil types (e.g. Moffat and Cope 1984) and thus soil should not be viewed as an homogeneous blanket over the surface. Climate and the occurrence of individual storm or drought events in prehistory are unrecorded variables, the evidence for which is sometimes inferred from the nature and composition of colluvial sequences (Allen 1991). It is because of this set of variables that it is not yet possible to produce accurate predictive models incorporating the complex relationships of slope morphology, pedological variation, climatic vagary and past land-use, though this has been achieved on the chalk landscape to some extent by Harris and Boardman (1990).

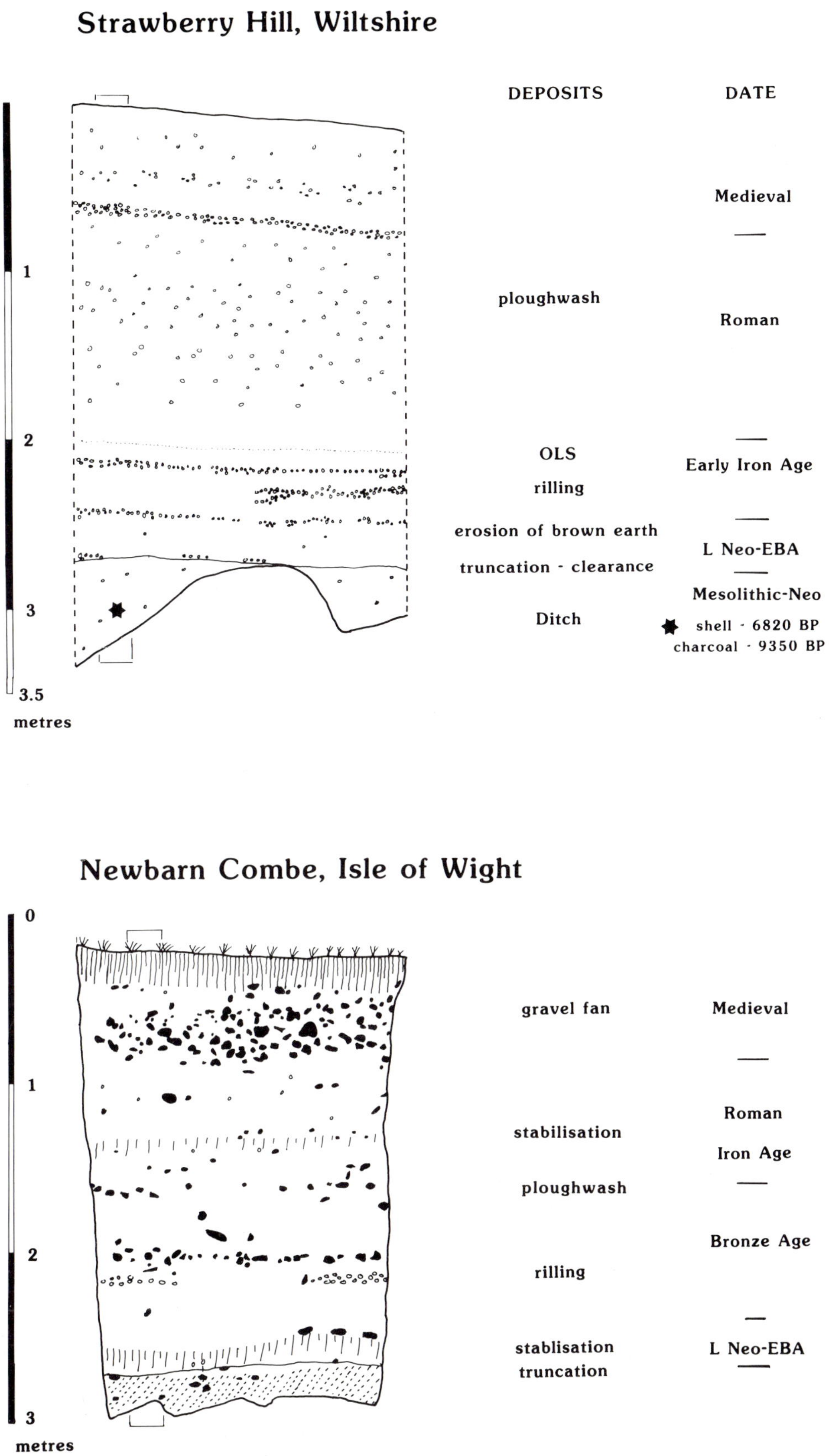


Figure 4.2 Colluvial sequences at Strawberry Hill and Newbarn Combe with dates of deposition and interpretation of erosion/deposition events. OLS = Old Land Surface; EBA = early Bronze Age; L.Neo = Late Neolithic.

| Site | Truncation of E. Holocene soil (latest date) | Buried soil/ stabilisation horizons | Flint horizons | | Chalk fans (? rill erosion) | Unsorted calcareous hillwash | Unsorted weakly calcareous hillwash | Well sorted, banded hillwash |
|---|---|---|---|---|---|---|---|---|
| | | | Gravel fans | Lavant deposits | | | | |
| **DORSET**<br>Middle Farm<br>(SY 672 902) | L Neo/Beaker | EBA | Mid-Late BA | - | - | - | EBA -Med | |
| Fordington Bottom<br>(SY 665 908) | L Neo/EBA | prob. BA | EBA | - | Roman | EBA -> Med | EBA and BA | IA and Roman |
| Hambledon environs<br>1 Combe Bottom<br>(ST 8565 1190)<br>2 Stepleton Lynchet<br>(ST 8579 1165)<br>Bell and Allen 1985 | Pre Iron Age<br>Pre Iron Age | -<br>- | -<br>- | -<br>- | -<br>- | Iron Age - Med<br>Iron Age - Med | -<br>- | -<br>- |
| **WILTSHIRE**<br>Strawberry Hill<br>(ST 996 528) | L Neo/EBA | Early Iron Age | - | - | Early Iron Age | L Neo -> Roman | - | - |
| Durrington Walls<br>(SU 152 436)<br>Wainwright 1971 | Neolithic | ??? Iron Age | - | - | - | L Neo -> | - | - |
| Figheldean<br>(SU 148 468) | Neolithic/EBA | relict<br>L Neo/EBA | - | - | - | BA -> Med | - | - |
| Heytesbury<br>(ST 939 418) | Neolithic/EBA | - | - | - | - | BA -> Med | - | - |
| Whitesheet Hill<br>(ST 802 352) | LBA/EIA | Mid Iron Age | - | - | - | IA -> Med<br>calcareous marl | - | - |
| Folly Bottom<br>(SU 173 424) | pre Roman | - | - | Roman/Med | - | - | - | - |

Table 4.1 Presence and date of recorded depositional and stabilisation events in colluvial sequences in Wessex; abbreviations are as follows: L = Late; E = Early; Neo = Neolithic; BA = Bronze Age; IA = Iron Age; Med = Medieval (continued overleaf).

| Site | Truncation of E. Holocene soil (latest date) | Buried soil/ stabilisation horizons | Flint horizons | | Chalk fans (? rill erosion) | Unsorted calcareous hillwash | Unsorted weakly calcareous hillwash | Well sorted, banded hillwash |
|---|---|---|---|---|---|---|---|---|
| | | | Gravel fans | Lavant deposits | | | | |
| **ISLE OF WIGHT**<br>Redcliff<br>(SZ 625 8555) | Neolithic/EBA | L Neo/EBA | - | - | - | L Neo/Beaker -> Roman/Med | - | - |
| Newbarn Combe<br>and<br>Newbarn Lynchet<br>(SZ 4355 8618) | Neolithic/EBA<br><br>LBA | ?1. EBA<br>2. IA/Roman<br>- | Medieval<br><br>- | -<br><br>- | BA/IA<br><br>- | EIA -> Med<br><br>LBA/EIA -> Med | EBA -> EIA<br><br>- | -<br><br>- |
| Duxmore<br>(SZ 563 874) | EBA | - | Roman | - | - | LBA -> Med | IA -> Roman | - |
| Gore Cliff<br>(SZ 493 763)<br>Preece 1980) | Roman | - | - | - | - | Roman -> Med | - | - |
| **HAMPSHIRE**<br>Compton<br>(SU 475 264) | BA | BA | - | - | - | BA -> | - | - |
| Bascombe<br>(SU 721 166)<br>Bell 1981b; 1983 | pre Iron Age | - | - | IA/Roman | - | - | Iron Age -> Med | |
| Chalton<br>(SU 729 160)<br>Bell 1981b; 1983 | pre BA | - | - | - | - | - | BA -> Med | |

Table 4.1 (continued) Presence and date of recorded depositional and stabilisation events in colluvial sequences in Wessex; abbreviations are as follows: L = Late; E = Early; Neo = Neolithic; BA = Bronze Age; IA = Iron Age; Med = Medieval.

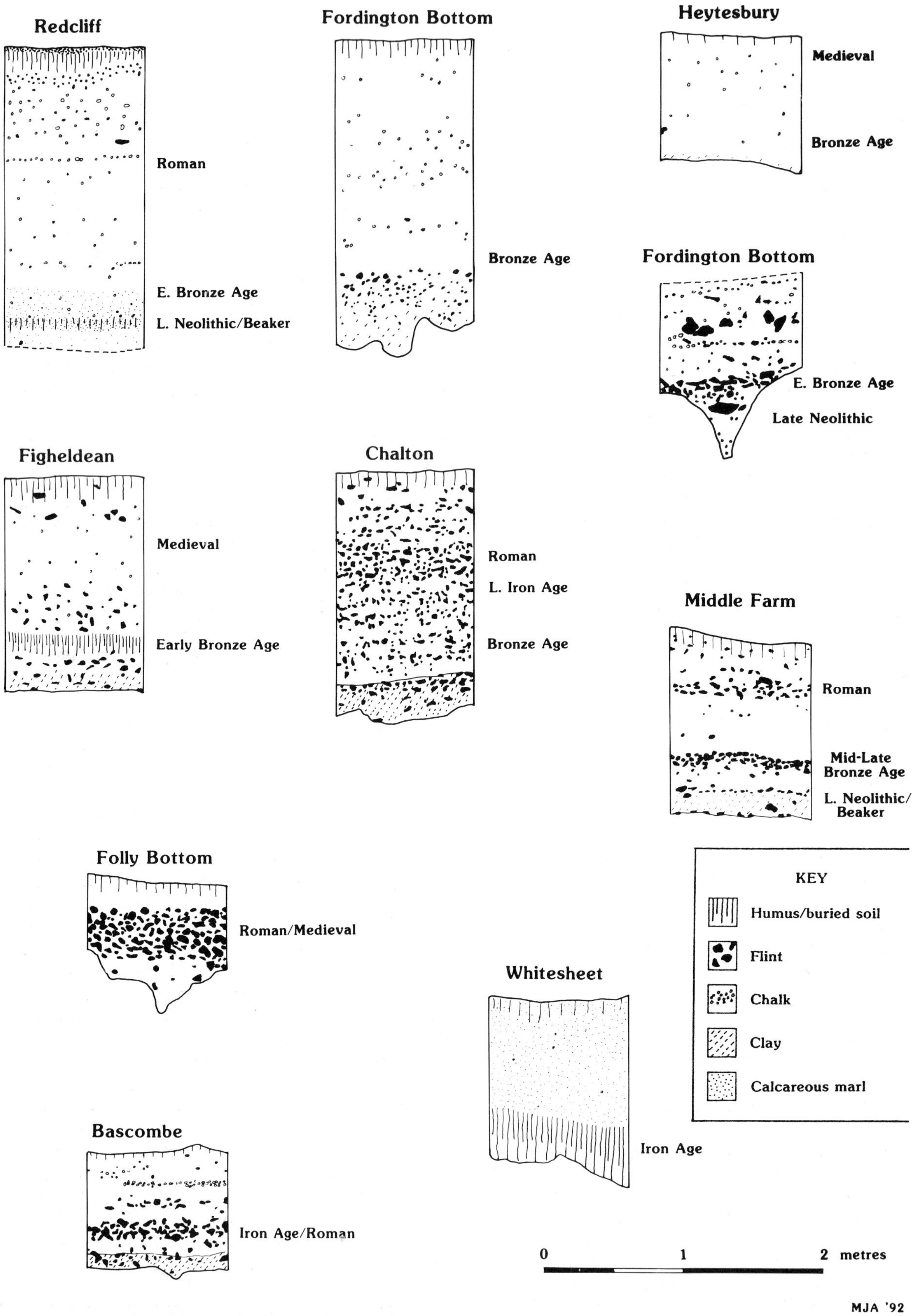


Figure 4.3 A representative selection of colluvial deposits investigated in Wessex.

## Products of erosion

It is not surprising that a number of different types of deposit can be recognised (see Figure 4.3), often with sequences varying considerably in a single valley system, let alone from area to area. Perhaps one of the main contributions to this study is the field record of a range of different types of deposits enabling description and characterisation of four main groups of colluvial deposit:

1a. Unsorted calcareous colluvium
1b. Unsorted weakly- or non-calcareous colluvium
2. Sorted or banded colluvium
3. Flint horizons (e.g. gravel fans)
4. Lenses of chalk pieces

Less common, but also recognised in the current studies, are sorted, or banded, usually weakly calcareous colluvium, stone-free (not worm worked) horizons and silty marls.

These categories can be divided into those which usually comprise most of the deposit and those which occur as discrete, relatively thin horizons within finer grained colluvial sequences. With the aid of modern analogues (Boardman and Robinson 1985; Stammers and Boardman 1984; Boardman 1984; 1988; 1992; Evans 1980; 1992; Allen 1988; 1991) and palaeo-environmental investigations, some interpretation and re-interpretation of the more regularly observed deposits can be suggested. The main deposits are usually fine grained material (silty clays to silty loams) with varying quantities of very small to medium-sized chalk pieces and small to medium flints. Organic matter is low (in 86% of the deposits measured by the writer organic matter was less than 2.1%) and magnetic susceptibility rarely exceeds 35 x $10^{-8}$ SI/Kg. Often these deposits are in excess of one metre in depth. They are generally unsorted although some horizonation can sometimes be detected and they often show an increase in chalk pieces and calcium carbonate content and decrease in magnetic susceptibility up profile.

These deposits are heavily reworked soil material and frequently occur in lynchets, for example Stepleton Lynchet, Hambledon Hill (Bell and Allen 1985), Fordington Bottom (Staines forthcoming) and the Newbarn lynchet, but have also been recorded in a number of dry valleys – Strawberry Hill (Figure 4.2), Newbarn Combe (Figure 4.2) and Redcliff (Figure 4.3) as well as footslope deposits such as Heytesbury (Barnes and Allen 1990). The unsorted nature of the deposits is thought to be a result of both erosion and post depositional reworking. Banded or well sorted deposits are less common (Table 4.1), but the distinction between one 'band' and a deposit containing sorted material may be largely a semantic one. Such sequences (e.g. Middle Farm, Dorchester; see Figure 4.3) indicate varying scales of erosional episodes and pulsed inputs of material rather than an almost continual accretion. The variation between highly calcareous and weakly calcareous, often more argillic, deposits is largely a product of the presence of former thicker soils and/or Clay-with-flints (Hodgson *et al.* 1967).

In general these deposits seem to produce molluscan assemblages dominated by *Helicella itala*, *Trichia hispida* and the Vallonias and are considered to be largely a product of tillage. The fact that they are originally very localised and probably washed from only one or two fields is borne out by their predominant occurrence in lynchets. The restricted origin of the eroded material is also seen at Middle Farm, Dorchester (Allen forthcoming a) where the fills of archaeological features (ditches and postholes) dug through Clay-with-flints on the plateau were almost devoid of land snails due to the local acidity. The valley deposits only 150 m away, which contained very few chalk pieces and superficially seemed to be non-calcareous, produced high mollusc numbers (over 600 shells per kilogram). Here colluviation is taking place largely from the thin soils on the chalk slope and not from the Clay-with-flints on the plateau. Although these unsorted deposits are very localised, there is evidence that they can be re-worked over longer distances (see for example, Duxmore, below).

The two most commonly observed discrete horizons are flint horizons and lenses of chalk pieces (see Table 4.1). The origin of these has been discussed previously (Allen 1991) and is summarised here. Flint horizons generally consist of medium to large flints often representing up to 70% of the sediment. They are predominantly nodular with a number of smaller fractured flint pieces. Although several scenarios have been invoked for deposits of flint nodules, including their accumulation as a result of rolling downslope aided by cattle or sheep, it is evident that the majority of these deposits are the result of gullying events (Stammers and Boardman 1984). Where such archaeological deposits have been excavated in plan (Ashcombe Bottom, Sussex), or orientation has been measured e.g. Ashcombe Bottom (Allen 1991) and Middle Farm (Allen forthcoming a), they show that they have originated from the valley side and can be compared with gravel fans (cf. Boardman 1984; Stammers and Boardman 1984). Gullying upslope results in large volumes of soil and stones being stripped from the slopes and washed away; the stones are dropped at breaks of slope while the finer material is transported further down the valley. These high energy events have a considerable erosive force and are known to remove large quantities of colluvium from valley floors (Boardman 1988). In Wessex such deposits have been noted predominantly in footslope locations (i.e. on the edge of the valley floor). No chronological hypothesis can be given for their occurrence: they have been recorded from earlier Bronze Age deposits at Fordington Bottom, Dorchester; middle Bronze Age at Middle Farm, Dorchester; Roman levels at Duxmore Combe, Isle of Wight and Medieval horizons at Newbarn Combe, also on the Isle of Wight (see Table 4.1).

At Folly Bottom, a large valley to the north-west of Amesbury, Wiltshire, the deposits comprised an undated, but probably later prehistoric, flint deposit in the centre of the valley (Figure 4.3). Little finer-grained colluvium was present. Augering showed that this deposit extended along the axis of the valley floor for almost 80 m. This may represent the coarse channel deposits of a temporary lavant or winterbourne; other examples are discussed by Bell (1992).

Thin (up to 4 cm) horizons of abundant, very small and small chalk nodules have been recorded from a number of sites (Table 4.1) and have been discussed by both Bell (1981a; 1992) and Allen (1991). When one was excavated in plan at Strawberry Hill, Wiltshire it showed a series of 'fingers' of chalk running downslope (Figure 4.4) and was similar to chalk fan deposits observed from modern erosion episodes (Figure 4.5). We can suggest these chalky lenses often represent chalk fan deposits, probably originating from rilling of the A and A/C horizons within an arable field (cf. Allen 1991, 44).

More problematic are the stone-free horizons that have been recorded from sites such as Strawberry Hill and Newbarn Combe (Figure 4.2). A good example of some of the problems of interpretation that these horizons can give is seen at Strawberry Hill. A series of stone-free and stony horizons was located near the base of the 3.4 m colluvial sequence (Figure 4.2). The upper stone-free horizon was associated with a quantity of sherds of an early Iron Age carinated bowl. Although in the field these sequences were interpreted as earthworm-worked horizons and seem to represent stabilisation and cessation of erosion, soil micromorphological evidence cast doubt on this. Richard Macphail reports that the stone-free horizon was 'a massive chalky "slurry" of weathered chalk, sand, silt, and mollusc fragments, arionid granules, fine charcoal and organic matter. It was a dense soil with closed vughy porosity. Juxtaposed varying chalky soil fragments occur'. He also suggests that this stone-free deposit originated from an intensively cultivated rendzina soil. The overlying chalky stone horizon is probably part of a similar erosion event and indicates sorting of the soil material prior to deposition rather than post-depositional reworking by worms. These deposits in turn were rapidly buried preventing *in situ* biological reworking or mixing by cultivation. In other cases (e.g. Bourne Valley, Sussex) Macphail *et al.* (1987) have shown that a stone-free horizon is a result of *in situ* biological reworking and pedogenesis.

If the chalky and stone-free horizons discussed above are to be preserved they must be buried rapidly and deep enough to prevent mixing by biological activity and cultivation. This is also true for the burial of soils. If each sediment increment is thin and the recurrence rate is low, then a soil horizon will be reworked biologically or by cultivation and not buried and sealed, only the lower portion of the soil profile being constantly removed from the biologically active or cultivation layer due to its depth. The consequence of total biological reworking and/or mixing by cultivation of such horizons would be to produce the uniform, typically homogeneous, calcareous colluvium, commonly observed. This then forces us to consider the origin of the main colluvial layers seen in valley locations. Are these a result of rilling and sorting, as described earlier, which have subsequently been humanly (i.e. by tillage) or biologically reworked, or are they a result of other processes of soil removal from the slopes without sorting, such as sheetwash (Boardman and Bell 1992)? In order to answer such questions, detailed soil micromorphological studies are required; such ideas cannot be tested through field description or molluscan analyses alone.

*Spatial variation and colluvial diversity*

As noted above a number of very different deposits have been observed, many occurring in a single sequence (see Figure 4.2). In a few instances multiple trenching of a single valley has revealed considerable spatial variation within the deposits. Eight sections through the centre of the 2.3 km long valley floor at Fordington Bottom, Dorchester, showed that gravel fans were highly localised and fine grained Post-glacial deposits varied between 1.2 m and 2.3 m deep (Figure 4.3) over a distance of less than 250 m. At Duxmore, Isle of Wight, five trenches were excavated by David Tomalin in a narrow scarp combe less than 60 m wide and only 200 m long (Figure 4.6). No two trenches were alike or produced the same sequence (Allen personal observation) even though the majority were less than 15 m apart. Deposits at the head of the combe were shallow and non-calcareous whereas those on the combe floor were deep and highly calcareous. Gravel fans were recognised in at least two of the five trenches. Two separate gravel fans were recorded, temporally separated and with the origin of their runoff at right angles to each other. Material in the younger and coarser fan had been washed down the valley axis, while the earlier fan originated from the valley side.

Within a single small combe, a complex and varied erosion history can be envisaged such that any particular location within the valley may have been both a receptor and transport site. There is no evidence to suggest that any particular deposit type (e.g. flint horizons) is confined to, or even occurs predominantly within, any particular chronological period. Nor can such deposits be consistently linked with specific episodes such as the first tillage of an area. However, a single event can also produce a number of spatially distinct deposits. At Duxmore the deeper deposits near the mouth of the combe were not only more calcareous than elsewhere, but the whole sequence was more homogeneous. This suggests, perhaps, greater amounts of reworked material and thus erosion of the colluvial deposits themselves. In contrast, higher up the valley more discrete horizons relating to specific events or periods could be discerned.

The lateral and temporal variation in the deposits is particularly noteworthy as they force one to question whether the single valley section is ever truly representative. Most of the writer's own work is only based on single sections; nevertheless at Ashcombe Bottom (Allen 1984; Macphail 1992) several soil types were recognised even within a single valley transect.

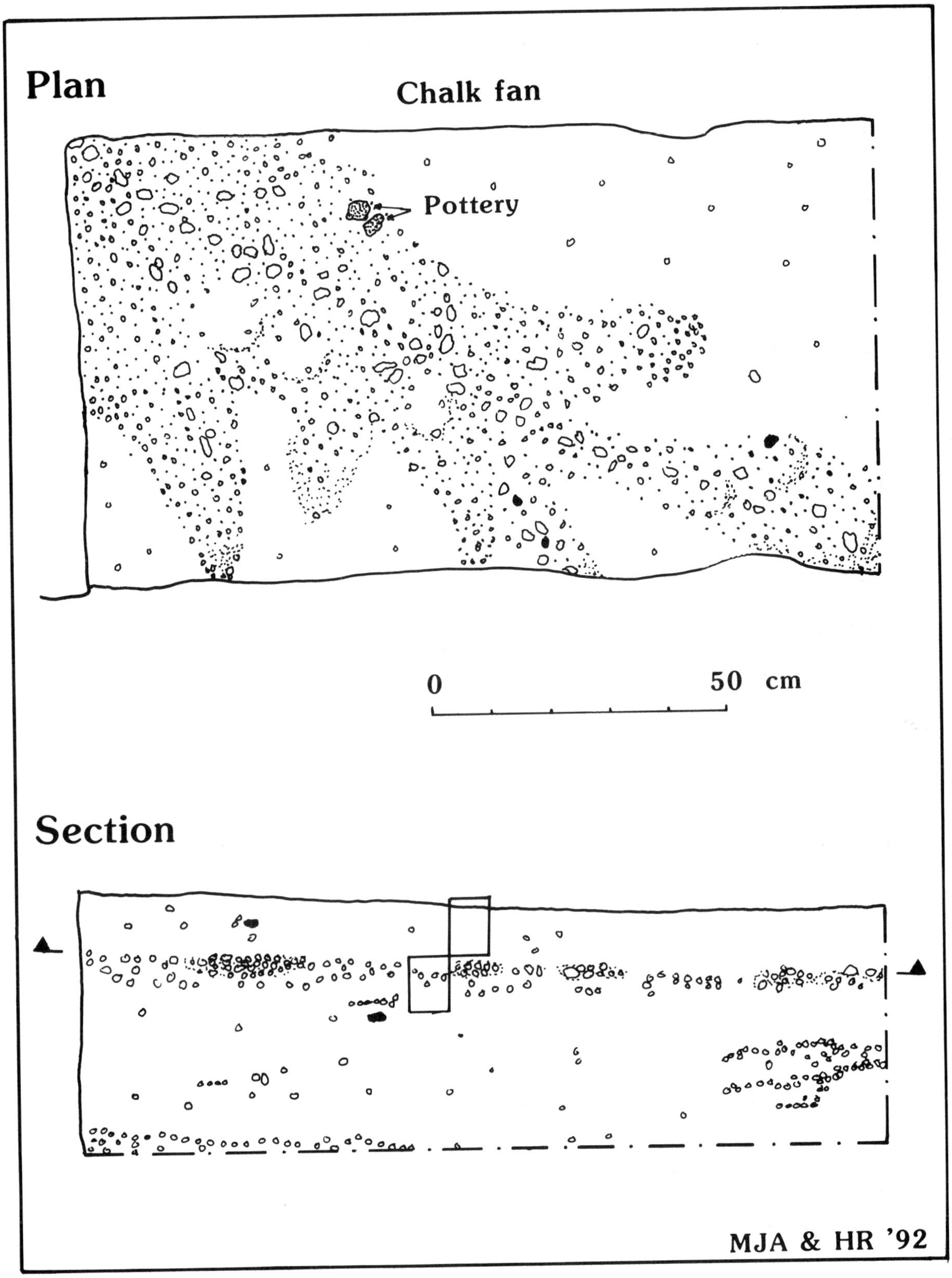


Figure 4.4 Plan and section of a chalk fan excavated at Strawberry Hill, rectangular boxes are the positions of micromorophological samples.

Figure 4.5 Chalk fan as a result of rill erosion from an arable field, Ashcombe Bottom, Sussex.

**Interpreting the past land-use using colluvial deposits**

Dated episodes of colluvium have been used to produce a generalised scheme for the evolution of the chalk landscape and soil patterning (cf. Bell 1981a; 1983). Recent studies summarised above may also help to explain episodes of erosion and pulsed inputs of colluvium. The main contribution of archaeology to erosion studies is the possibility of dating erosion episodes and linking them with human activities. Specific spatial relationship with archaeological sites and past activity can also be seen, as Bell demonstrated at Itford (Bell 1983).

*Former Holocene soils*

An understanding of the former Holocene soils on the Wessex chalkland is not only required for the reconstruction of past human activity, but is also pertinent to the erodibility, vegetational and agricultural potential of the landscape. In some cases buried soils from which palaeoenvironmental information has been obtained have been found in dry valleys (e.g. Pitstone, Buckinghamshire; Evans 1966; Evans and Valentine 1974). These occurrences are, however, rare as many sites show truncation of the former early Holocene soils (see Figures 4.2 and 4.3 and Bell 1992). Where they survive, they do so only in subsoil hollows or as pedo-relicts as indicated by soil micromorphological investigations (e.g. Ashcombe Bottom, Macphail 1992).

Soil micromorphological data (Macphail *et al.* 1987; Macphail 1992) and field records indicate that the early Holocene soils may have been brown earths or calcareous brown earths. In locations where field interpretations have suggested the existence of basal old land surfaces this was not always confirmed by subsequent analysis. Sections of the Newbarn Combe, a large dry valley on the Isle of Wight, produced a weakly calcareous, non-snail-bearing, deposit in the base of the valley (Figure 4.2). It was evident, however, that it was not *in situ* but was derived from the severely eroded former Holocene soil. Similarly the basal deposits at Strawberry Hill, Wiltshire, which were sealed by over 3 m of colluvium, proved to be of colluvial origin.

Evidence for former Holocene soils is therefore scarce and this in itself indicates a significant erosion and truncation of these horizons. Most colluvial sequences began in the Bronze Age or earlier and thus the removal of the probably loessic brown earth soils

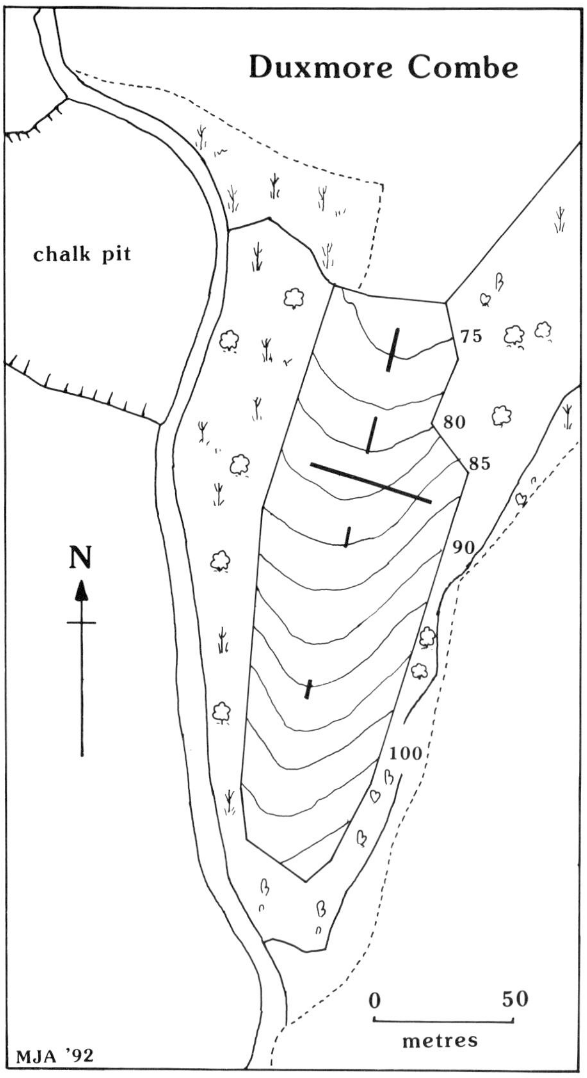


Figure 4.6 Plan of Duxmore Combe, Isle of Wight, showing the location of the machine-cut trenches; the valley floor is contoured at 2.5 m intervals.

(Macphail 1992) was evidently prior to this date. Truncation of the former basal soils can be seen to be severe and extensive. A number of sections at Fordington Bottom show that the basal deposits frequently comprised extensive gravel fans which sealed a ditch of at least early Bronze Age date (Figure 4.3). A humic basal buried soil at the foot of Whitesheet Hill (Figure 4.3) proved to be of Iron Age date indicating total removal of all former soils and deposits before stabilisation and pedogenesis. The lack of these early Holocene soils provides evidence of an extensive episode of erosion but, to date, the re-deposition of this material has not been evident, except perhaps in the larger riverine deposits such as those reported by Scaife and Burrin (1883; 1985; forthcoming). The cause of this erosion is laid predictably at the door of early forest clearance, but the reason for such continued and wholesale erosion is often not fully explained although it might have required only one major episode of erosion to truncate the early Holocene soil from any one site.

*Evidence of prehistoric land-use*

Molluscan analysis has been used to define different localised environments within a single valley catchment and recognise varying, possibly cyclical land-use regimes. Two main observations can now be made. (1) The typically unsorted, calcareous, colluvium is usually equated with arable activity and the mollusc assemblages reflect this (e.g. Thomas 1977; Bell 1983). (2) Temporally limited grassland episodes can be identified. Scaife tentatively speculated a cyclical pattern of rotational arable and pasture from pollen evidence on the Isle of Wight (Scaife 1984) but there was no corroborative evidence and molluscan studies by Bell (1983) on the South Downs did not indicate this. Now, however, molluscan analysis of a shallow calcareous footslope deposit below Knook Horse Hill, Heytesbury, Wiltshire (Allen in Barnes and Allen 1990) has demonstrated this possibility. The fluctuation of two groups of species, those typical of disturbed arable contexts (*Pomatias elegans, Helicella itala* and *Vallonia excentrica*) and those associated with grassland areas (*Carychium tridentatum, Pupilla muscorum* and *Vallonia costata*), suggest alternating arable and pasture, perhaps of a rotational farming regime. The composition of such assemblages would repay further ecological investigation.

Likewise at Figheldean in the Avon Valley, Wiltshire (Allen and Wyles forthcoming), Fordington Bottom, Dorset (Allen forthcoming b) and Newbarn Combe, Isle of Wight (Allen in preparation) molluscan analysis of colluvial sequences indicates that two different environments may have existed contemporaneously in the valley. In these cases it has been suggested that pasture existed in the valley bottom, contemporaneously with arable land on the adjacent slopes, from which most of the soil material is derived. If this is so, then the products of erosion have been carried from, and through, arable fields and transported over grassland before being deposited in the centre of the valley. This system seems to have been reversed by the Medieval period, for often field systems are seen on valley floors and pasture occurred on the valley sides. This is also often the case today. Here we can suggest that erosion and deposition has influenced the location of human settlement and farming. Early farming communities seem to have located settlement in the valley bottoms or at the base of slopes which afford some shelter as well as the valley providing a routeway. Tillage was conducted on the gentler slopes and hilltops. However, by later prehistory, extensive soil tillage and erosion resulted in thin rendzinas which were less profitable to till on the valley sides, while thicker soils existed in the valley bottoms. As a consequence settlement and farming locations may have been reversed; settlement was relocated onto the gentle slopes and plateaux and tillage conducted on the valley floors.

*Environmental systems, thresholds and landscape sensitivity*

The response of soils to pressures exerted by human activities is complex. In an attempt to understand at what point erosion might occur it may be useful to consider briefly the concepts of environmental thresholds and landscape sensitivity which have been successfully deployed by Burrin and Scaife (1988) for alluvial deposits. This may enable us to address a number of questions. An environmental system can withstand a series of pressures caused by external changes. However, eventually these changes will approach the environmental threshold, which once breached, may result in erosion. Erosion and soil thinning result in a changed environmental system (Figure 4.7). Vegetation clearance and the onset of new activities are probably greater contributing factors to erosion than continued ongoing processes such as tillage.

In Figure 4.7 a hypothetical series of events are depicted based upon much of the data summarised in Table 4.1. A variety of erosional events are depicted to illustrate a hypothetical history of sedimentation and soil degradation. Stable conditions are portrayed by the flat portions of the graph, and destabilisation, caused by either a single activity (or events) or the continuation of, for instance, tillage, is depicted by the steeper portions of the graph. As a sequence of events occur, so a hypothetical colluvial sequence is postulated and a change from thicker brown earth soils to thin grey rendzinas mapped. What is now required is full evaluation of this hypothetical model using empirical data (see Table 4.1 for examples) and analysis of environmental system thresholds.

*Chronology of erosion*

Initial clearance of the assumed climax Atlantic forest on the chalk would result in a dramatic change in the environmental system. Soil is not only disturbed but lies unprotected from precipitation and wind. The lack of vegetation cover will decrease evapotranspiration, increase the local groundwater levels and increase runoff making the soil particularly susceptible to erosion. This may explain the truncation of the early Holocene soils as observed in many valley sequences (see Table 4.1 and Figure 4.3) which seems to have occurred in, or by, the Bronze Age. Truncation may, however, have been associated with the preparation of ground for cultivation and subsequent tillage which was widespread by the Bronze Age. It is, therefore, still not certain whether we have to infer catastrophic events as a result of woodland clearance or repeated smaller scale erosional episodes associated with tillage to explain the removal of the early Holocene soils from valley locations.

Despite the potential for the *onset* of tillage to result in severe erosion no major deposits have been attributable to this event, but many have been interpreted as having eroded from arable fields. Colluvium is fairly widespread in the Bronze Age (Table 4.1), and the lack of basal soils indicates earlier (? Neolithic) erosion. The major reason for this occurrence is the virtual wholesale clearances for settlement and farming in the Bronze Age. In the middle to later Bronze Age further erosion episodes have been recognised which have been attributed to winter sowing (Allen 1988; 1991) as Jones (1981) and Reynolds (1981) have detailed. This evidence for Bronze Age and earlier erosion is contrary to Bradley's initial thoughts that colluviation was largely an Iron Age or later phenomenon (Bradley 1978, 41 and 123). Colluvium does, however, become significantly more calcareous in Roman and post-Roman deposits indicating thinning of the soils, and also, in some sequences, such as Strawberry Hill, Newbarn Combe (Figure 4.2) and Gore Cliff (Preece 1980), the Roman and post-Roman deposits are appreciably thicker as well as more calcareous.

The occurrence of chalk lenses and gravel fans, however, seem to have no chronological significance (Table 4.1). it has been argued that winter sowing, in particular, is more likely to result in gravel fans (Boardman and Robinson 1985) and Allen (1988) has suggested that such practices may have occurred as early as the Bronze Age.

Recent work in the Dorchester area (Smith *et al.* forthcoming; Allen forthcoming c) has enabled the writer to suggest that soil erosion can be seen to influence human action. Soil erosion has been causally linked to changes in agricultural regimes. Furthermore, the consequence of soil erosion may have had profound effects upon farming practices and the use of whole landscapes. Thus to a large extent the present-day soil pattern and archaeological landscape is a legacy of prehistoric mismanagement.

## Conclusions

In this paper I have detailed a number of investigations which have been based upon archaeological themes directed at determining land-use history in relation to colluviation. However, I have also attempted to fit this approach within the broader remit of soil studies and the conceptual issues of landscape thresholds and equilibria. In so doing I hope that I have shown how closely related all these disciplines are within the context of soil erosion studies. Studies of soil erosion from an archaeological perspective enable some basic statements to be made.

(1) Chalkland colluvium is largely a result of fluvial (i.e. sheetwash, rill and gully) erosion episodes rather than a continual process of soil creep.

(2) Former early Holocene woodland soils in Wessex were highly erodible and suffered severe erosion and truncation upon vegetation clearance by past human populations.

(3) The establishment of tillage led to episodes of erosion; *continued* tillage, which resulted in soil thinning, ultimately played a role in the pattern of past settlement and farming location, if not practices.

(4) We are still not in a position to determine the precise mechanisms of prehistoric erosion nor have we yet been able to determine whether many of the

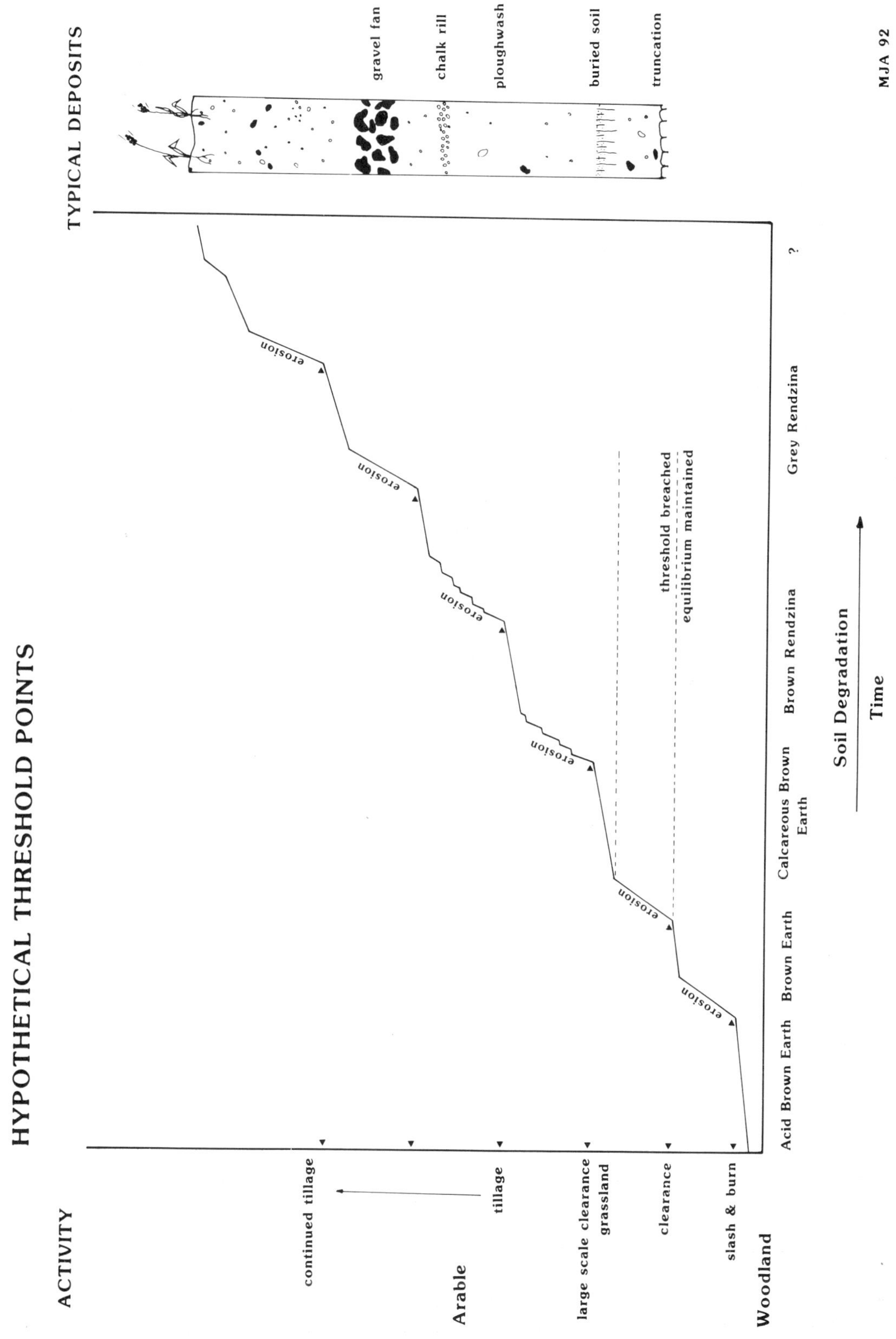


Figure 4.7 Thresholds; schematic figure of hypothetical environmental thresholds and resultant deposits.

homogeneous prehistoric colluvial deposits are the result of gradual incremental erosion or the biological reworking and physical mixing by cultivation of a pulsed, and thus layered, colluvial sequence.

(5) Past erosion events are highly complex even within a single landscape unit and result in significant local diversity, and it is perhaps worth reiterating a statement made by Burrin and Scaife (1988, 212) in relation to alluvial deposits: 'It can be suggested from this diverse pattern and magnitude of valley response through space and time that caution needs to be exerted in archaeological assessment, interpretation and significance of such environmental change'.

**Acknowledgements**

I am indebted to Wessex Archaeology for enabling me to draw freely from their data for this paper and to my research supervisor Arthur ApSimon for his support as much of the fieldwork and analysis was conducted whilst the writer was in receipt of an SERC award. It is a pleasure to acknowledge the continued collaboration, discussions with, and information provided by, Richard Macphail. The comments of a number of people have influenced the content of this paper; these include Martin Bell, John Boardman, John Evans and Rob Scaife, to whom my warm thanks are given. Rosie Braithwaite, Les Allen and Sarah Wyles kindly read previous versions of this paper and their improvements on my original text are gratefully acknowledged.

The soil micromorphological data from Strawberry Hill was provided by Richard Macphail and the work on the Isle of Wight was largely arranged through the kind offices of David Tomalin of the Isle of Wight Archaeological Committee.

**References**

Allen, M. J. 1984 *Excavations at Ashcombe Bottom: first interim report, August 1984.* Lewes Archaeological Group.

Allen, M. J. 1986 Magnetic susceptibility as a potential palaeoenvironmental determinant. *Circaea* 4, 18–21.

Allen, M. J. 1988 Archaeological and environmental aspects of colluviation in south-east England. In: Groenman-van Waateringe, W. and Robinson, M. (eds.) *Man-made Soils.* British Archaeological Reports, International Series 410, Oxford, 67–92.

Allen, M. J. 1991 Analysing the landscape: a geographical approach to archaeological problems. In: Schofield, A. J. (ed.) *Interpreting Artefact Scatters: contributions to ploughzone archaeology.* Oxbow Books, Oxford, 39–57.

Allen, M. J. forthcoming a The Mollusca. In: Smith, R. J. C., *Excavations on the Southern Dorchester By-pass.* English Heritage Monograph.

Allen, M. J. forthcoming b. Land-use history as evidenced by the molluscan and colluvial sequences at Fordington Bottom. In Excavations at Fordington Bottom, Western Link. In: Smith, R. J. C., *Excavations on the Southern Dorchester By-pass.* English Heritage Monograph.

Allen, M. J. forthcoming c. Environment, fields and land management. In: Smith, R. J. C., *Excavations on the Southern Dorchester By-pass.* English Heritage Monograph.

Allen, M. J. in prep. *The land-use history of the Southern English chalklands with an evaluation of the Beaker Period using environmental data; colluvial deposits as environmental and cultural indicators.* Ph.D. thesis in preparation for Department of Archaeology, University of Southampton.

Allen, M. J. and Macphail, R. I. 1987 Micromorphology and magnetic susceptibility: their combined role in interpreting archaeological soils and sediments. In: Fedoroff, N., Bresson, L. M. and Courty, M. A. (eds.) *Soil Micromorphology.* Plaisir: L'Association Français pour l'Etude du Sol, 669–676.

Allen, M. J. and Wyles, S. F. forthcoming The molluscan evidence. In: Newman, C. Excavations at a Romano-British site at Figheldean, Avon Valley. *Wiltshire Archaeological Magazine.*

Barnes, I. and Allen, M. J. 1990 *Codford-Heytesbury A36 road improvement: archaeological evaluation.* Wessex Archaeology Internal Report.

Bell, M. G. 1981a Valley sediments and environmental change. In: Jones, M. and Dimbleby, G. (eds.) *The Environment of Man: the Iron Age to Anglo-Saxon Period.* British Archaeological Reports, British Series 87, Oxford, 75–91.

Bell, M. G. 1981b *Valley sediments as evidence of prehistoric land-use.* Unpub. Ph.D. thesis, Institute of Archaeology, University of London.

Bell, M. G. 1982 The effects of land-use and climate on valley sedimentation. In: Harding, A. F. (ed.) *Climatic Change in Later Prehistory.* Edinburgh University Press, 127–142.

Bell, M. G. 1983 Valley sediments as evidence of prehistoric land-use on the South Downs. *Proceedings of the Prehistoric Society* 49, 119–150.

Bell, M. G. 1986 Archaeological evidence for the date, cause and extent of soil erosion on the chalk. *SEESOIL* 3, 72–83.

Bell, M. G. 1992 The Prehistory of soil erosion. In: Bell, M. G. and Boardman, J. (eds.) *Past and Present Soil Erosion.* Oxbow Books, Oxford, 21–35.

Bell, M. G. and Allen, M. J. 1985 *Valley sediments and molluscs from around Hambledon Hill, Dorset.* Unpublished manuscript for R. Mercer.

Boardman, J. 1984 A morphometric approach to soil erosion on agricultural land near Lewes, East Sussex. In: Lukehurst, C. T. and Gant, R. L. (eds.) *Issues in Countryside Research,* Brighton Polytechnic, Kingston/Brighton Research Papers, 1–10.

Boardman, J. 1988 Severe erosion on agricultural

land in East Sussex, UK, October 1987. *Soil Technology* 1, 333–348.
Boardman, J. 1992 Current erosion on the South Downs: implications for the past. In: Bell, M. G. and Boardman, J. (eds.), *Past and Present Soil Erosion*. Oxbow Books, Oxford, 9–19.
Boardman, J. and Bell, M. G. 1992 Past and present soil erosion: linking archaeology and geomorphology. In: Bell, M. and Boardman, J. (eds.) *Past and Present Soil Erosion*. Oxbow Books, Oxford, 1–8.
Boardman, J. and Robinson, D. A. 1985 Soil erosion, climatic vagary and agricultural change on the Downs around Lewes and Brighton, autumn 1982. *Applied Geography* 5, 243–258.
Bradley, R. 1978 *The Prehistoric Settlement of Britain*. Routledge and Kegan Paul, London.
Burrin, P. J. and Scaife, R. G. 1988 Environmental thresholds, catastrophe theory and landscape sensitivity: their relevance to the impact of man on valley alluviation. In: Bintliff, J. L., Davidson, D. A. and Grant, E. G. (eds.) *Conceptual Issues in Environmental Archaeology*. Edinburgh University Press, 211–232.
Evans, J. G. 1966 Late glacial and post-glacial subaerial deposits at Pitstone, Bucks. *Proceedings of the Geologists' Association* 77, 347–364.
Evans, J. G. 1968 Periglacial deposits on the Chalk of Wiltshire. *Wiltshire Archaeological Magazine* 63, 12–26.
Evans, J. G. 1972 *Land Snails in Archaeology*. Seminar Press, London.
Evans, J. G. and Valentine, K. W. G. 1974 Ecological changes induced by prehistoric man at Pitstone, Buckinghamshire. *Journal of Archaeological Science* 1, 343–351.
Evans, R. 1980 Mechanics of water erosion and their spatial and temporal controls: an empirical viewpoint. In: Kirkby, M. J. and Morgan, R. P. C. (eds.) *Soil Erosion*. Chichester, 109–128.
Evans, R. 1992 Erosion in England and Wales. In: Bell, M. G. and Boardman, J. (eds.) *Past and Present Soil Erosion*. Oxbow Books, Oxford, 53–66.
Fisher, P. F. and Macphail, R. I. 1985 Studies of archaeological soils and deposits by micromorphological techniques. In: Fieller, N. J. R., Gilbertson, D. D. and Ralph, N. G. A. (eds.) *Palaeoenvironmental investigations*. British Archaeological Reports, International Series 258, Oxford, 93–112.
Harris, T. and Boardman, J. 1990 A rule-based expert system approach to predicting waterborne soil erosion. In: Boardman, J., Foster, I. D. L. and Dearing, J. (eds.) *Soil Erosion on Agricultural Land*. Wiley, Chichester, 401–412.
Hodgson, J. M., Catt, J. A. and Weir, A. H. 1967 The origin and development of clay-with-flints and associated horizons on the South Downs. *Journal of Soil Science* 18, 85–102.
Jones, M. 1981 The development of crop husbandry. In: Jones, M. and Dimbleby, G. (eds.) *The Environment of Man: the Iron Age to Anglo-Saxon Period*. British Archaeological Reports, British Series 87, Oxford, 93–127.
Kerney, M. P., Brown, E. H. and Chandler, T. J. 1964 The late glacial and post-glacial history of the chalk escarpment near Brook, Kent. *Philosophical Transactions of the Royal Society, London* B. 248, 135–204.
Kwaad, F. J. P. M. and Mücher, H. 1979 The formation and evolution of colluvium on arable land in northern Luxembourg. *Geoderma* 22, 173–192.
Macphail, R. I., Romans, J. C. C. and Robertson, L. 1987 The application of micromorphology to the understanding of Holocene soil development in the British Isles; with special reference to Holocene cultivation in the British Isles. In: Fedoroff, N., Bresson, L. M. and Courty, M. A. (eds.) *Soil Micromorphology*. L'Association Français pour l'Etude du Sol, Plaisir, 647–656.
Macphail, R. I. 1992 Soil micromorphological evidence of ancient soil erosion. In: Bell, M. G. and Boardman, J. (eds.) *Past and Present Soil Erosion*. Oxbow Books, Oxford, 197–215.
Moffat, A. J. and Cope, D. W. 1984 The Hampshire Chalklands. In: Jarvis, M. G. and Findlay, D. C. (eds.) *Soils of the Southampton District*. British Society of Soil Science, Southampton, 82–116.
Preece, R. C. 1980 The biostratigraphy and dating of a Postglacial slope deposit at Gore Cliff, near Blackgang, Isle of Wight. *Journal of Archaeological Science* 7, 255–265.
Quine, T. and Walling, D. 1992 Patterns and rates of contemporary soil erosion derived using caesium-137 measurement, analysis and archaeological significance. In: Bell, M. G. and Boardman, J. (eds.) *Past and Present Soil Erosion*. Oxbow Books, Oxford, 185–196.
Reynolds, P. J. 1981 New approaches to familiar problems. In: Jones, M. and Dimbleby, G. (eds.) *The Environment of Man: the Iron Age to Anglo-Saxon Period*, British Archaeological Reports, British Series 87, Oxford, 19–49.
Richards, J. C. 1990 *The Stonehenge Environs Project*. English Heritage Archaeological Report 16, London.
Scaife, R. G. 1984 Gallibury Down, Isle of Wight – pollen analysis of a Bronze Age downland palaeosol. *Ancient Monuments Laboratory Report 4240*.
Scaife, R. G. and Burrin, P. J. 1983 Floodplain development and vegetational history of the Sussex High Weald and some archaeological implications. *Sussex Archaeological Collections* 121, 1–10.
Scaife, R. G. and Burrin, P. J. 1985 The environmental impact of prehistoric man as recorded in the upper Cuckmere valley at Stream Farm, Chiddingly. *Sussex Archaeological Collections* 123, 27–34.
Scaife, R. G. and Burrin, P. J. forthcoming Archaeological inferences from fluvial responses; some findings from southern England. In:

Needham, S. P. and Macklin, M. G. (eds.) *Archaeology under Alluvium; archaeology and the river environment in Britain.*

Stammers, R. L. and Boardman, J. 1984 Soil erosion and flooding on Downland areas. *The Surveyor* 164, 8–11.

Thomas, K. D. 1977 A preliminary report on the Mollusca from the lynchet section. In: Bell, M. G. *Excavations at Bishopstone. Sussex Archaeological Collections* 115, 258–264.

Wainwright, G. J. 1971 *Durrington Walls: Excavations 1966–1968.* Society of Antiquaries Research Report 29, London.

# CHAPTER 5

# EROSION IN ENGLAND AND WALES – THE PRESENT THE KEY TO THE PAST

R. EVANS

*Soil erosion induced by misuse of the land is widespread in Britain. In the lowlands water and wind thin the soil of tilled fields; in the uplands peat moorland is spectacularly hagged and gullied and on overgrazed steep slopes bare subsoil and scree are expanding in area. The reasons for present-day erosion are briefly described. In the lowlands valley floor deposits testify that water erosion was widespread in the past, as long ago as the Neolithic period, but in the uplands it seems likely that except during forest clearance erosion did not occur widely until the late eighteenth century. In the lowlands erosion often occurs now where it did in the past, primarily because these landscapes are inherently erodible when cultivated. As land use changes through time, so the risk of erosion changes. The expansion over the last few decades of the area of land sown to winter cereals has led to an increase in the extent and severity of water erosion. Changes in land use in the past could have had similar impacts. What is happening now, therefore, is a key which helps explain the distribution and severity of erosion in the past.*

## Introduction

Soil erosion induced by the misuse of land is widespread in Britain. Much work has been done in the last three decades to assess how erosion has been initiated, to measure its rates and to assess its impacts. The knowledge gained from this work can help explain why erosion occurred where it did in the past. Erosion resulting from forest clearance is discussed elsewhere (Evans 1992), in this paper the erosional impacts of industrial and agricultural activities are described.

Erosion is most obvious now in the uplands, especially the peat moors, which are often deeply hagged and gullied and appear to be eroding rapidly. Grazed steep slopes in the uplands and lowlands also appear vulnerable to erosion in many localities. Paradoxically, there is presently less evidence for erosion in the past in the uplands than there is in the lowlands (Evans 1992).

Rarely is erosion of arable land sufficiently spectacular for people to take note of it. Water and wind erosion of arable land are generally obliterated by the farmer within a short time of their occurrence, so that the mass of the population now, and in times past, has not realised that erosion of arable land has been occurring with significant impacts.

## Erosion of peat moors

Wind and sheep undercut the turf mat of the upstanding hag or peat-face, which is continually trimmed back (Evans 1990a) so exposing an ever-widening band of peat. The peat, on drying out, is susceptible to wind erosion, especially if the surface is disturbed by grazing animals, usually sheep. Peat is also susceptible to frost action, which prepares the surface for erosion by water and wind. At high altitudes, above *c*. 450 m OD, the interaction of climate and the grazing animal is such that once exposed, peat will continue to erode (Evans 1990a). Vegetation cannot get a hold until the mineral subsoil becomes exposed to give a more stable and erosion-resistant surface. But how is the peat first exposed?

Around the fringes of peat moors, at their convex margins, sheep can initiate bare soil scars. Burning, started accidentally or deliberately by the gamekeeper, often leaves a low step at the margin of the burnt heather and unburnt cotton grass; sheep can create scars in this step. In other places, in the centres of the moors or on watershed cols, pools have dried out and the edges of these are now retreating. The drying out of these pools could be related to a series of warmer summers and drier winters, such as those we have witnessed recently, or to the draining of the pools because the water-table has been lowered. Channels cutting back to the pool complex, could release water from the pools or, if they do not reach the pools, lower the water-table sufficiently for the pools to dry out. But what causes channel incision and/or headward retreat?

Present-day evidence suggests that gullies do not retreat rapidly in cotton grass covered peat moors, even in rare storms, as the tussocks are resistant to channel incision and head-cutting (Evans, unpublished). However, as can be seen at present, runoff in storms can strip mosses growing in depressions.

Peak District moors started eroding widely about 250 years ago (Tallis 1990) and present-day rates of erosion could account over that time for much of the peat removed (Burt and Labadz 1990). Defoe's description in 1726 of the summit of the Cheviot (Defoe 1927), is of a smooth grassy non-eroding plateau. It is not now. Palynological evidence shows that for long periods much upland moorland was, as it is now, covered by heather and cotton grass, along with mosses (Simmons 1990). In the Peak District about the turn of the nineteenth century moorlands appeared dominantly mossy (Farey 1811), although shortly after they were mostly covered in heather (Rhodes 1824). In the Cheviots in the seventeenth century valley mires became dominated by mosses (Tipping, pers. comm.). Perhaps the cooler wetter conditions of the Little Ice Age gave mosses a competitive edge over heather and cotton grasses. The mosses were susceptible to sulphate pollution turned out by the factories and mines of the Industrial Revolution (Tallis 1990). Incision and headward retreat into the dying mosses, once started, could have been rapid.

Tallis (1990) noted that mosses declined around the time of the climatic optimum of the thirteenth century, and erosion was initiated in the Berwyn Mountains and the Peak District at that time. However, in this instance, erosion stabilised later when mosses began to grow again.

In the eighteenth century, especially toward the end of that period, sheep began to appear on the hills in large numbers (Prince 1989). They could have initiated a change in vegetation on the drier moors, so that heather became less dominant. Heather is vulnerable to grazing (Macdonald 1990). Compaction of the turf could have led to greater runoff (Evans 1990a). Burning of the moors, started accidentally by lightning or, with the arrival of the gamekeeper to the moors in Victorian times, deliberately, would also increase runoff and erosion (Evans 1992).

The coincidence then of a change in vegetation related to a change in climate, the onset of the Industrial Revolution and the increase in numbers of sheep on the moors, seems to have led in the second half of the eighteenth century to large storms triggering massive erosion. Since then, the increasing number of sheep on the moors, and the burning of the moors, have ensured that in many localities gullies and hags have continued to erode.

**Erosion of steeply sloping rough grazings**

In the 1970s sheep numbers in England and Wales reached levels similar to those pertaining in the Victorian High Farming era, and were probably slightly less than those found at the height of the fold-course system in the late eighteenth/early nineteenth century (Table 5.1). Many hundreds of thousands of sheep were kept in those counties where fold-course agriculture was supreme in the eighteenth and nineteenth centuries (Luccock 1805; Prince 1989), such as Norfolk, and those counties containing large proportions of chalk downland (Table 5.2). With the onset of the severe agricultural depression after the thriving High Farming period of the Victorian 1870s, fold-course agriculture declined. Since 1920 (Table 5.1), the low point, the number of sheep in England and Wales has risen, but now the distribution of sheep is markedly different to what it was in the 1870s. The decline in the number of flocks in the lowlands has been more than matched by the increase on the rough grazings in the uplands (Table 5.2).

Since the Second World War sheep have initiated erosion of mineral soil on steep grassy slopes in many localities in the uplands (Evans 1990a; Carr and Evans, unpublished). Sheepscars are formed which comprise a backwall and scar apron (Evans 1977). The backwall continually retreats because sheep rub themselves upon it or it is broken down by their hooves, whilst the apron extends because the sheep disturb the surface and frost, wind and rain act upon it. Stones are also kicked downslope from the apron onto the grass, so covering and killing it. In places large areas of thin soil are being stripped from the underlying weathered rock.

Grazing intensities which initiate erosion and, coincidentally, cause heather moor to be replaced by grasses, are about one sheep per 0.5 ha of grazing land (Carr and Evans, unpublished). Historically, these grazing pressures are very high (Table 5.1), especially in the uplands (Table 5.2). However, it is not just the

| Period | Mid 14th century | Mid 16th century | 1805 | 1874 | 1920 | 1973 | 1988 |
|---|---|---|---|---|---|---|---|
| Sheep | 7–8 | 8.5 | 25.6 | 22.9 | 13.4 | 19.4 | 29.7 |
| Source | Thirsk 1967a | Thirsk 1967a | Luccock 1805 | MAFF 1874 | MAFF 1920 | MAFF 1974 | MAFF 1989 |

*Notes*

1. During the Victorian High Farming era numbers of sheep were highest in 1874.
2. The county structure of England and Wales was altered in 1974

Table 5.1 Numbers (millions) of sheep in England and Wales - from the mid fourteenth century to the present.

| | 1086 | Mid 14th century | 1805 | 1874 | 1973 |
|---|---|---|---|---|---|
| Lowland Counties: | | | | | |
| Norfolk | 46.2 | - | 722.2 | 795.4 | 75.5 |
| Sussex | - | 110.0 | 863.8 | 576.6 | 252.2 |
| Hampshire | - | - | 516.6 | 612.7 | 91.3 |
| Dorset | 22.3 | - | 632.2 | 524.3 | 95.2 |
| Wiltshire | - | - | 701.0 | 765.3 | 128.1 |
| Upland Counties: | | | | | |
| Cumberland | - | - | 378.4 | 562.8 | 728.1 |
| Westmorland | - | - | 223.7 | 348.5 | 597.5 |
| Yorkshire - | | | | | |
| North Riding | - | - | 365.3 | 731.5 | 880.8 |
| West Riding | - | - | 383.1 | 770.6 | 896.0 |
| Wales | - | - | 1254.2 | 3064.7 | 6551.7 |
| Sources | Darby 1977 | Glasscock 1973 | Luccock 1805 | MAFF 1874 | MAFF 1974 |

*Notes*

1. During the Victorian High Farming era numbers of sheep were highest in 1874.
2. The county structure of England and Wales was altered in 1974.

Table 5.2 Numbers (thousands) of sheep in selected lowland and upland counties and in Wales at different periods.

total number of sheep on a hillside which is important, but their distribution in relation to the vegetation on that slope.

On a slope which is dominantly covered by less palatable vegetation, such as heather, bracken or mat grass, but in which there are patches of palatable grass species, sheep will preferentially graze and shelter where the grass is. So a small number of sheep within a large extent of heather, for example, will graze the grass at high stocking densities. Erosion by overgrazing could be triggered, therefore, at what appear overall to be low stocking densities.

In the uplands, the present-day impacts of herding large flocks of sheep out onto the open fell are often severe, for trampling creates not just footpaths along which water can flow, but over large areas exposes the underlying subsoil, scree and rock.

Fences which restrict the ranging of animals can also lead to breakdown of the turf along the side of the fence because of overgrazing as well as simply the impact of too many hooves.

In the last few years, because of the present-day system of agricultural subsidies, sheep numbers have also increased in the British lowlands. Large flocks of sheep are present once again on the chalk downlands, for example. These flocks are often grazing the steep slopes of the escarpments and valley sides, creating scars on irregular slopes, and elsewhere disrupting the surface vegetation, especially along sheep-paths. Scree formation in the underlying chalk rubble on these thinly soil-covered slopes can then occur.

Rabbits are also initiating erosion now, especially where the vegetation has been disturbed by sheep. Subsoil and chalk debris from rabbit burrows and scrapes quickly overwhelm the vegetation, and the bare soil is then easily eroded by the combined actions of water (rilling) and gravity (slumping). There has recently been a dramatic increase in the number of rabbits (Long 1990), because of a run of warm dry winters. This has resulted in some slopes on the chalk of southern England and in the Yorkshire Dales, for example, becoming severely denuded.

In times past when sheep numbers were high in a locality, soil erosion could have been initiated on steep slopes. In the Scottish Highlands formerly soil-covered screes have been dated to the late eighteenth century, about the time when sheep grazing began to displace crofting in the Highlands (Innes 1983).

In the lowlands sheep have been found in large numbers since at least Roman times (Trow-Smith 1957) especially on chalk and limestone hills (e.g. Defoe 1927; Young 1813a). They were grown for their wool and to fertilize the land and, before the eighteenth century were probably most numerous (Table 5.1) in Tudor times (Thirsk 1967a) and the Medieval period (Donkin 1973; Glasscock 1973) prior to the Black Death, when monasteries, among other large landlords, made a good living from sheep. In the lowlands large flocks of sheep, up to 1000 per 0.4 ha in Tudor times (Thirsk 1967b), were folded on fields of stubble, ley grasses or turnips. These flocks would often be driven along valley sides and escarpments to the fields. As Luccock (1805) noted, sheep had to be able 'to climb without difficulty the most abrupt slopes of a billowy district' of Wiltshire downland. Such flocks could have caused severe disruption of the turf.

Until the late eighteenth century rabbits were generally confined within warrens, so it was not until after this time that widespread erosion would take place on sandlands and the chalk outcrop. Rabbits were especially numerous on the southern chalk (Sheail 1971).

Cattle ranching, often associated with boundary banks, was widespread in the second millennium BC on the southern chalk downlands (Fowler 1981; Ryder 1981). In Medieval times there were the vaccaries of the monasteries in northern England (Donkin 1962; 1973; Winchester 1987) and cattle widely grazed the British uplands in the sixteenth to eighteenth centuries (Trow-Smith 1957; 1959). The impacts on the land of a driven herd could have been severe locally, especially along boundary lines.

Between the slopes which are susceptible to overgrazing and the arable fields is a zone of (now improved) grassland where erosion is rare (Evans 1990b) except around gates or along tracks. In the past this grazed grassland was also probably at little risk of erosion, only if it was converted to crops would it become vulnerable. In a lowland landscape water and wind erosion are common only where arable crops are widely grown.

## Present-day erosion of arable land - factors relevant to erosion in the past

### *Water erosion*

Much more arable land is at risk of erosion by water than by wind (Evans 1990b). Today, rilling of arable land is more widespread and occurs most frequently on sandy soils, for example, in the West Midlands and Nottinghamshire; it is least common on clayey soils (Evans 1988a). Erosion is common on light loams, for example, in Norfolk, and on loamy soils with a high silt content, for example, on the chalk downlands and on Devonian rocks in the county of Hereford and Worcester. Sands are inherently more erodible than clays, being less cohesive and structurally more unstable soils. They are also easily worked soils, so a wide range of crops can be grown on them and hence they are vulnerable to erosion over a much longer period than clays, for example, which are drilled dominantly to autumn-sown crops. Soils with a high silt content are also often structurally unstable and easily slake and cap so that rain runs off rather than soaks into the ground.

Modern technology, particularly the use of large tractors, allows the farmer to work the land mostly when he wants, and to cover large areas of ground quickly. The drawback is that the compacted wheelings of tractors and other agricultural machinery (and most farmers drive up and downslope), form an efficient means of transporting rainfall downslope. Most erosion, therefore, is down wheelings or is associated with them, for example, rills 'break out' from wheelings when the direction of steepest slope diverges from that down which the tractor was driven. To facilitate the use of large machines many field boundaries have been removed since the Second World War (Barr *et al.* 1986), and this has increased the risk of erosion (Evans 1988a).

Soils on chalk landscapes, especially the southern downlands, erode extensively (Boardman 1992). Although field boundaries have been removed and the silty soils are structurally unstable, the major reason that erosion is extensive is the land use on the downs.

Large areas of the southern downlands are now sown to winter cereals, in contrast to only about 20 years ago, when land use was dominantly spring-sown barley, grass for cattle and sheep, and some winter wheat. This change in land use is exemplified by the Sompting basin in Sussex, which covers 10 km² of which 61% is now drilled to winter cereals in contrast to the early 1960s and in 1975 when the figure was about 15% (Boardman and Evans 1991). The increase in area drilled in autumn is the response of farmers who are being paid for the amount they produce; autumn-sown cereals out-yield spring-grown cereals. Until recently the dominant autumn-sown cereal was wheat, but within the last 15 years or so, new varieties of autumn-sown barley have come onto the market, and spring-grown barleys have often been replaced by autumn-sown varieties.

In winter, soils are saturated for a long period when the crop cover, especially if the land was drilled late, is often thin (<30%), runoff is little impeded therefore (Evans 1990c). As more land in a drainage basin goes under winter cereals so flooding occurs more frequently (Boardman and Evans 1991).

The expansion in area of land sown to winter cereals has been country wide, in England and Wales by a factor of about three since the late 1960s (Evans and Cook 1986). Much of this expansion has taken place in: (1) the wetter west where grass has been converted to arable, and where weeds are no longer a problem thanks to modern herbicides, and (2) on the chalk and sands and light loams where the major cereal crop used to be spring-sown barley. Spring barley fields are at risk of erosion for only a short period as the crop is quick to grow in spring and protects the ground surface.

When sheep are folded in autumn and winter on sloping fields of turnips or kale, in wet weather the soil between the rows of crop becomes puddled and runoff and erosion can occur even when the rainfall is gentle (Evans 1988b).

Rates of erosion are lower in fields sown to cereals, than they are in fields sown to sugar beet, potatoes, many horticultural and vegetable crops, and grass (Evans 1988a). All these, except potatoes where the soil is ridged, need smooth seedbeds with the result that little rainfall is absorbed and stored on the surface.

The period when they are sown, often late spring or summer, often coincides with a period of more intense storms.

Erosion is most severe where soils contain a high proportion of coarse silt or fine sand, for example, soils on the Yeovil Sands in Somerset, where on average >3 $m^3$ per ha per field erode in a year (Evans 1988a). Volumes of soil moved are high on sands (>1 $m^3$) and least on clays (<0.3 $m^3$).

Rills and gullies are often confined to small parts of fields (Evans 1988a). On heavy textured soils most rills are in depressions and valley floors, not on valley sides. This is because the combined flow from both valley sides is needed to overcome erosional thresholds. Even on sands and other erodible soils, rills are found only in depressions and valley floors in a large number of fields. Erosion becomes more severe as the relief and slope gradient of a landscape increase (Evans 1990c).

The amount of sediment deposited within a field is highest where soils contain much sand or coarse silt and, conversely, most sediment is transported out of a field where eroded soils are clayey (Evans 1990c). When light textured soils erode, if rainfall is not intense sandy deposits are laid down on the footslope, and below this, often with a gap between, are found silty deposits which become finer textured and more organic-rich towards the centre of the valley floor. When erosion is severe the finer deposits are usually carried out of the field and the coarser sediments laid down in the valley floor, especially against field boundaries. When rills and gullies, the latter too big to be traversed by tractors, have incised into stony subsoils, large gravel fans are deposited on slope concavities or in valley floors. Fans such as these are frequently seen on the South Downs, for example.

Most fields do not erode every year. On average they probably erode about one or two years in ten, although rilling may occur more frequently on erodible soils.

*Wind erosion*

Nowadays, wind mostly erodes sandy soils, especially fine sands, and peats (Evans and Cook 1986). Both these soils have low densities, especially the organic soils, and little cohesion, so that when they are bare of crop particles are easily rolled, saltated or borne by the wind. Areas with soils particularly prone to blowing are the East Anglian Fens, the Lancashire Mosses, the Vales of Trent and York, and sandlands in Suffolk, Nottinghamshire and the West Midlands. Soils of heavier texture are only attacked by the wind after they have been dried out and the clods of earth broken down into small particles by freeze-thaw cycles acting over a period of days. Wind erosion of loamy and clayey soils is rare, and is most obvious when soil particles are deposited on snow. In two decades I have seen only two such examples of erosion, in the Chiltern Hills, near Tring, Hertfordshire and near Strathaven, south of Glasgow, Strathclyde.

Wind erosion happens mostly in spring-time, and although it can occur in fields sown to barley, it is more common in fields drilled with sugar beet, carrots or onions. The seedbeds for these crops are made as smooth as possible, the seeds also take a long time to germinate and grow so the ground is unprotected for a considerable period. Blowing can also occur in autumn after preparation of seedbeds for winter crops.

Blown soil is deposited against field boundaries and in field ditches.

**Erosion of arable land in the past**

*Water erosion - factors which would have exacerbated erosion*

Erosion would be most widespread when population pressures were greatest as more land was sown to crops in these periods, for example, in Romano-British and Medieval times (Evans 1990d). Erosion would also become more prevalent when land was sown to crops vulnerable to erosion, for example, the growing of autumn-sown wheat in the Iron Age (Helbaek 1952 and Romano-British periods (Helbaek 1952; Applebaum 1972). In the Middle Ages, as pressure on land increased, there gradually developed a two and three course rotation system with one year of fallow, and by the thirteenth century such a rotation was an essential of good farming (Thirsk 1964). There was probably much erosion of the fallow land in this period (Evans 1990d). Agricultural improvement, which gathered strength in the second part of the eighteenth century (Prince 1989), probably reduced the incidence of erosion because grassland often replaced the fallow. However, fields down to turnips which were fed to sheep as part of a fold-course agricultural system would be susceptible to washing in winter.

Many crops which were, and still are, mostly grown on light land exacerbate erosion, primarily because they are spring-sown and take a long time to germinate and cover the ground at a time of year when heavy storms are not uncommon. Carrots, for instance, which were grown on the Suffolk Sandlings in the late eighteenth century (Young 1813b). More recently, sugar beet was introduced into eastern England in the 1920s (Evans 1990b). Potatoes are often grown on light land, and have become widespread as a field crop since the beginning of the nineteenth century (Salaman 1949). Compared with the present, rilling of sandland fields sown to spring barley was probably more common in the past. It was recommended the seedbed should be as fine as dust and rolled (Tusser (1580) in Payne and Herrtage 1878), which would produce a surface smoother than that drilled nowadays. Also, the crop should not be sown until May (Markham 1635), much later than now, to avoid the competition of weeds.

In times past, land would mostly be worked downslope, as it is now. The large heavy turnwrest (Kentish) plough could be taken along slope (Blith 1653) but was probably not widely used for that. Manoeuvring a team of oxen or horses pulling a heavy plough along a slope steeper than about 5 degrees must have been difficult, especially when the plough had to be turned round at the end of the field. Ridge-and-furrow of whatever age, is mostly aligned downslope to improve surface drainage, particularly on heavier soils. Even on freely drained soils, such as those on the chalk of the Lincolnshire Wolds and Sussex South Downs, traces of ridge-and-furrow aligned down slope can still be seen on air photos or from across the valley. The risk of soil erosion would be increased by working the land in these ways.

Medieval ridge-and-furrow was often *c.* 200 m long, a more than adequate length for water to gather, flow and erode (Evans 1980). However, the small fields of prehistoric and Romano-British times, mostly less than 1.0 ha, usually smaller than 0.5 ha, with maximum downslope lengths generally considerably less than 100 m (Bowen 1962; Bowen and Fowler 1978; Fowler 1981) would have reduced the severity of erosion.

Prior to the twentieth century there is little in the British literature which tells explicitly of erosion. Fisher (1868, quoted in Douglas 1970) noted erosion in Norfolk and Young (1804a) referred to a dell in Hertfordshire which filled with water and soil. Winter rains and melting snows which washed and scoured the soil were also noted in Devon (Vancouver 1808). In early nineteenth century Worcestershire, mill ponds filled with soil washed from adjacent counties (Pitt 1813). In the late eighteenth and early nineteenth centuries it was noted that sandy soils (Stone 1794; Young 1804b), and those on chalk downland (Vancouver 1810) were liable to erode by water, as they are now.

Blith (1653) seemed unconcerned about winter water 'running over lands' but considered noteworthy 'Sands brought forth by the violence of strong land floods'. Summer storms could severely erode recently drilled turnips (Kilvert 1870-79) and fallows (White 1788). Dung (Blith 1653; Fitzherbert (1534) in Skeat 1882), manure (Blith 1653; Tusser (1580) in Payne and Herrtage 1878; Young 1813c), 'mucilage' (Stone 1794) or 'fatness' (Plattes 1639) was washed off land; by implication soil particles too must have been transported out of the field. Ways were devised to work the land to stop this happening (Fitzherbert (1534) in Skeat 1882) and to catch the transported dung (Blith 1653; Plattes 1639).

From the evidence noted above it seems likely that water erosion was widespread in the past. This erosion was accompanied by the laying down of deposits in dells and valley floors within fields, and the transporting of soil particles into streams to be later deposited in mill ponds and as alluvium on valley floors or coastal marshes.

### *Valley floor deposits as indicators of past erosion - some general comments*

The presence of alluvium and colluvium in valley floors is evidence of water erosion in the past; erosion and deposition started when the land was cleared of woodland (Bell 1981; 1982; 1983; Evans 1990d). Alluvium, often but not always fine textured, flanks permanent streams and is waterlogged for much of the year. Colluvium is found in dry valley floors and on footslopes, as well as banked against field boundaries; it has not travelled far. In some instances, where adjacent valley sides are too steep for agriculture, colluvium could have been derived from erosion triggered by overgrazing, probably by sheep, or it could have come from fields higher up the valley.

Colluvium has been widely mapped in England by soil surveyors. By definition (Avery 1980) it is material washed down from fields; it often contains higher but fluctuating organic matter contents down the soil profile than do adjacent similar soils on slopes, and may contain charcoal fragments and pottery artifacts, as well as frequently burying an earlier-formed soil.

Colluvium occurs especially on the chalk (Bell 1981; 1982; 1983), in places on sandlands and light loams in the Midlands and East Anglia, on the Devonian outcrop in Herefordshire, on the Jurassic and Permian Limestones in southern and central England (Evans 1990d), and on the Jurassic Yeovil Sands in Somerset (Colborne and Staines 1987). Soils on most of these rocks, other than on the limestones where they are heavier textured, contain much silt or sand. Colluvium is occasionally found over clayey till or bedrock.

The distribution of colluvium bears a marked similarity to the distribution of present-day erosion (ENDS 1984; Evans 1990b), except on the lowland limestones where rilling is not often recorded at present. The general lack of colluvium on clays is not surprising for much sediment is carried out of eroded clay fields nowadays.

Except on sandlands, these eroded areas have for long periods been used for mixed though dominantly arable farming (Applebaum 1972; Fowler 1981; Hallam 1988; Piggott 1981; Prince 1989; Thirsk 1967a; 1984), not pastoral farming. The southern chalklands are the prime example of this, in places settled since Neolithic times. Sandlands were often under heath until the late eighteenth century when great efforts were made to improve and reclaim this 'waste' land, to provide food for a fast expanding population (Evans 1990d; Mingay 1989). Reclamation has continued intermittently up to the present, for example, since the Second World War, parts of the Suffolk Sandlings.

The estimated volume of colluvium deposited in a valley floor can be equated with the depth of soil stripped from the adjacent valley slopes. Values of surface lowering are 170–200 mm on sandstones and siltstones; generally between 90–200 mm on the chalk, values being higher where soils are more silty; mostly less than 100 mm on other loamy soil parent materials; and lowest (*c.* 50 mm or less) where soils are heaviest in texture (Evans 1990d). The relativities of these values compare well to present-day erosion rates which are higher on soils containing much sand or silt and lowest on clays.

Erosion of arable land on chalk has taken place at intervals over 5000 years, whereas much of the sandlands has only been in agricultural use for the last 250 years or so. The discrepancies in rates of erosion are greater than they first appear therefore, and we do know that at present erosion is more rapid, occurs more frequently and is more extensive on sandlands. The present-day low rates of erosion on clays accompanied by the transport of soil out of the eroded fields can explain why on much of the outcrop of Jurassic and Permian limestones with their dominantly fine-textured soils, rates of colluviation were lower.

*Valley floor deposits as indicators of past erosional processes - a detailed study*

Recently, a soil map was published of 200 km² of land near Yeovil in Somerset (Colborne and Staines 1987). Colluvium and alluvium occur extensively in this locality. Plant debris or charcoal from the base of three colluvial soil profiles were dated by radiocarbon methods (Staines, unpublished). Erosion at the three sites post-dates 290±80 BP (I–14111; *c.* AD 1695), 1460±80 BP (I–14110; *c.* AD 525) and 3300±90 BP (I–14295 *c.* 1300 BC). Bronze Age and Romano-British pottery was found at another colluvial site. These dates imply that erosion began at different times at each site.

Two colluvial soils were mapped in valley floors, on footslopes and occasionally banked-up against field boundaries: the Yeovil Series, a light silty soil (sandy silt loam to very fine sandy loam), and the Conway Series (colluvial phase), generally slightly heavier in texture (clay loams to silty clay loams). The soils are dominantly found on the Upper Lias Yeovil Sands, comprising soft and hard sandstones, siltstone and limestone, and rarely on the Middle Lias Pennard Sands of siltstone and soft sandstone. They were not mapped on mudstones and limestones of other parts of the Jurassic sequence nor on rocks of the Cretaceous period. The soils on the slopes adjacent to the Yeovil and Conway (colluvial phase) Series comprise dominantly stoneless fine sandy silt loams, silty clay loams or clay loams; soils on slopes of other lithologies are mostly more stony and of clay loam or clay texture.

The position in the landscape of these two colluvial soils is different. The Yeovil Series is often higher in the landscape, and may be within steep sided valleys etched in escarpments; 16% of the basins containing Yeovil Series lie directly upslope of the colluvial phase of the Conway Series. Two soil series, the Bridport and South Petherton Series, dominantly of fine sandy silt loam topsoil texture, comprise 80% of the soils on slopes adjacent to the Yeovil Series whereas they comprise only 59% of the slopes draining to the Conway (colluvial phase); the remaining soils in the basins are dominantly silty clay loams. The heavier texture of the Conway (colluvial phase) Series is related therefore to its lower position in the landscape and that a greater proportion of the adjacent soils are heavier textured.

Where colluvium was mapped erosion nowadays is extensive on land tilled for crops, though much of the area is under grass, and rates of erosion are high (Colborne and Staines 1985; Evans 1988a). Estimated amounts of drainage basin surface lowering in this locality are high, the mean being 133 mm (Table 5.3).

Staines (unpublished) estimated that of the 34 km² of erodible soils, 25 km² had eroded and 11 km² was covered by colluvium. Assuming a mean depth of colluvium of 1 m, 11 M m³ of soil have been eroded. Comparing the depth of eroded and little-eroded soils Staines considered that on average the surface had been lowered by 400 mm. Using this figure of surface lowering he estimated that 10.057 M m³ had been removed. The two calculations equate well. For the whole of the area at risk of erosion Staines estimated a surface lowering of 170 mm. This is slightly higher than the 146 mm calculated from Table 5.3.

Yeovil Series colluvial soils

Surface lowering of drainage basins containing colluvium in valley floors:

| Number | Mean | Median | Range |
|---|---|---|---|
| 79 | 133 | 121 | 7–386 |

Surface lowering of slopes above colluvial footslopes:

| Number | Mean | Median | Range |
|---|---|---|---|
| 38 | 292 | 235 | 60–789 |

Surface lowering all slopes:

| Number | Mean | Median | Range |
|---|---|---|---|
| 117 | 184 | 159 | 7–789 |

Conway Series colluvial soils

Surface lowering of drainage basins containing colluvium in valley floors:

| Number | Mean | Median | Range |
|---|---|---|---|
| 87 | 85 | 74 | 11–286 |

Surface lowering of slopes above colluvial footslopes:

| Number | Mean | Median | Range |
|---|---|---|---|
| 8 | 261 | 283 | 45–590 |

Surface lowering all slopes:

| Number | Mean | Median | Range |
|---|---|---|---|
| 95 | 100 | 76 | 11–590 |

Table 5.3 Surface lowering (mm) of the Yeovil Sands, Somerset

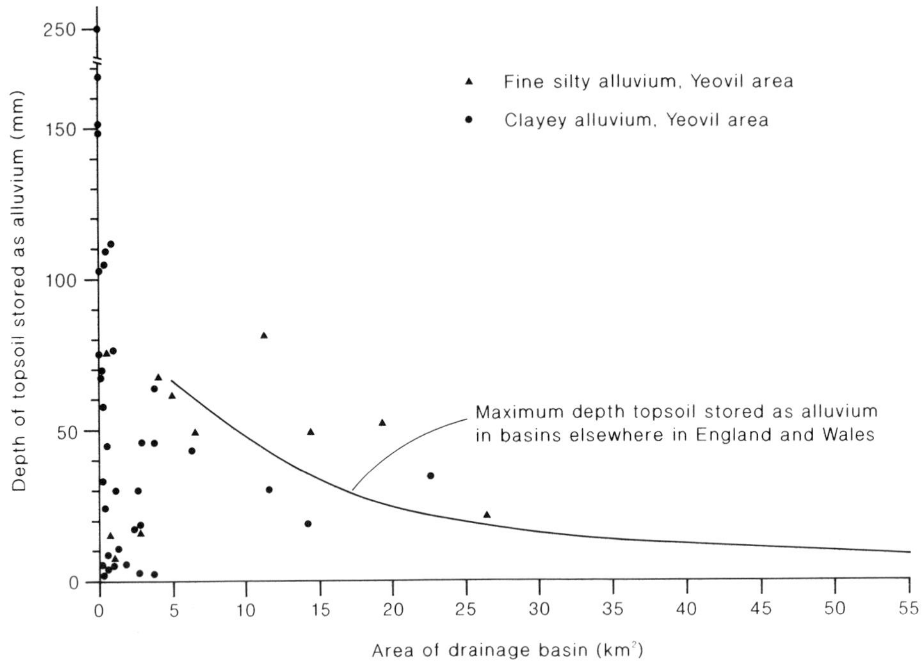


Figure 5.1 Depth of eroded soil stored as alluvium in the Yeovil area, Somerset, and elsewhere in England and Wales (Evans 1990d).

Depths of surface stripping are higher, to be expected from present-day evidence, of the generally coarser textured soils flanking the Yeovil Series (Table 5.3). Surface lowering appears much higher where colluvium is found along footslopes rather than in valley floors. This is probably not so, but relates to the transporting of soil out of those fields which contain valleys. Sediment carried in water flowing through rills debouching onto footslopes is mostly deposited within a short distance as slope angle declines and flow velocity decreases, deposition is especially marked if a field boundary holds up flow. However, the converging in a valley floor of water flowing from adjacent slopes results in sediment being transported along the valley floor; deposition may take place where the gradient declines, but mostly sediment is deposited only where the water is ponded back against a field boundary. And even then, much water, with its sediment load, often flows over and down valley of the field boundary. The sediment transported out of fields may be deposited downvalley as alluvium.

In those drainage basins where alluvium is found down valley, the equivalent of a further 32 mm of stripped surface soil is stored in the valley floor between the merging of the colluvium with the alluvium and the confluence with the alluvium of the next valley system. In 12 larger drainage basins where both colluvium and alluvium were mapped, ranging in size from 39–2646 ha (mean 689 ha), the volumes of soil stored in the valley floor are equivalent to a basin-wide topsoil stripping on average of 93 mm (range 50–153 mm).

In those valleys where no colluvium was mapped but alluvium was, the depth of surface lowering of the 27 valleys contributing silty clay loam deposits is 66 mm, but only 41 mm for the nine valleys where finer textured alluvium is derived from clays.

The depth of topsoil stored in alluvium can be greater than that generally found in England and Wales (Figure 5.1), reflecting not only that the soils on the Yeovil Sands are more erodible than most, but that clayey soils here appear more erodible than in other soil-mapped localities. Clays in nearby Dorset also erode more rapidly than, for example, they do in Cambridgeshire and Bedfordshire, which was partly attributed to the much higher rainfall in south western England (Evans 1988a).

The colluvial drainage basins eroded are generally of smaller size than those found elsewhere, reflecting the greater erodibility of these soils. For instance, over 50% of the basins are smaller than 20 ha in extent (Figure 5.2), whereas for basins elsewhere in England and Wales the figure is below 40% (Evans 1990d). As drainage basin area increases in size rates of surface lowering decline exponentially (Figure 5.3).

An analogy can be drawn between the distribution of rills and the severity of erosion in present-day fields with the distribution and severity of erosion in the landscape of the past. At the field scale rilling generally affects only small parts of the field, so though rates of soil loss are high in those localities they are much lower when averaged over the whole field (Evans 1988a). At the landscape scale the

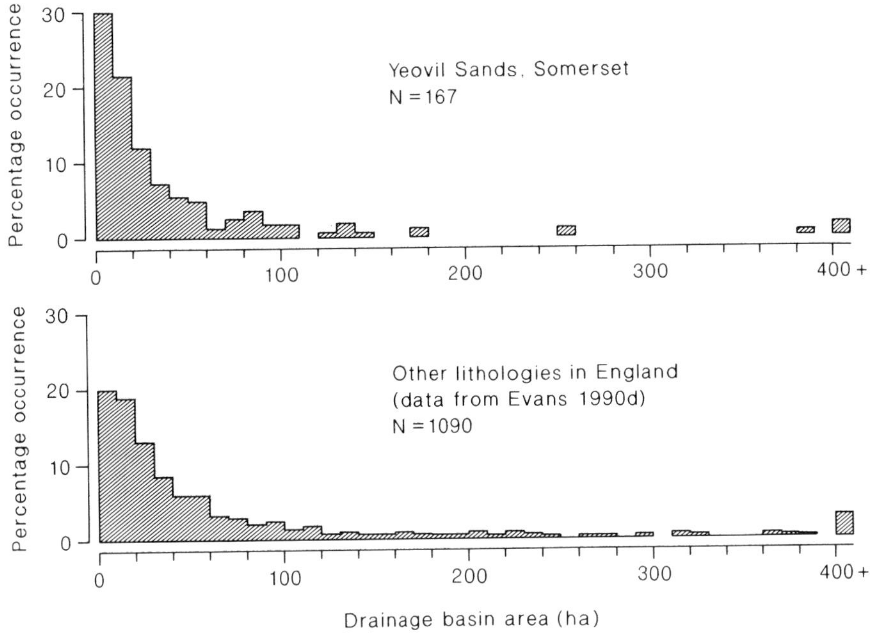


Figure 5.2 Areas of colluvial drainage basins on Yeovil Sands, Somerset, and elsewhere in England.

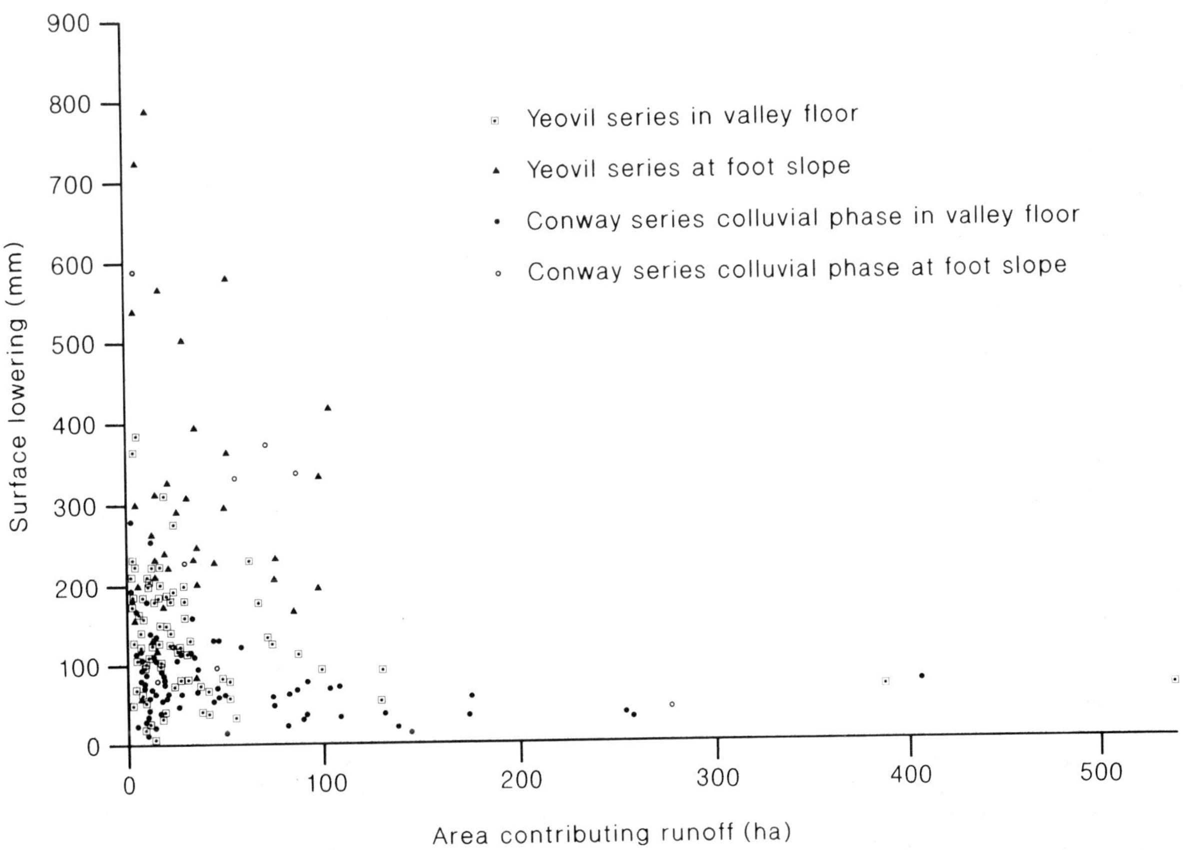


Figure 5.3 Drainage area and surface lowering on Yeovil Sands

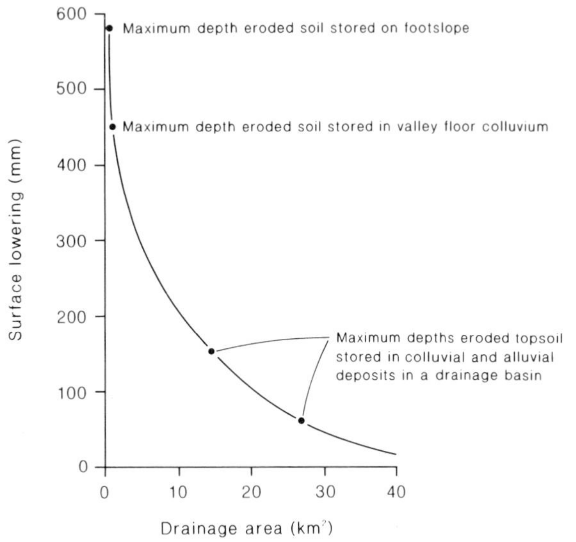


Figure 5.4 Drainage basin area and surface lowering in the Yeovil area, Somerset.

smaller the eroded drainage basin the higher the rate of topsoil stripping (Figure 5.4).

*Soil textures of colluvium and adjacent valley side topsoils as indicators of erosional processes*

The colluvial soils in the Yeovil area are not very different in texture from the soils on adjacent slopes. This is not always so, although often it is on chalk. Examination of 24 Soil Survey publications (for list contact author) which give particle-size analyses of both colluvial soils and soils on adjacent slopes, suggests that colluvium derived from soils on slopes with more than *c.* 25% clay has a lower clay content (Figure 5.5). Conversely, colluvium derived from soils on slopes containing less than *c.* 25% clay has a higher clay content. These changes can be related to the erosional processes acting now.

Soils with a low clay / high sand content are easily eroded. When rainfall is heavy, sand is deposited on footslopes and in valley floors, and finer particles may either be transported out of the field or come to rest where water is ponded back, especially above field boundaries. However, when rainfall is light, sand is not moved far and finer particles come to rest on the lower part of the footslope and in the valley floor. Small erosional events occur more frequently and so the coarser deposits laid down in valley floors at infrequent intervals gradually become replenished by fines, and ploughing mixes the deposits so slowly the soil becomes more loamy.

Soils like those on Yeovil Sands or loessially derived soils on chalk, have a high silt or fine sand content, and often contain only a small fraction coarser than fine sand (>200µ) size. It is the coarser fraction which separates out as the velocity of runoff slows. The clay fraction (<2µ) remains bound with the silt fraction as aggregates. Hence, the colluvium is similar in texture to the uneroded topsoil.

Heavier textured soils, especially on slopes, are less easily eroded. Rain sufficiently intense to initiate rilling on slopes undoubtedly leads to fine particles being carried out of the field in a valley floor rill (Evans 1990c). However, the coarser sand particles and soil aggregates often are deposited in the valley floor, so lightening the texture of the topsoil.

In many present-day instances of severe erosion on chalk downland, massive gravel fans are deposited on footslopes and valley floors (Boardman 1992). Similar features are occasionally recorded in archaeological excavations (Bell 1983 and 1992, p.27). The present-day soils on the chalk are often very thin (<200 mm), much thinner than they used to be. Because of this rills are now more frequently incising through the soil into the chalk rubble beneath; the soils are also more stony now because ploughing incorporates chalky rubble into the topsoil.

*Wind erosion of lowland soils*

What archaeological evidence there is for wind erosion in the past (Cornwall 1953; Evans and Jones 1979) refers to soils mostly on the southern chalk outcrop and on the Thames terraces in Oxfordshire. It seems to have occurred mostly in the Bronze Age, a time when the climate was warmer and drier than now. Soils in these localities are considered to be at no risk of blowing now. Perhaps wind erosion was a winter phenomenon? Alternatively, the soil surface could have been bare of vegetation, because of overgrazing for instance, and the surface could have been constantly disturbed by grazing herds.

Sand dunes in the Suffolk Breckland, still visible in fields near Brandon, are reminders that misuse of arable land and of overgrazing of heath by rabbits (Sheail 1971: Young 1813b) in the past, did lead to wind erosion. Defoe (1927) in 1726 described Bagshot Heath as a 'desert', implying blowing sand. Young (1804b) noted blowing sand in parts of Norfolk, and the coastal sandlings in Suffolk (1813b). He also commented that rolling sandy soils in Sussex (1813a) made them blow, and that there were 'deserts' in parts of Lincolnshire (1813c).

However, large parts of the sandlands in Lincolnshire, Nottinghamshire, the West Midlands and Suffolk, for example, for which there is present-day evidence of wind erosion (Evans 1990b), have only been at risk since they were enclosed for arable in the last 250 years (Evans 1990d). Blowing of peat and/or fine sand in the Lancashire Mosses, the Vales of York and Trent (Evans 1990d), and the East Anglian Fens (Seale 1975), has only come about since these areas were drained and reclaimed on a large scale for arable agriculture. Drainage began in the 1630s in the Fens, the late eighteenth century in the Mosses, and more recently still, much of it this century, in the Vale of York.

Prior to the seventeenth century wind erosion of arable soils was probably uncommon, except in the

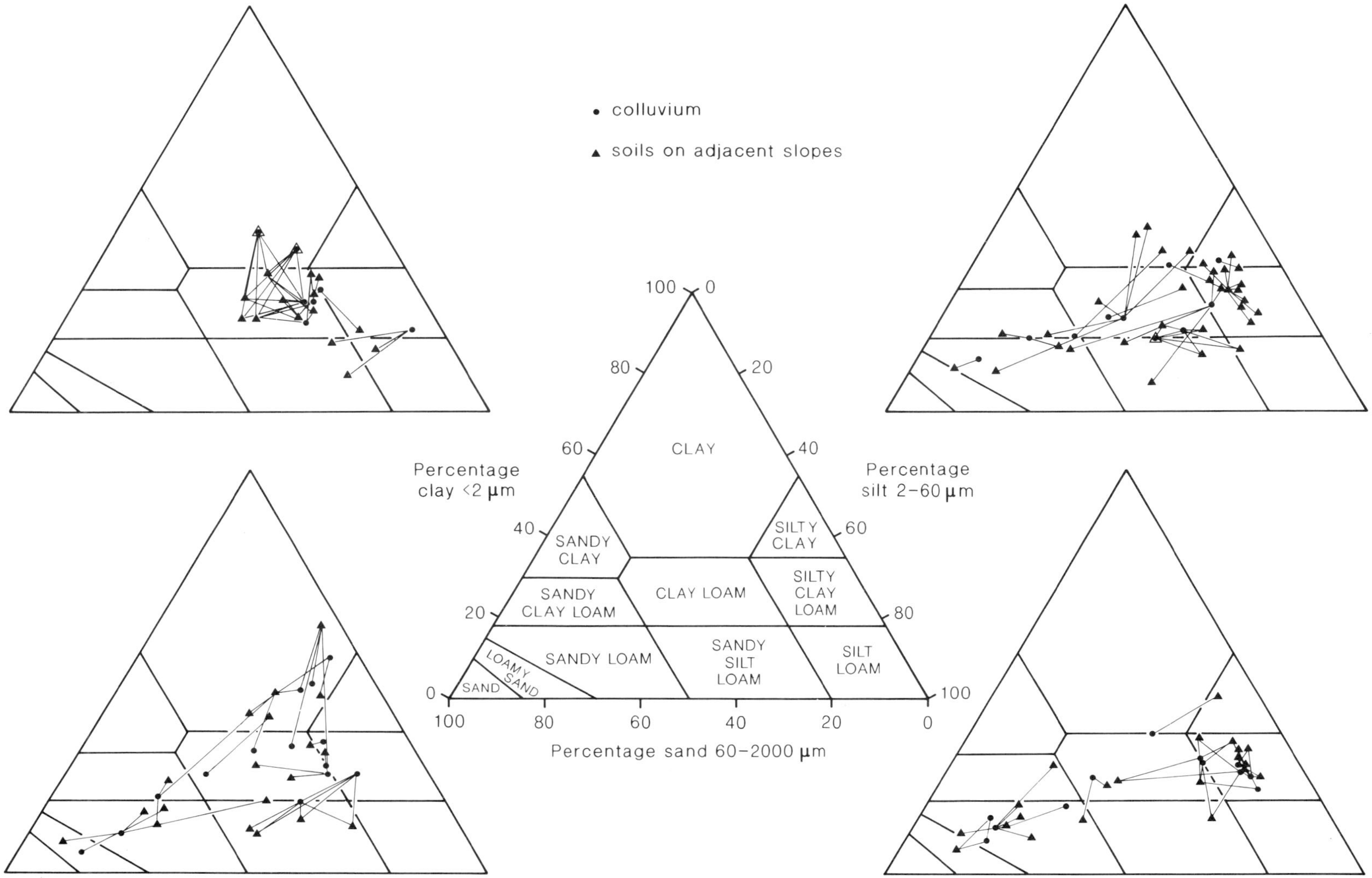


Figure 5.5 Particle-size analyses of topsoils of colluvium and soils on adjacent slopes in 24 localities in England and Wales. To avoid overlap and confusion the 24 localities have been divided up and plotted onto four triangular diagrams.

Breckland, and had little impact on either the farmer or the landscape. Water erosion was far more important, as it is now.

## Conclusions

The present does provide a key to the past. The factors which control erosion now: land use, landform, soil and crop types, the direction the land is worked, and economic and population pressures, did so in the past. Both literature sources and stratigraphical sequences show that erosion in the past often occurred where it is most frequently found now, that is, on the more erodible landscapes of the arable lowlands. It is fortunate that in the prehistoric past fields were often smaller in size then they were later, otherwise erosion could have been more severe. More land is probably at risk of erosion at present than at any time since the Medieval period.

When population pressure became intense in the seventeenth and particularly the eighteenth centuries, the response was to enclose the communal arable open-fields and replace the dunged fallow by grazed grass (Thirsk 1967b; 1984; Prince 1989). Arable land was also increased at the expense of the waste (Prince 1989). Improving pastures and introducing better ways of feeding cattle and particularly sheep, especially the use of root crops either fed directly to folded sheep on light soils or harvested and stall-fed where soils were heavier, improved the carrying capacity of the land and allowed it to be more heavily fertilized (Mingay 1989; Prince 1989). This not only improved yields of cereals, so helping to feed the rapidly expanding population, but the increasing area of land sown to grass reduced the risk of erosion.

With the coming of artificial fertilizers in the middle of the nineteenth century the chance came to shift to continuous arable. However, the collapse of the agricultural economy in the late 1870s which lasted, except for a brief period during the First War, to 1939, forestalled this development until after the Second War. Since then, the policy of the British Government, and latterly the European Community, has been to increase the intensity of food production leading both to more water and wind erosion of arable land and erosion of upland slopes due to overgrazing.

Modern techniques of farming have exacerbated erosion and probably have increased its severity. However, present-day rates of erosion and frequency of occurrences cannot account for the depths of soil stripped in the past and now stored as colluvium and alluvium. Even on long-settled land such as the southern chalk, if erosion was acting at present-day rates it would have had to have occurred much more often than it does now to have deposited the amounts of topsoil presently stored as colluvium. It is also assumed that most erosion occurred on slopes, where presently it mostly occurs in valley floors. Similar calculations for those localities where erosion is more severe now, the Yeovil Sands area in Somerset or the Midlands sandlands, for instance, yield a similar answer. In the past, rates of erosion were either higher than they are now, or, and more likely, rilling occurred more often.

Until grass began to replace fallow land, from the sixteenth century on (Thirsk 1967b; 1984; Prince 1989), much bare land was at risk of erosion in wet winters or in spring or summer storms. Weeds were probably rampant in those days (Applebaum 1972; Fitzherbert (1534), in Skeat 1882; Mate 1985; Markham 1635; Tusser (1580) in Payne and Herrtage 1878) and could have protected the soil, but soils were ploughed up to four times during the fallow to kill the weeds (Thirsk 1967b). Continuous stirring of the soil, and the use of the land for growing crops every year when it was not fallow (often one year in two or three), would lead to low soil organic matter contents and structurally unstable soils. Such soils slaked to give smooth capped surfaces vulnerable to rilling.

Fitzherbert ((1534) in Skeat 1882) noted that land should not be left fallow in winter as the rain would wash the dung off the land, and also beat the soil flat so that it would bake hard. It is noteworthy that those areas where soil erosion was widespread in the past are areas where soils are structurally unstable now. The devastating impacts of summer storms on fallows (White 1788), and even fields down to root crops (Kilvert 1870-79), give some idea of the impacts of what must have been quite frequent events in earlier times.

As evidence grows that erosion was widespread and frequent in the past, it becomes puzzling why it was rarely explicitly mentioned (Blith 1653) in the literature of the sixteenth to nineteenth centuries which dealt with improving agricultural techniques. That erosion occurred was implicit in the warnings that light land had to be worked in certain ways to stop manure from being washed away (Blith 1653; Fitzherbert (1534) in Skeat 1882; Tusser (1580) in Payne and Herrtage 1878), and that this transported manure (and soil) could be trapped in downslope fields (Blith 1653; Plattes 1639). Perhaps the impacts of erosion, over the short term – flooding and silting, over the long term – declining agricultural productivity, were not comprehended. Or, erosion was so widespread and frequent that it was considered nothing could be done about it, particularly in a subsistence economy where food production, not sustainability of land use, was the prime concern.

## Acknowledgements

I wish to thank an unknown referee and Dr. J. Boardman for their helpful comments on earlier drafts of this paper. S. J. Staines and Dr. R. Tipping kindly gave permission to quote from their work.

## References

Applebaum, S. 1972 Roman Britain. In: Finberg, H. P. R. (ed.) *The Agrarian History of England and Wales. Vol. I. II. AD 43–1042*. Cambridge University Press, Cambridge, 1–277.

Avery, B. W. 1980 *Soil Classification for England and Wales*. Soil Survey Technical Monograph No. 14, Harpenden.

Barr, C., Benefield, C., Bunce, B., Ridsdale, H. and Whittaker, M. 1986 *Landscape Changes in Britain*. Institute of Terrestrial Ecology, Abbots Ripton.

Bell, M. 1981 Valley sediments and environmental change. In: Jones, M. and Dimbleby, G. W. (eds.) *The Environment of Man: the Iron Age to the Anglo-Saxon Period.* British Archaeological Reports 87, Oxford, 75–91.

Bell, M. 1982 The effects of land-use and climate on valley sedimentation. In: Harding, A. F. (ed.) *Climatic Change in Later Prehistory*. Edinburgh University Press, Edinburgh, 127–42.

Bell, M. 1983 Valley sediments as evidence of Prehistoric land-use on the South Downs. *Proceedings of the Prehistoric Society* 49, 119–50.

Blith, W. 1653 *The English Improver Improved.* John Wright, London.

Boardman, J. 1992 Current erosion on the South Downs and implications for the past. In: Bell, M. and Boardman, J. (eds.) *Past and Present Soil Erosion*. Oxbow Books, Oxford, 9–19.

Boardman, J. and Evans, R. 1991 *Flooding at Steepdown.* Report to Adur District Council.

Bowen, H. C. 1962 *Ancient Fields*. British Association for the Advancement of Science, London.

Bowen, H. C. and Fowler, P. J. (eds.) 1978 *Early Land Allotment in the British Isles*. British Archaeological Reports 48, Oxford.

Burt, T. and Labadz, J. 1990 Blanket peat erosion in the southern Pennines. *Geography Review* 3, 31–5.

Carr, S. C. and Evans, R. Overgrazing and erosion in a Lake District valley, unpublished manuscript.

Colborne, G. J. N. and Staines, S. J. 1985 Soil erosion in south Somerset. *Journal of Agricultural Science, Cambridge* 104, 107–12.

Colborne, G. J. N. and Staines, S. J. 1987 *Soils in Somerset I. Sheet ST41/51.* Soil Survey Record No. 111, Harpenden.

Cornwall, I. W. 1953 Soil science and archaeology with illustrations from some British Bronze Age monuments. *Proceedings of the Prehistoric Society* 19, 129–47.

Darby, H. C. 1977 *Domesday England.* Cambridge University Press, Cambridge.

Defoe, D. 1927 *A Tour thro' the whole Island of Great Britain* Vol. 2. Peter Davies, London.

Donkin, R. A. 1962 Cattle on the estates of Medieval Cistercian Monasteries in England and Wales. *Economic History Review* 15, 31–53.

Donkin, R. A. 1973 Changes in the early Middle Ages. In: Darby, H. C. (ed.) *A New Historical Geography of England.* Cambridge University Press, Cambridge, 75–135.

Douglas, I. 1970 Sediment yield from forested and agricultural lands. In: Taylor, J. A. (ed.) *The Role of Water in Agriculture.* Pergamon Press, Oxford, 57–88.

ENDS 1984 Soil erosion: an unsustainable face of modern farming, *Environmental Data Services (ENDS) Report* 115, August, 9–10.

Evans, J. G. and Jones, H. 1979 Mount Pleasant and Woodhenge: the land molluscs. In: Wainwright, G. J. (ed.) *Mount Pleasant, Dorset: Excavations 1970–71.* Society of Antiquaries of London, Research Report 37, 190–213.

Evans, R. 1977 Overgrazing and soil erosion on hill pastures with particular reference to the Peak District. *Journal of the British Grassland Society* 32, 65–76.

Evans, R. 1980 Characteristics of water-eroded fields in lowland Britain. In: De Boodt, M. and Gabriels, D. (eds.) *Assessment of Erosion.* John Wiley, Chichester, 77–87.

Evans, R. 1988a *Water Erosion in England and Wales 1982–1984*. Report for Soil Survey and Land Research Centre, Silsoe.

Evans, R. 1988b Some effects of soil erosion in Britain. *Geography Review* 1, 34–7.

Evans, R. 1990a Erosion studies in the Dark Peak. *Proceedings of the North of England Soils Discussion Group* 24, 39–61.

Evans, R. 1990b Soils at risk of accelerated erosion in England and Wales. *Soil Use and Management* 6, 125–31.

Evans, R. 1990c Water erosion in British farmers' fields – some causes, impacts, predictions, *Progress in Physical Geography* 14, 199–219.

Evans, R. 1990d Soil erosion: its impact on the English and Welsh landscapes since woodland clearance. In: Boardman, J., Foster, I. D. L. and Dearing, J. A. (eds.) *Soil Erosion on Agricultural Land.* John Wiley, Chichester, 231–54.

Evans, R. 1992 Sensitivity of the British landscape to erosion. In: Thomas, D. and Allison, R. (eds.) *Landscape Sensitivity*. John Wiley, Chichester, in press.

Evans, R. (unpublished) Impact of a large storm on a stream network in the British Pennines, unpublished manuscript.

Evans, R. and Cook, S. 1986 Soil erosion in Britain, *SEESOIL* 3, 28–59.

Farey, J. 1811 *A General View of the Agriculture and Minerals of Derbyshire.* Macmillan, London.

Fowler, P. J. 1981 Later Prehistory. In: Piggott, S. (ed.) *The Agrarian History of England and Wales. Vol. I. I. Prehistory.* Cambridge University Press, Cambridge, 61–298.

Glasscock, R. E. 1973 England circa 1334. In: Darby, H. C. (ed.) *A New Historical Geography of England*. Cambridge University Press, Cambridge, 136–85.

Hallam, H. E. (ed.) 1988 *The Agrarian History of England and Wales. Vol. II. 1042–1350.* Cambridge University Press, Cambridge.

Helbaek, H. 1952 Early crops in southern England, *Proceedings of the Prehistoric Society* 18, 194–233.

Innes, J. L. 1983 Lichenometric dating of debris flow deposits in the Scottish Highlands. *Earth Surface Processes and Landforms* 8, 579–88.

Kilvert, F. 1870-79 *Kilvert's Diary 1870-1879.* Plomer, W. (ed.) reprinted 1977. Jonathan Cape,

London.
Long, E. 1990 Rabbits. Revivalists run riot. *Farmers Weekly* 6 July, 65–8.
Luccock, J. 1805 *The Nature and Properties of Wool.* Edward Baines, Leeds.
Macdonald, A. 1990 Heather damage: a guide to types of damage and their causes. *Research and Survey in Nature Conservation. Conservation Management in the Uplands* 28, Nature Conservancy Council, Peterborough.
MAFF 1874 *Agricultural Returns of Great Britain.* HMSO, London.
MAFF 1920 *Agricultural Returns of Great Britain.* HMSO, London.
MAFF 1974 *Agricultural Statistics UK 1973.* HMSO, London.
MAFF 1989 *Agricultural Statistics UK 1988.* HMSO, London.
Markham, G. 1635 *The English Husbandman.* William Sheares, London.
Mate, M. 1985 Medieval agrarian practices: the determining factors. *Agricultural History Review* 33, 22–31.
Mingay, G. E. 1989 Farming techniques. In: Mingay, G. E. (ed.) *The Agrarian History of England and Wales. Vol. VI. 1750–1850.* Cambridge University Press, Cambridge, 275–383.
Payne, W. and Herrtage, S. (eds.) 1878 *Five Hundred Pointes of Good Husbandrie* by T. Tusser, reprint of 1580 edition. English Dialect Society, London.
Plattes, G. 1639 *Discovery of Infinite Treasure*, facsimile of first edition, reprinted 1974. Walter J. Johnson Inc., and Theatrum Orbis Terrarum Ltd., Amsterdam.
Piggott, S. (ed.) 1981 *The Agrarian History of England and Wales. Vol. I. I. Prehistory.* Cambridge University Press, Cambridge.
Pitt, W. 1813 *General view of the Agriculture of Worcester*, reprinted 1969. David and Charles, Newton Abbot.
Prince, H. C. 1989 The changing rural landscape 1750–1850. In: Mingay, G. E. (ed.) *The Agrarian History of England and Wales. Vol. VI. 1750–1850.* Cambridge University Press, Cambridge, 7–83.
Rhodes, E. 1824 *Peak Scenery.* London.
Ryder, M. L. 1981 Livestock. In: Piggott, S. (ed.) *The Agrarian History of England and Wales. Vol. I.I. Prehistory.* Cambridge University Press, Cambridge, 299–410.
Salaman, R. N. 1949 *The History and Social Influence of the Potato.* Cambridge University Press, Cambridge.
Seale, R. S. 1975 *Soils of the Ely District.* Memoir, Soil Survey, Harpenden.
Sheail, J. 1971 *Rabbits and their History.* David and Charles, Newton Abbot.
Simmons, I. 1990 Upland peats. Human impact on an ecosystem. *Geography Review* 3, 27–30.
Skeat, W. W. (ed.) 1882 *The Book of Husbandry* by Master Fitzherbert, reprint of 1598 edition, English Dialect Society, London.
Staines, S. J. Age and significance of soil truncation and colluviation in south Somerset, unpublished manuscript.
Stone, T. 1794 *General View of the Agriculture of the County of Bedford.* Hodson, London.
Tallis, J. H. 1990 Aspects of blanket peat erosion. *Proceedings of the North of England Soils Discussion Group* 24, 27–38.
Thirsk, J. 1964 The common fields. *Past and Present* 29, 3–25.
Thirsk, J. 1967a Enclosing and Engrossing. In: Thirsk, J. (ed.) *The Agrarian History of England and Wales. IV. 1500–1640.* Cambridge University Press, Cambridge, 200–55.
Thirsk, J. 1967b Farming techniques. In: Thirsk, J. (ed.) *The Agrarian History of England and Wales. Vol. IV. 1500–1640.* Cambridge University Press, Cambridge, 161–99.
Thirsk, J. (ed.) 1984 *The Agrarian History of England and Wales. Vol. V. 1640–1750.* Cambridge University Press, Cambridge.
Trow-Smith, R. 1957 *A History of British Livestock Husbandry to 1700.* Routledge and Kegan Paul, London.
Trow-Smith, R. 1959 *A History of British Livestock Husbandry 1700–1900.* Routledge and Kegan Paul, London.
Vancouver, C. 1808 *General View of the Agriculture of the County of Devon*, reprinted 1969. David and Charles, Newton Abbot.
Vancouver, C. 1810 *General View of the Agriculture of Hampshire, Including the Isle of White*, Richard Phillips, London.
White, G. 1788 *The Natural History of Selborne*, reprinted 1951. Lutterworth Press, London.
Winchester, A. J. L. 1987 *Landscape and Society in Medieval Cumbria.* John Donald, Edinburgh.
Young, A. 1804a *General View of the Agriculture of Hertfordshire*, reprinted, David and Charles, Newton Abbot, 1971.
Young, A. 1804b *General View of the Agriculture of the County of Norfolk*, reprinted, David and Charles, Newton Abbot, 1969.
Young, A. 1813a *General View of the Agriculture of the County of Sussex*, reprinted, David and Charles, Newton Abbot, 1970.
Young, A. 1813b *General View of the Agriculture of the County of Suffolk*, reprinted, David and Charles, Newton Abbot, 1969.
Young, A. 1813c *General View of the Agriculture of Lincolnshire*, reprinted, David and Charles, Newton Abbot, 1970.

# CHAPTER 6

# SOIL EROSION ON THE LOWER GREENSAND AT WOBURN EXPERIMENTAL FARM, BEDFORDSHIRE - EVIDENCE, HISTORY AND CAUSES

J. A. CATT

*Detailed records at Woburn Experimental Farm, which start soon after the farm was opened in 1876, show that water erosion on the sandy soils over the Lower Greensand caused no problems with field experiments before World War II. However, it has increased considerably since then, even on very gentle slopes. This is attributed to use of herbicides and heavy farm machinery and to an increase in root crops which have replaced pasture and winter cereals. However, the postwar erosion does not entirely explain the extent of a widespread loamy colluvium, which is derived partly from loess and Chalky Boulder Clay occurring on the Lower Greensand plateau south-east of the farm. Very little colluvium overlies deposits of lakes which dried up more than 100 years ago, and it is likely that most of the mapped colluvium originated by Medieval or earlier soil erosion.*

## Introduction and history of the experimental farm

Woburn Experimental Farm is situated in the village of Husborne Crawley, near Woburn, Bedfordshire. It lies astride the junction of the Lower Greensand or Woburn Sands (Wyatt *et al.* 1988) (Lower Cretaceous) and Oxford Clay (Upper Jurassic) at an elevation of 79–110 m O.D., and has a total area of 77 ha.

The Agricultural Holdings Act (England) of 1875 provided for payment of compensation to outgoing tenant farmers for soil improvements, the benefits of which are unexhausted on the date the tenant departs. The original field experiments at Woburn were designed to evaluate these unexhausted benefits so that departing tenant farmers could be appropriately compensated (Johnston 1977). From earlier experiments at Rothamsted Experimental Station (Harpenden, Hertfordshire), Lawes (1875) had suggested certain formulae for calculating the residual 'manure values' of purchased animal foods, but these were disputed in part by J. C. A. Voelcker (1876), consulting chemist to the Royal Agricultural Society of England (RASE). The 9th Duke of Bedford, then Vice-President of the society, consequently offered the society possession of a farm (Crawley Mill Farm) on his Woburn Estate and money to pay for experiments to settle the matter. Together Lawes and Voelcker began experiments in 1876 on not only the residual manurial value of animal feedstuffs and artificial fertilizers but also on the continuous growing of winter wheat and spring barley, to see whether the success that Lawes had previously achieved growing cereals continuously on the heavier (Clay-with-flints) soils at Rothamsted could be repeated on light sandy soils.

J. C. A. Voelcker died in 1884 and his son (J. A. Voelcker) succeeded him as both consultant chemist to the RASE and Director of Woburn Experimental Farm. Tables for the manurial values of animal foods, artificial fertilizers and lime were later published by Voelcker and Hall (1902; 1913). When the society decided to quit the tenancy of the farm in 1921 Voelcker himself took it on for five years until Rothamsted's governing body (The Lawes Agricultural Trust) assumed responsibility for both the tenancy and scientific programme in 1926.

The RASE withdrew from the farm in 1921 partly because current opinion was that the land was too light and unfit for arable farming. However, the success of later research at Woburn on various problems of farming light sandy soils, such as use of organic and green manures, liming, trace element deficiencies, irrigation, and control of weeds, nematodes and root fungal diseases, has brought large increases in the potential yields of many arable crops on sandy soils. As a result many sandy areas in the English Midlands, East Anglia, Lincolnshire, eastern Yorkshire, Lancashire, Somerset and the Welsh Borderlands have come under cultivation and now successfully produce a range of arable and vegetable crops. However, this expansion onto lighter soils has also increased certain national environmental problems, notably soil movement by wind and runoff (Reed 1979; Evans 1990).

Erosion has caused several types of problem with the field experiments at Woburn Experimental Farm; they include irregular loss of crops, transfer of specially treated soil across plot boundaries and yield variation resulting from irregular distribution of colluvium by earlier erosion. Many of the fields originally made available for experiments on Crawley Mill Farm are irregularly sloping, and were recognized from the start as potentially unsuitable for careful field trials, with the result that flatter fields (originally Stackyard and later Far Field) were added to the experimental farm. The steeper slopes are typical of the irregular rounded topography of the Lower Greensand outcrop, especially near its dissected north-west facing scarp.

Although the supposed uniformity of the sandy soil of Stackyard Field was often quoted as a feature

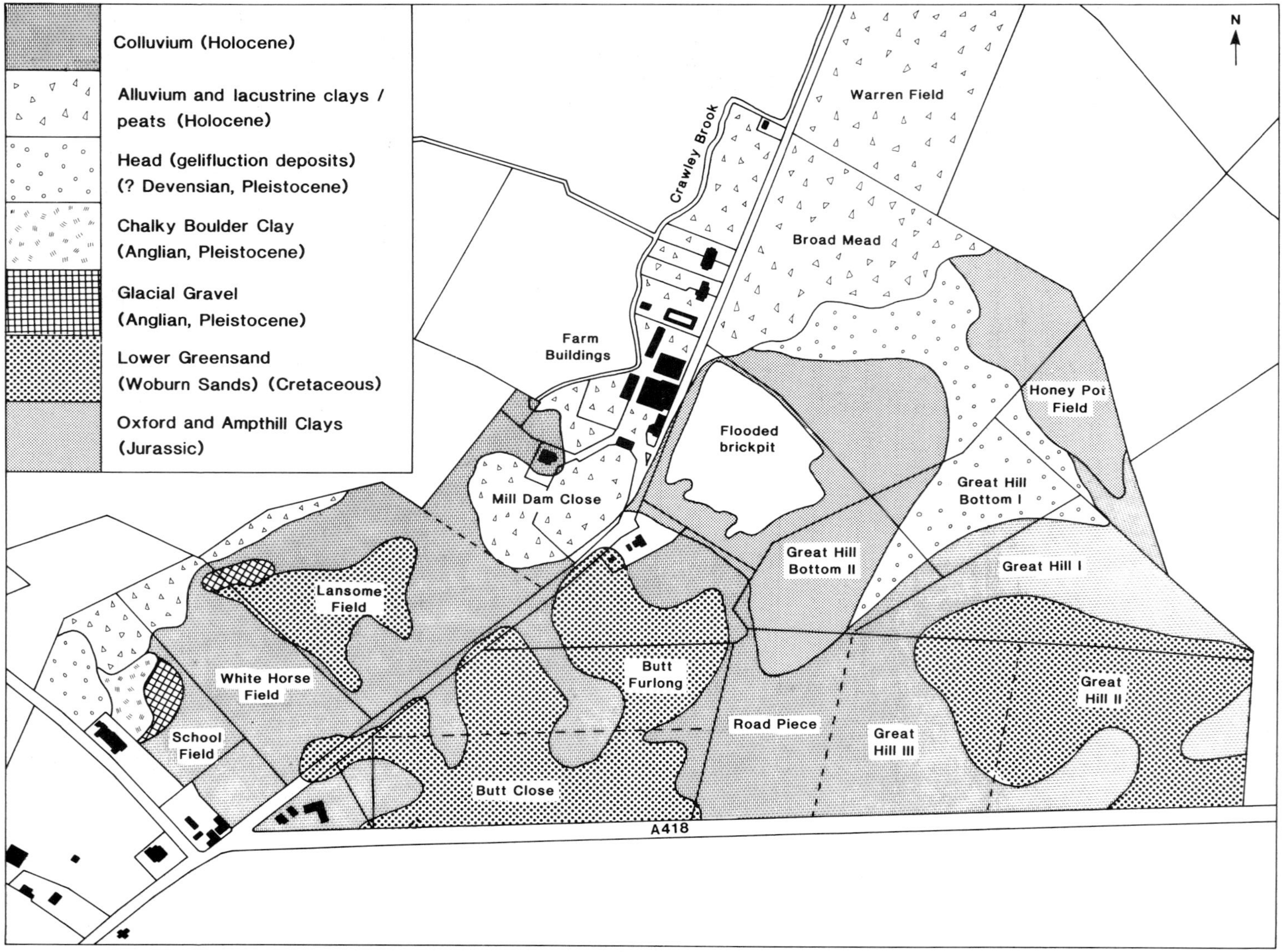


Figure 6.1 Map of Crawley Mill Farm (part of Woburn Experimental Farm) to show distribution of colluvium and other parent materials.

making it more suitable for experiments than others on the farm, the effect of soil variation resulting from earlier erosion (i.e. presence or absence of loamy colluvium) was recognized from the start as a cause of yield variation. Commenting on barley yields for the first decade on Stackyard, Lawes (1888) wrote 'from some cause, possibly some slight difference in the texture of the soil, the two unmanured plots differ considerably, one yielding close upon 27 bushels per acre and the other 23 bushels'. As explained by Catt *et al.* (1980), the main effect of the colluvium on crop yields is through its large available water capacity compared with the Lower Greensand; in dry years the increased water supply improves yields, but in wet years the greater water retention can encourage nematodes and other pests and diseases, thus decreasing yield.

**Soils of the experimental farm**

The main parent materials of soils on the farm are Upper Jurassic (Oxford and Ampthill) Clays, Lower Greensand, Chalky Boulder Clay and glacial gravels, head (gelifluction) deposits, alluvial deposits of the Crawley Brook (an east bank tributary of the River Ouzel), lacustrine clay and peat deposited in shallow lakes occupying minor valleys tributary to the Crawley Brook, and Holocene colluvium formed by erosion of the soils on the Lower Greensand and incorporation of silt and clay from other sources. Figures 6.1 and 6.2, based mainly on soil mapping by Catt *et al.* (1975; 1977; 1980), show the distribution of these parent materials on part of the original (Crawley Mill) farm and on Stackyard Field. Their wider distribution in the surrounding region was described by Wyatt *et al.* (1988), though these authors mapped the colluvium as part of the head (slope) deposits.

The colluvium occurs mainly on low-lying parts of the Lower Greensand outcrop. On Crawley Mill Farm (Figure 6.1) it mantles the lower slopes of Great Hill and the low knoll on Lansome and White Horse Fields; it also obscures the base of the Lower Greensand where it rests on Upper Jurassic Clays south of the farm buildings. To the north of Great Hill it spreads over head deposits, and on School Field it partially covers an outlier of glacial gravels and the Holocene alluvial and lacustrine deposits. On Butt Close and Butt Furlong colluvium mainly occupies the floors of shallow dry valleys cut into basal parts of the Lower Greensand. A similar relationship exists on Stackyard Field, and is shown very clearly in Figure 6.2, which has land surface contours surveyed at 20 cm vertical intervals. On these three fields the Lower Greensand emerges from beneath the colluvium mainly on ridges and knolls between the valleys.

Throughout the farm the colluvium was distinguished from Lower Greensand on the basis of particle size distribution over the upper 80 cm of the soil profile; colluvium was mapped where the sum of % silt plus twice the % clay exceeds 30 over at least half the top 80 cm, i.e. where more than half the upper part of the soil profile is loam containing more silt and/or clay than the (sand or loamy sand) soil derived purely from the Lower Greensand. On most of the fields the distribution of colluvium was based upon analyses of samples taken at the intersections of either 20 m or 40 m square grids (Catt *et al.* 1975; 1977). However, on Stackyard the distribution was based on analyses of subsoil samples (40–60 cm depth) taken on a 10 m grid (Catt *et al.* 1980), so here the colluvium is more precisely mapped than on other fields.

Two soil series were recognized in the colluvium. Lowlands series (previously Stackyard series), a colluvial brown earth (Avery 1980; Clayden and Hollis 1984), is well-drained and lacks greyish mottling within the uppermost 70 cm of the profile. Flitwick series, a gleyic ferritic brown earth, has greyish and rusty mottling appearing between 40 and 70 cm depth, and occurs on the floors of deeper valleys where the groundwater table is close to the surface. In these situations the colluvium is usually thicker (1–3 m) and heavier in texture (% silt + 2 (% clay) = 50 – 85) than where it is well drained.

Soils derived from Chalky Boulder Clay and glacial gravel occupy only small isolated areas on School and Stackyard Fields (Figures 6.1 and 6.2), but are more extensive on intervening fields as yet not mapped in detail. Chalky till occurs widely in Woburn Park (Wyatt *et al.* 1988), which lies south and east of the farm generally at greater heights (110–130 m O.D.) on the crest of the Lower Greensand escarpment. The small outliers of till and gravel on and below the dissected Lower Greensand scarp show that this area was over-ridden by the Chalky Boulder Clay (Anglian?) glacier but has since been more severely eroded than the escarpment crest and dipslope.

Evidence for periglacial erosion probably during the Devensian Stage is provided by the head deposits on School Field and some of the unmapped fields to the south. These deposits are mainly grey or greyish brown, firm stony clays or sandy clay loams; they are more clayey, stony and compact than the colluvium. On the farm they usually overlie chalky till at 1–2.5 m depth, and are derived partly from the till because the stones are mainly flint fragments and other glacial erratics. The head deposit mantling slopes below the junction of the Lower Greensand and Jurassic clays on the northern flank of Great Hill (mainly Great Hill Bottom I and Broad Mead Fields) has a different origin, however. It is a variable mixture of sand and clay with only a few stones, which are entirely carstone (sandy ironstone) fragments and small quartzite pebbles from the Lower Greensand. Catt *et al.* (1975) suggested that it was ejected from the junction of the clay and sand by water compressed between the impervious clay and permafrost in the overlying sand.

**Composition and origin of the colluvium**

The sand which is the main component of the fine earth (<2 mm) in the colluvium is similar in particle size distribution and mineralogical composition to the Lower Greensand, and is obviously derived from it. The seven samples of Lower Greensand from the

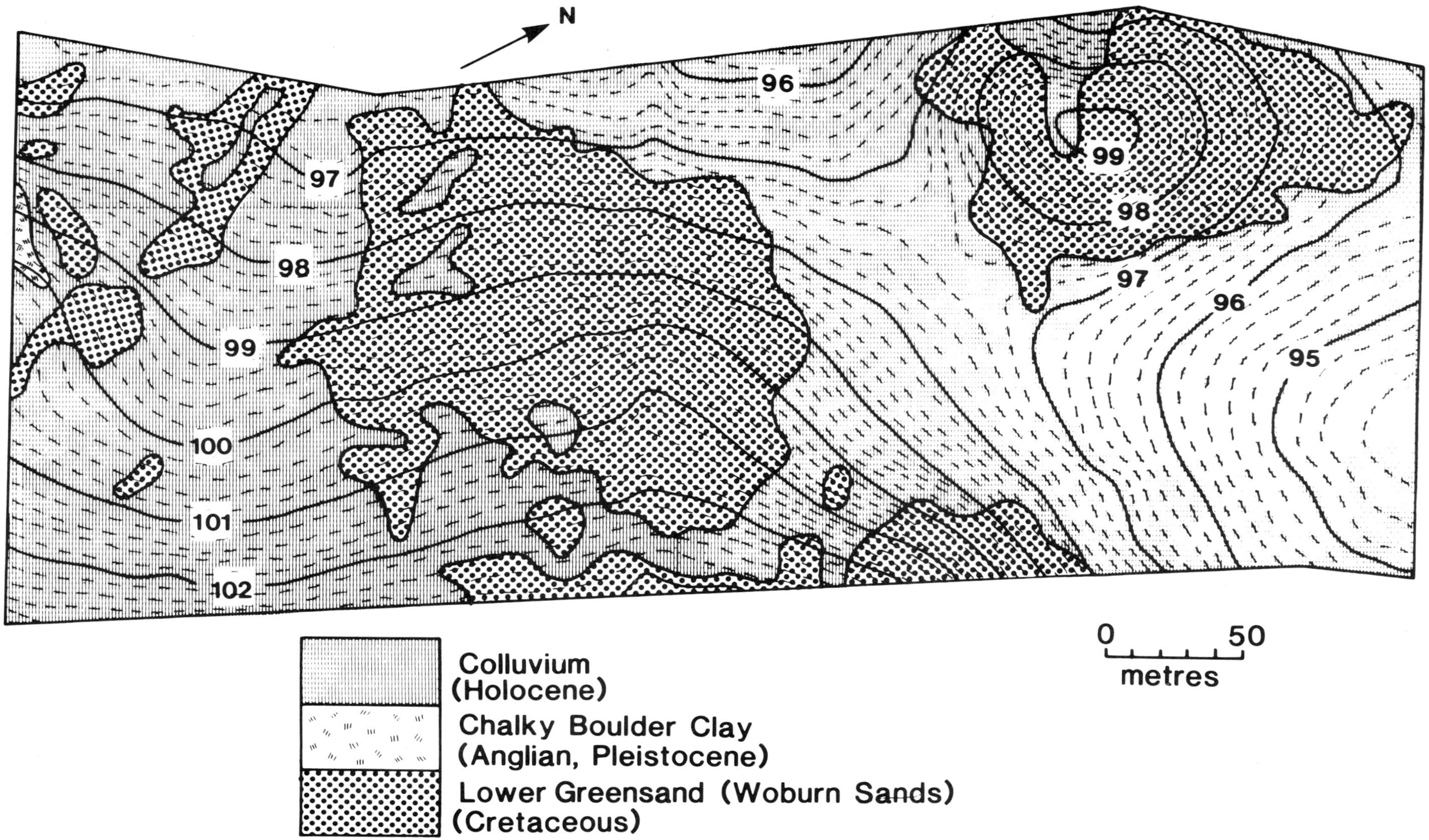


Figure 6.2 Map of Stackyard Field, Woburn Experimental Farm, to show distribution of colluvium and other parent materials in relation to surface contours at 20 cm vertical intervals.

Woburn area analysed by Catt *et al.* (1975) contained 0.3–12.9% clay (mean 7.4%) and 0.2–1.9% total silt (2–62 μm) (mean 1.3%). Twenty-seven samples of colluvium from seven profiles (Catt *et al.* 1975; 1977), spread over a total depth of 147 cm, contained 5.5–21.0% clay (mean 11.7%) and 8.4-41.1% silt (mean 17.9%). Compared with the Lower Greensand, the colluvium is therefore slightly enriched in clay and greatly enriched in silt. Both of these additional components could be derived from several possible sources:

(1) Concentrated by the colluviation processes from the small clay and silt contents of the Lower Greensand.
(2) Incorporation of Jurassic clay material, though this is unlikely because it would involve some uphill movement of the fine material from the clay outcrops, which lie downslope of most of the colluvium.
(3) Incorporation of fine material from the Chalky Boulder Clay; as the colluvium often contains occasional flint fragments and other erratics, this is a likely source.
(4) Incorporation of loess, which was deposited very widely over southeast England during the Late Devensian (Catt 1978).

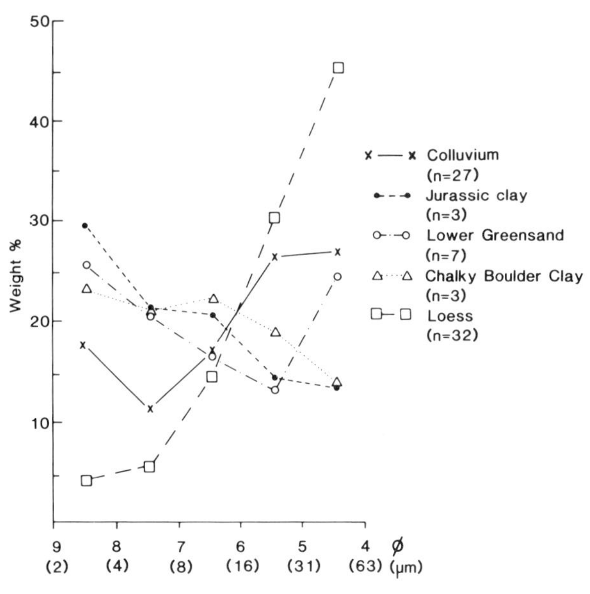


Figure 6.3 Mean particle size distribution at ø intervals of silt fractions from Woburn colluvium and four possible sources of the silt.

These possible sources can be distinguished by mineralogical composition and, for the silt component, by detailed particle size distribution as well.

Figure 6.3 shows the mean particle size distribution at ø intervals (ø = $-\log_2$ mm) of silt from the 27 samples of colluvium and of silt from the four possible sources. None of the possible sources has silt which strongly resembles that in the colluvium. The finer part of the silt could be derived from Jurassic clay, Chalky Boulder Clay or Lower Greensand, but the coarser silt seems to have a size distribution more like loess. On this evidence it seems likely that the silt in the colluvium is derived from two or more of the possible sources.

Table 6.1 shows the mean mineralogical composition of coarse silt (6-4ø) from 21 samples of the colluvium and from samples of the same four possible source sediments. The occurrence of flint fragments, tremolite/actinolite, brown hornblende and fairly large amounts of green hornblende in the colluvium shows that most of the coarse silt was derived either from the Chalky Boulder Clay or from loess. However, the similar amounts of epidote in the colluvium and loess and the presence of staurolite and kyanite in both these deposits but their absence from the Chalky Boulder Clay suggest that the coarse silt is essentially from the loess. This conclusion does not apply to the fine silt (9–6ø), for which there is at present no mineralogical information.

Almost all the 21 colluvium samples analysed contained small amounts of opal in the light fraction of the coarse silt. Many of these particles have the characteristic shapes of grass phytoliths, which indicate that the colluvium is at least partly derived from topsoil in which grass grew for a fairly long period. This could be as recent as the mid-twentieth century, because many of the fields both on and close to the experimental farm have been cultivated only since the Second World War. However, there were probably many earlier periods in the later Holocene when grassland was extensive in the area.

Table 6.2 shows the mean mineralogical composition of clay (<2 μm) fractions from the colluvium and from possible sources of the clay. The relative amounts of layer silicate minerals (kaolinite, illite, interstratified smectite-illite and chlorite) suggest that most of the clay in the colluvium is derived from the Chalky Boulder Clay. Unweathered Chalky Boulder Clay cannot have provided the hydrated iron oxides (goethite and lepidocrocite) in the colluvium, but these could have been formed by weathering processes in soils developed on the till. They could also be derived from the Lower Greensand or from loess, but the absence of smectite-illite and the range of other layer silicate clay minerals in loess suggest that little or none of the clay in the colluvium came from loess.

To summarize, the particle size and mineralogical analyses of the colluvium suggest it is a mixture of sand derived mainly from the Lower Greensand, coarse silt essentially from loess and stones and clay derived principally from the Chalky Boulder Clay.

| | Colluvium, mean of 21 samples | Lower Greensand, mean of 7 samples | Chalky Boulder Clay, mean of 3 samples | Jurassic Clay, mean of 3 samples | Late Devensian loess, mean of 32 samples |
|---|---|---|---|---|---|
| *(a) Light fraction* | | | | | |
| Quartz % | 81.8 | 91.4 | 79.6 | 87.9 | 82.0 |
| Alkali feldspar % | 13.9 | 5.1 | 14.7 | 11.0 | 15.0 |
| Flint % | 0.9 | - | 1.0 | - | 0.9 |
| Muscovite % | 1.1 | 0.5 | 4.2 | 1.0 | 1.4 |
| Glauconite % | 1.6 | 3.0 | 0.5 | 0.1 | 0.7 |
| Opal % | 0.7 | - | - | - | - |
| *(b) Heavy fraction* | | | | | |
| Zircon ‰ | 209 | 423 | 99 | 520 | 126 |
| Tourmaline ‰ | 45 | 158 | 30 | 59 | 31 |
| Epidote ‰ | 315 | 152 | 178 | 80 | 387 |
| Zoisite ‰ | 29 | 12 | 28 | 20 | 19 |
| Garnet ‰ | 34 | 11 | 41 | 58 | 42 |
| Green hornblende ‰ | 104 | 10 | 158 | 5 | 109 |
| Tremolite/actinolite ‰ | 35 | - | 21 | - | 28 |
| Brown hornblende ‰ | 5 | - | 9 | - | 7 |
| Red rutile ‰ | - | 1 | 10 | - | 3 |
| Yellow rutile ‰ | 61 | 84 | 32 | 93 | 33 |
| Brown rutile ‰ | 14 | 3 | 8 | 34 | 12 |
| Anatase ‰ | 48 | 70 | 21 | 59 | 18 |
| Brookite ‰ | 4 | 17 | 9 | 4 | 3 |
| Staurolite ‰ | 8 | 13 | - | 17 | 12 |
| Kyanite ‰ | 7 | 10 | - | 10 | 6 |
| Andalusite ‰ | - | 1 | - | - | - |
| Apatite ‰ | - | - | 16 | - | 3 |
| Collophane ‰ | - | - | 4 | 8 | 2 |
| Augite ‰ | 1 | - | 10 | 1 | 1 |
| Chlorite ‰ | 78 | 32 | 316 | 29 | 140 |
| Biotite ‰ | 2 | 3 | 10 | - | 18 |
| Spinel ‰ | 1 | - | - | - | - |
| Sphene ‰ | - | - | - | 3 | - |

Table 6.1 Mineralogical composition of coarse silt (16–63 µm) fractions of Woburn colluvium and possible source sediments

| | Colluvium, mean of 24 samples | Lower Greensand, mean of 6 samples | Chalky Boulder Clay, mean of 3 samples | Jurassic Clay, mean of 3 samples | Late Devensian loess, mean of 7 samples |
|---|---|---|---|---|---|
| Kaolinite % | 12 | 2 | 15 | 2 | 6 |
| Illite % | 16 | 33 | 30 | 15 | 18 |
| Smectite-illite % | 72 | 65 | 55 | 83 | - |
| Smectite-vermiculite % | - | - | - | - | 8 |
| Vermiculite-illite % | - | - | - | - | 10 |
| Smectite % | - | - | - | - | 35 |
| Vermiculite % | - | - | - | - | 12 |
| Chlorite % | tr | - | tr | - | tr |
| Goethite % | tr | tr | - | - | tr |
| Lepidocrocite % | tr | tr | - | - | tr |
| Quartz % | - | - | - | - | 8 |
| Feldspar % | - | - | - | - | 3 |

Table 6.2 Clay (<2 μm) mineral composition of Woburn colluvium and four possible source materials

The presence of opaline phytoliths throughout several profiles of the colluvium indicates incorporation of surface material from grassland soils.

**History of soil erosion and deposition of colluvium**

Erosion at Woburn Farm was first mentioned in print by Moffatt (1951), who reported that on 21 May 1950 a storm with 1.75 inches (44 mm) of rain in under 3 hours caused erosion in fields not under grass or corn crops, and in some places produced gullies 1 m wide and 10 cm deep; potato tubers were exposed by washing away of ridges, and in low-lying areas sugar-beet plants were buried under downwashed soil.

By the late 1960s erosion was becoming a fairly frequent problem on several of the experimental fields. A field experiment on long-term phosphorus nutrition of barley, potatoes and sugar-beet started on Stackyard in 1968 (Johnston *et al.* 1976) was abandoned in 1972 and sown to a grass-clover ley in 1973, because erosion events in March 1969 and October 1971 (Williams 1978) had produced deep gullies across some plots and deposited soil with absorbed phosphate in a temporary lake on other plots. Mattingly *et al.* (1975) suggested that erosion on Stackyard accounted for a sharp decrease in the organic nitrogen contents of certain plots between 1959 and 1972; they also reported observations of runoff on Stackyard in the early 1950s.

A more recent example of the problems in field experiments resulting from soil movement across plot boundaries occurred on a nematicide dosage trial on Stackyard in 1983. Six of the plots previously treated with nematicide were covered by 8–10 cm of untreated soil washed from a large area of bare soil upslope; this occurred through several erosion events shortly before the test crop of potatoes was sown in early May. The depth of untreated soil was considered enough to decrease severely the effectiveness of the treatment, and the year's results for the affected plots were consequently abandoned.

Other fields on which erosion has often caused problems are Butt Close, Butt Furlong, Lansome and Great Hill. Catt *et al.* (1975) reported sheet erosion on Butt Close during a storm in May 1973 when over 50 mm of rain fell in one hour. They suggested that runoff was caused by a plough pan which severely limited infiltration of water into the normally well-drained soil; the situation was aggravated by runoff from the main road (A418) which is adjacent to the field and 1–2 m above it. The erosion on Stackyard could also be increased by runoff from the same road. On both these fields quite severe erosion has occurred on slopes less than 1°.

Great Hill has steeper slopes (5–7°) than most of the other experimental fields. It has been taken in and out of cultivation several times during the history of the experimental farm. After an initial period in pasture it was sown to spring barley in 1892, and then carried various cereal and root crops (also at times gorse as an experimental forage crop) until autumn 1931. It was then put down to grass-legume mixtures but ploughed and sown to barley again in 1942. The postwar experiments on this field were then frequently damaged by erosion; potatoes were especially susceptible and often had to be re-ridged after heavy rain. The experiments were consequently abandoned again in the mid-1980s, and some of the steepest slopes are now used for erosion plots designed to measure the effects of soil cultivation and crop residue incorporation on erosion and to provide runoff and soil loss data for validation of the European Soil Erosion Model (EUROSEM) currently being developed by Silsoe College and various EC collaborators. Since the erosion experiment was started in spring 1989, 24 erosion events have been measured on one or more of the eight plots. The first was on 7 July 1989, 17 occurred between 20 October 1989 and 25 February 1990 and 6 between 25 December 1990 and 8 January 1991. Soil losses from the 875 $m^2$ plots ranged from 29 to 1347 kg (equivalent to 0.33–15.39 t $ha^{-1}$) for individual events in 1989 and early 1990, but were all <10 kg per plot in the winter of 1990–1991.

Despite the increasing frequency of recorded erosion in various parts of the farm over the last 40 years, there is no evidence for it occurring at all between 1876 and 1950. The detailed diaries of the farm recorders for the period 1882–1947, which are archived in Rothamsted library, report many other factors influencing crop performance, including bad weather (frost, wind, hailstorms, heavy rain), weeds, diseases, insect pests, soil pH, damage by game, birds, vermin or rabbits, shading effects of trees and cultivation techniques. Yet the visually obvious and experimentally damaging effects of erosion were never reported, even though the farm staff seem to have been aware of the danger; for example, Miller (1930) commented that 'the light soil at Woburn must be continuously covered with crops'. This suggests that any erosion which occurred before 1950 was slight and passed unnoticed because it did not interfere with farmwork or result in a loss of yield. Even on the steep slopes of Great Hill, which were often cultivated and sometimes carried root crops or gorse during this period, there was never sufficient erosion to warrant a note in the diaries.

Although the rate of soil erosion at Woburn has increased since World War II, it is unlikely that the total erosion over this period has been enough to explain the extensive deposits of colluvium on Stackyard and other fields (Figures 6.1 and 6.2). As Lawes (1888) was probably aware of the textural variation in the soil of Stackyard, much of the deposition of colluvium there had probably occurred before the farm was transferred from the ducal estate in 1876. Also, the mineralogical composition of the colluvium on Stackyard and other fields suggests that much of it originated on the higher ground in Woburn Park, and therefore predates building of the long, continuous, high estate wall on the southern side of the A418 road. The present section of the wall between Husborne Crawley and Ridgmont, adjacent to Great Hill, Road Piece and Butt Close (Figure 6.1) was erected in 1899–1901, but this was a replacement of an earlier wall built between 1805 and 1817 when the Woburn Park was extended northwards (M. Gwilliam, pers. comm.). The Woburn-Husborne Crawley section of wall adjacent to Stackyard probably also dates from this earlier period. The wall would have prevented any colluvium from the park reaching the experimental farm, though it is also unlikely that any erosion has occurred in the essentially wooded and grassed parkland since construction of the wall, because there is no evidence for accumulation of colluvium on the upslope (park) side of the wall.

Early maps of the Woburn estate and stratigraphic evidence also suggest that most of the colluvium is at least several centuries old. On parts of School, White Horse and Lansome Fields the colluvium partly overlies the Holocene lacustrine and alluvial deposits, so some of it must have accumulated after the lakes and stream dried up. The distribution of lacustrine deposits shows that two lakes previously existed on this part of the farm, one at the northwestern ends of School and White Horse Fields and the other in the northeastern part of Lansome Field (previously known as Mill Dam Close). In both areas the lacustrine deposits comprise grey organic silty clay with fine ochreous mottles over a thin loamy peat or humose clay over gravelly sand, which in turn rests on Lower Greensand.

The lake on School and White Horse Fields lay in the head of a short valley tributary to the nearby Crawley Brook, and was either maintained by the high groundwater table close to the stream or originated as a temporary ox-bow lake in an abandoned meander of the stream. Loamy peat (containing 11.6% organic carbon) from a depth of 105–118 cm near the boundary between the two fields (National Grid Reference SP 960359) gave a radiocarbon age of 3085±85 years B.P. (Birm.–761). As this sample was from the uppermost layer of the peat, the date suggests that the lake existed in the Bronze Age; the overlying 105 cm of organic clay loam (27–32% clay) therefore probably accumulated in the lake after about 3000 B.P. It is not known when this lake disappeared. The only evidence is that no lake is shown here on the earliest detailed map of the ducal estate (*c*. 1760 A.D.), so the 0–25 cm of sandy loam colluvium overlying part of the lacustrine clay has taken at least two centuries to accumulate. The nearby slopes from which this must have been eroded range from 1 to 2°, which is typical of many of the source slopes for the much greater thicknesses (1–2 m) of colluvium found elsewhere on the farm. This suggests that most of the colluvium on the farm is older than the disappearance of the lake.

The other lake (the mill pond on Lansome Field) is shown on estate maps up to 1850 A.D. but does not appear on the first (1884) edition of the Ordnance Survey. It was therefore drained at some time between these two dates. The lake may have been fed partly by springs issuing from the base of the Lower Greensand, which must lie almost under the old lake site, but early maps show that it was also supplied by a stream diverted from the Crawley Brook and originally entering the lake on its northwestern side. Early in the nineteenth century this feeder stream was diverted along the present northwestern margin of the newly enclosed Lansome Field and then along the boundary between Lansome and Mill Dam Close (dashed line in Figure 6.1), thus entering a somewhat enlarged mill pond from the south. By 1850 A.D. a new feeder stream from the northwest was supplying a second smaller pond northwest of the original, but both ponds were drained before 1884 possibly when the RASE took possession of the farm. The 0–25 cm of colluvium overlying the lake deposits near their southern margin have therefore accumulated since about 1875, probably mainly since 1950. They have been derived from parts of Lansome Field to the south which slope at 1.25–2.5°, so evidence here again suggests that the much greater thicknesses of colluvium elsewhere represent an earlier and presumably longer or more severe period of erosion.

Unfortunately we have no archaeological evidence on the farm for the age of the colluvium, and there is little evidence in the area for early farming activities which could have resulted in extensive soil erosion and deposition of colluvium. The Bedfordshire County Council's Sites, Monuments and Buildings Record held in the County Planning Department shows that 'evidence for prehistoric, Romano-British and Saxon occupation or activity in the Woburn and Husborne Crawley areas is very limited. It is difficult to say whether this is due to a real absence of settlement or to other factors, such as land-use history of the area; in particular the extensive non-arable land of Woburn Park is not conducive to archaeological discovery. However, former arable cultivation is indicated by ridge and furrow earthworks in several parts of Woburn Park. Soil erosion and colluvial deposition is thus likely to have occurred during the early Medieval period, but whether this was a new phenomenon locally or merely the continuation of a well established situation is not really known' (M. Gwilliam, pers. comm.). From evidence in other parts of South East England the colluvium at Woburn Farm could date from any period from Neolithic to late Medieval.

### Causes of postwar erosion

The increase in erosion at Woburn Experimental Farm since World War II has probably resulted from several factors. First, there has been an increase in the area of experimental root crops, especially potatoes; these replaced prewar pasture and winter cereals, which would have offered the soil more protection from erosion. Second, heavy modern soil-cultivating machinery compacts the subsoil if it is wet, thus decreasing infiltration rates and encouraging runoff; the bulk density of cultivation pans at Woburn is often >1.8 g $cm^{-3}$ (Williams 1978). Third, herbicides have been used increasingly over the last 20–30 years to eliminate weeds, which previously helped stabilize the soil, even though the farm staff then spent much time hoeing by horse and hand.

A progressive decline in organic matter has occurred in the sandy soils at Woburn under prolonged cultivation and arable cropping. For example, on plots of the Continuous Barley Experiment (Stackyard Field) given inorganic or no fertilizers, the mean soil carbon and nitrogen contents decreased from 1.49% and 0.156% respectively in 1876 to 0.64% and 0.084% in 1959 (Mattingly *et al.* 1975). Loss of organic matter under these conditions is another factor that could have encouraged erosion by weakening bonds between soil particles, though experiments reported by Johnston and Warren (1964) suggested that even large dressings of peat and some other organic materials do little to strengthen the structure of sandy Woburn soil. Williams (1978) suggested that the type of organic matter is more important than the total amount in determining the stability of Woburn soil.

### Conclusions

The detailed records kept at Woburn Experimental Farm show very clearly that serious soil erosion causing problems with the field experiments began about 1950. Since then it has increased in frequency, even on quite gentle slopes (less than 1°). In important experimental areas it has been minimized by establishing grass strips between plots, but on the steeper (5–7°) plots of the recently established erosion experiment up to 17 erosion events have occurred per year. The postwar increase in erosion is attributed to greater areas of root crops, use of herbicides and subsoil compaction by heavy agricultural machinery.

Holocene colluvium derived from grassland soils on the Lower Greensand, mixed with loess and Chalky Boulder Clay derived from the Lower Greensand dipslope, is widespread on the farm and reaches 3 m in thickness in some of the dry valleys. The postwar soil erosion accounts for very little of this, as deposits of lakes which dried up a century or more ago are covered by 25 cm or less of colluvium, and the farm, situated on the Lower Greensand scarpslope, is separated from the dipslope by the long, continuous, high wall of the Woburn Estate built 100–200 years ago. The exact age of the main colluvial deposits is uncertain; they could have accumulated during any earlier period when deforestation and soil cultivation occurred in the area.

The success of agricultural experiments earlier this century at Woburn led to a considerable expansion of arable agriculture onto sandy soils in many parts of England. Indirectly it was therefore partly responsible for the currently expanding problem of water erosion on these soils. Current trends in United Kingdom agriculture may resolve this problem by encouraging the return of sandy soils to non-arable uses. Alternatively the soil management practices incorporated into the Woburn and other erosion experiments may show how erosion can be decreased to a level which is acceptable in the long term.

### Acknowledgement

I thank Mr Michael Gwilliam, County Planning Officer for Bedfordshire, and staff of the County Conservation and Archaeology Section for information on the history of Woburn Estate and archaeological records for the Woburn area.

### References

Avery, B. W. 1980 Soil classification for England and Wales (higher categories). *Soil Survey Technical Monograph* 14. Rothamsted Experimental Station, Harpenden.

Catt, J. A. 1978 The contribution of loess to soils in

lowland Britain. In, Limbrey, S. and Evans, J. G. (eds.) *The effect of man on the landscape: the lowland zone.* Council for British Archaeology Research Report 21, London, 12–20.

Catt, J. A., King, D. W. and Weir, A. H. 1975 The soils of Woburn Experimental Farm. I. Great Hill, Road Piece and Butt Close. *Report Rothamsted Experimental Station for 1974*, Part 2, 5–28.

Catt, J. A., Weir, A. H., King, D. W., Le Riche, H. H., Pruden, G. and Norrish, R. E. 1977 The soils of Woburn Experimental Farm. II. Lansome, White Horse and School Fields. *Report Rothamsted Experimental Station for 1976*, Part 2, 5–32.

Catt, J. A., Weir, A. H., Norrish, R. E., Rayner, J. H., King, D. W., Hall, D. G. M. and Murphy, C. P. 1980 The soils of Woburn Experimental Farm III. Stackyard. *Report Rothamsted Experimental Station for 1979*, Part 2, 5–39.

Clayden, B. and Hollis, J. M. 1984 Criteria for differentiating soil series. *Soil Survey Technical Monograph* 17. Rothamsted Experimental Station, Harpenden.

Evans, R. 1990 Soils at risk of accelerated erosion in England and Wales. *Soil Use and Management* 6, 125–131.

Johnston, A. E. 1977 *Woburn Experimental Farm. A hundred years of agricultural research devoted to improving productivity of light land.* Rothamsted Subject Day Booklet 4, Rothamsted Experimental Station, Harpenden.

Johnston, A. E., Mattingly, G. E. G. and Poulton, P. R. 1976 Effect of phosphate residues on soil P values and crop yields. I. Experiments on barley, potatoes and sugar beet on sandy loam soils at Woburn. *Report Rothamsted Experimental Station for 1975*, Part 2, 5–35.

Johnston, A. E. and Warren, R. G. 1964 Improving soil structure at Woburn. *Rothamsted Experimental Station Report for 1963*, 39–41.

Lawes, J. B. 1875 On the valuation of unexhausted manures. *Journal of the Royal Agricultural Society* 36, 3–38.

Lawes, J. B. 1888 The permanent wheat and barley experiments in Stackyard Field, Woburn. *Journal of the Royal Agricultural Society* 49, 1–8.

Mattingly, G. E. G., Chater, M. and Johnston, A. E. 1975 Experiments made on Stackyard Field, Woburn, 1876–1974. III. Effects of NPK fertilizers and farmyard manure on soil carbon, nitrogen and organic phosphorus. *Report Rothamsted Experimental Station for 1974*, Part 2, 61–77.

Miller, H. G. 1930 The Woburn Farm. *Report Rothamsted Experimental Station for 1930*, 104.

Moffat, J. R. 1951 The farms. *Report Rothamsted Experimental Station for 1950*, 121–127.

Reed, A. H. 1979 Accelerated erosion of arable soils in the United Kingdom by rainfall and run-off. *Outlook on Agriculture* 10, 41–48.

Voelcker, J. A. and Hall, A. D. 1902 The value of unexhausted manures obtained by the consumption of foods by stock. *Journal of the Royal Agricultural Society* 63, 76–114.

Voelcker, J. A. and Hall, A. D. 1913 Compensation for the unexhausted manurial values of feedingstuffs and fertilizers. *Journal of the Royal Agricultural Society* 74, 104–119.

Voelcker, J. C. A. 1876 On the theoretical and practical value of purchased food and of its residue as a manure. *Journal of the Royal Agricultural Society* 37, 203–230.

Williams, R. J. B. 1978 Effects of management and manuring on physical properties of some Rothamsted and Woburn soils. *Rothamsted Experimental Station Report for 1977*, Part 2, 37–54.

Wyatt, R. J., Moorlock, B. S. P., Lake, R. D. and Shephard-Thorn, E. R. 1988 Geology of the Leighton Buzzard-Ampthill district. *British Geological Survey Technical Report* WA/88/1.

# CHAPTER 7

# SLOPE EROSION AND COLLUVIATION AT THE FLOODPLAIN EDGE

A. G. BROWN

*The break of slope at the floodplain edge provides optimum conditions for curvature dependent deposition in the form of sheets, fans or inwashes. In some cases this material may interdigitate with alluvial sediments, or overlie a palaeochannel infill at the floodplain edge. Under ideal conditions it may be possible to trace a colluvial band, probably produced by one storm, from the slope edge onto the floodplain, to date it and to relate it to human activity on that slope. Magnetic susceptibility of colluvial / alluvial profiles may be indicative of the processes of hillslope erosion. Land boundaries trap substantial colluvial deposits providing evidence of historical slope erosion. Stratigraphic evidence of colluviation onto floodplains and the use of magnetic susceptibility measurements are illustrated in this paper using sites in North Germany and the Midlands of the United Kingdom.*

## The floodplain edge and curvature dependent colluvial deposition

The sudden break in slope which generally exists between the floodplain and terraces or valley slopes provides optimum conditions for curvature-dependent deposition of colluvium. Curvature-dependent deposition is deposition resulting from a downslope decrease in gradient, typical of the lower parts of convexo-concave slopes. Although the term colluvium presents difficulties, it is generally regarded as a diamicton (i.e. non-sorted terrigenous deposit) which can be produced by a variety of gravity-driven slope processes including both Newtonian and non-Newtonian flows. Although it is frequently differentiated from alluvial sediment transported in well-defined channels (Ritter and Miles 1973), this distinction is difficult to make in practice, and therefore colluvium is often identified on the basis of its geomorphological position and sedimentary characteristics rather than its origin; for example the United Kingdom Soil Survey definition of colluvium is based upon its sedimentary and pedological characteristics. In this paper the term 'colluvium' will be used for all deposits stratigraphically and morphologically linked to a source slope irrespective of sorting and fabric.

In relatively low-energy environments colluvial deposits at floodplain edges are most commonly sheets or thin fans deposited by overland flow. These occur below critical slope angles where previously confined water spreads-out at the downslope end of rills and gullies, losing both depth and velocity (and sometimes discharge due to percolation), which reduces its power to transport the load supplied to it. It is the change in the hydraulic geometry of flow that occurs at the end of the rill, and not a sudden change in gradient (rare in any case), which is the cause of deposition (Bull 1976). There is, however, an indirect relationship with slope as rills are formed on the steepest part of convexo-concave slopes and the intersection point (the point where flow changes from being erosive and confined to depositional and unconfined) will often be found at the base of this segment and above the lower gradient toe segment.

In the absence of post-depositional mixing, which is often the cause of poor sorting, inwashes may be identifiable in alluvial contexts due to differences in fabric, texture or mineralogy. These inwashes will only be preserved where the sedimentation is cumulative, as is the case in floodplain backswamps or topogenous mires. Where there has been lateral reworking of the floodplain by meander migration, or floodplain stripping (Nanson 1986) caused by high energy floods, these deposits will only be preserved until the next flood when they will be mixed into alluvial deposits downstream (Figure 7.1). Depending upon the size of the floodplain and the erodibility of its slopes, colluvial deposits vary in their relative contribution to the floodplain fill. In one of the few studies which has quantified the colluvial input, Lattman (1960) calculated that it made-up about 20% of the total valley fill. The valley, which was in the Appalachian plateau, Pennsylvania, was about 335 m wide and the colluvium was built-out from the valley side as the stream shifted laterally to the opposite edge of the floodplain. Lattman (1960) was able to differentiate the colluvium on stratigraphic grounds and because it contained slabs of local siltstone dipping back into the slope; however, in many cases it may be extremely difficult to distinguish colluvial from alluvial sediments (Ritter and Miles 1973). Plots of cumulative colluvial soil area against cumulative basin area show that colluvium is preferentially found in small catchments (Brown 1987) and is probably more important in smaller rather than larger valleys. The texture and sorting of undisturbed colluvium can be related to flow competence and slope angle. From experimental data de Ploey (1984; 1984a) has related critical slope angle ($S_{cr}$ in m $m^{-1}$) below which colluviation occurs to the mean grain diameter (D in cm), the load concentration (C in g $l^{-1}$) and unit

Figure 7.1 Colluvial fan on floodplain gravels in Malaga Province, Southern Spain. The fan resulted from a severe rainstorm which occurred around 14 November 1989.

discharge (q in $cm^2s^{-1}$) in the form:

$$S_{cr} = D.C^{0.8}/q^{0.5}$$

As can be seen from Figure 7.2 greater grain diameters require greater critical slopes (i.e. low slope angles will produce fine colluvium) and for a given grain size the greater the slope the higher the load concentration will be. This implies that deposition on lower slopes or further downslope in concavities will produce finer colluvial deposits. Stratified colluvium and fining upwards may occur as local slope angle decreases. Studies of the micro-aggregates that result from raindrop impact have shown that they are very similar to the micro-aggregates in undisturbed soil (Imeson and Jungerius 1976) suggesting that most silt and clay colluvium is transported as micro-aggregates rather than individual mineral grains. The stratification and micro-aggregates are typical of Mücher's (1973) Type C colluvium as identified using soil micromorphology, and which he associates with 'accelerated soil erosion influenced by human activity'. Most of this work is derived from field observations and experiments in Belgium and Luxembourg, where the conditions of slope, soil and climate are not dissimilar to those found in Southern Britain. This work has also shown how the pattern of soil is largely determined by human alterations to natural slope processes during prehistoric and historic periods (Kwaad and Mücher 1977; 1979; Imeson *et al.* 1980). Additionally these studies have shown that colluvium may be stratified and that pollen and volcanic ash may be used to erect chronologies and estimate relative rates of accumulation.

In mountain environments the colluvial system is often dominated by the infilling and evacuation by mass-movements (landslides, mudslides and debris avalanches) of dells (Mills 1987) or hillslope hollows (Reneau *et al.* 1989), the dating of which has yielded

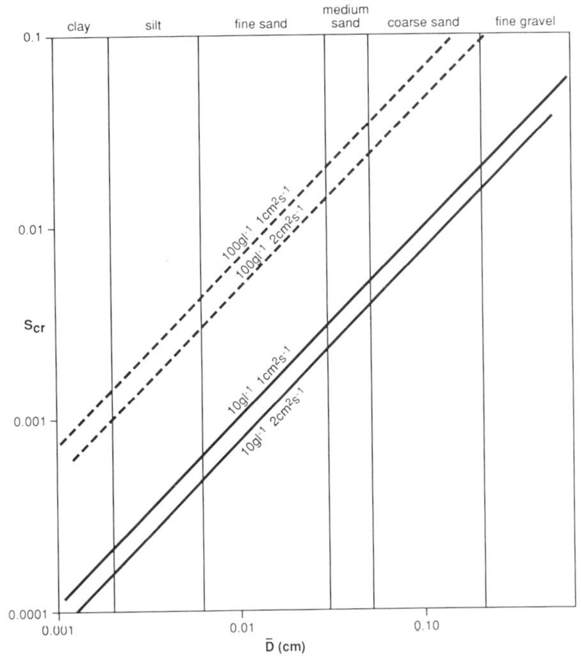


Figure 7.2 Graph of the relationship between critical slope angle for colluviation ($S_{cr}$) and mean grain diameter (D) for two different load concentrations (100 $gl^{-1}$ and 10 $gl^{-1}$) and two different unit discharges (1 $cm^2s^{-1}$ and 2 $cm^2s^{-1}$). Plotted from the equation given by de Ploey (1984).

hillslope erosion rates (Dietrich and Dorn 1984). The trapping of sediment by hillslope hollows has been modelled by Benda and Dunne (1987), and by Reneau and Dietrich (1991) by using a geometrical factor and the colluvial deposition rate. Benda and Dunne (1987) assumed transport perpendicular to the hollow axes, while the approach of Reneau and Deitrich (1991) uses a convergence angle to model volumetric colluvial transport rates ($q_v$) oblique to hollow axes as below:

$$q_v = \frac{1}{2p_s \mathrm{Sin}\beta} \frac{dM_{1h}}{dt}$$

where $dM_{1h}/dt$ is the average mass deposition rate of colluvium, $p_s$ is the average soil density and β is the convergence angle (with straight contours β=O).

While the rates are much lower the conveyance and transport of sediment into and out of colluvial hollows also occurs in non-mountainous regions, especially where gullying periodically strips-out hollow or swale colluvium (Thorne and Zevenbergen 1990) which may be subsequently redeposited on floodplains. However, in many valleys, especially dry-valleys, there is little, if any, evacuation of colluvial sediments (Bell 1981) and colluvial storage equals the distributed slope input less losses in solution.

This paper presents United Kingdom (Callow End from the West Midlands) and North German (North Saxony) examples of colluvial units within floodplain sections, ranging from a single event inwash to substantial cumulative units. Also, using data from the East Midlands, United Kingdom (Raunds in the Nene valley), an example is given of the relationships between cultural features and deposition at the floodplain boundary. Simple measurements of magnetic properties are used in an attempt to differentiate colluvial units from alluvial units and relate intra-unit variation to hillslope erosional processes.

**Callow End: a single event?**

This site which lies at the Teme-Severn junction in the West Midlands, is a palaeochannel with a partly organic infill. The bluff slope is cut into the Worcester terrace, the surface of which is about one to two metres above the floodplain surface (Figure 7.3). The stratigraphy includes an inwash from this terrace slope above the channel bank, which is a well sorted coarse sand, on average, only 5 mm or approximately 5 grains thick. The inwash thickens slightly towards the terrace edge. Considering its geometry and lack of internal structure it was most probably deposited by a single event, which on the basis of radiocarbon dating occurred sometime between *c.* 3,000 and 2,600 years BP. This was a rare event as it only occurred once between 4,000 years BP and 1,200 years BP. The lack of significant silt and clay in the inwash may have been caused by deposition in shallow water which may have drained from the site. This site allowed the investigation of the local vegetation history from a pollen core through the fill (Figure 7.3). Because of the small size of this, and other palaeochannel sites, the pollen source area is small and of the same order of magnitude as the sediment source area (Brown and Barber 1985). The extremely high lime pollen percentage and its decrease away from the terrace edge (Brown 1988) clearly indicates that the terrace was covered by lime dominated woodland. From this diagram it can be seen that the event occurred after most of the lime dominated terrace had been deforested and also after a decrease in hazel. The site at this time was still covered by alder woodland. The event also coincides with a rise in grass pollen, but pre-dates the full removal of the alder woodland from the floodplain itself. Charcoal fragments were counted but the event is not associated with particularly high values. The diagram suggests that we have the coincidence of a terrace-slope cleared of its natural lime and hazel cover, with possibly some exposed soil and/or cultivation, and a particularly severe storm. If the vegetation cover had remained constant one could estimate a recurrence interval for the erosional event from this depositional event at about 1 in 2,800 years,

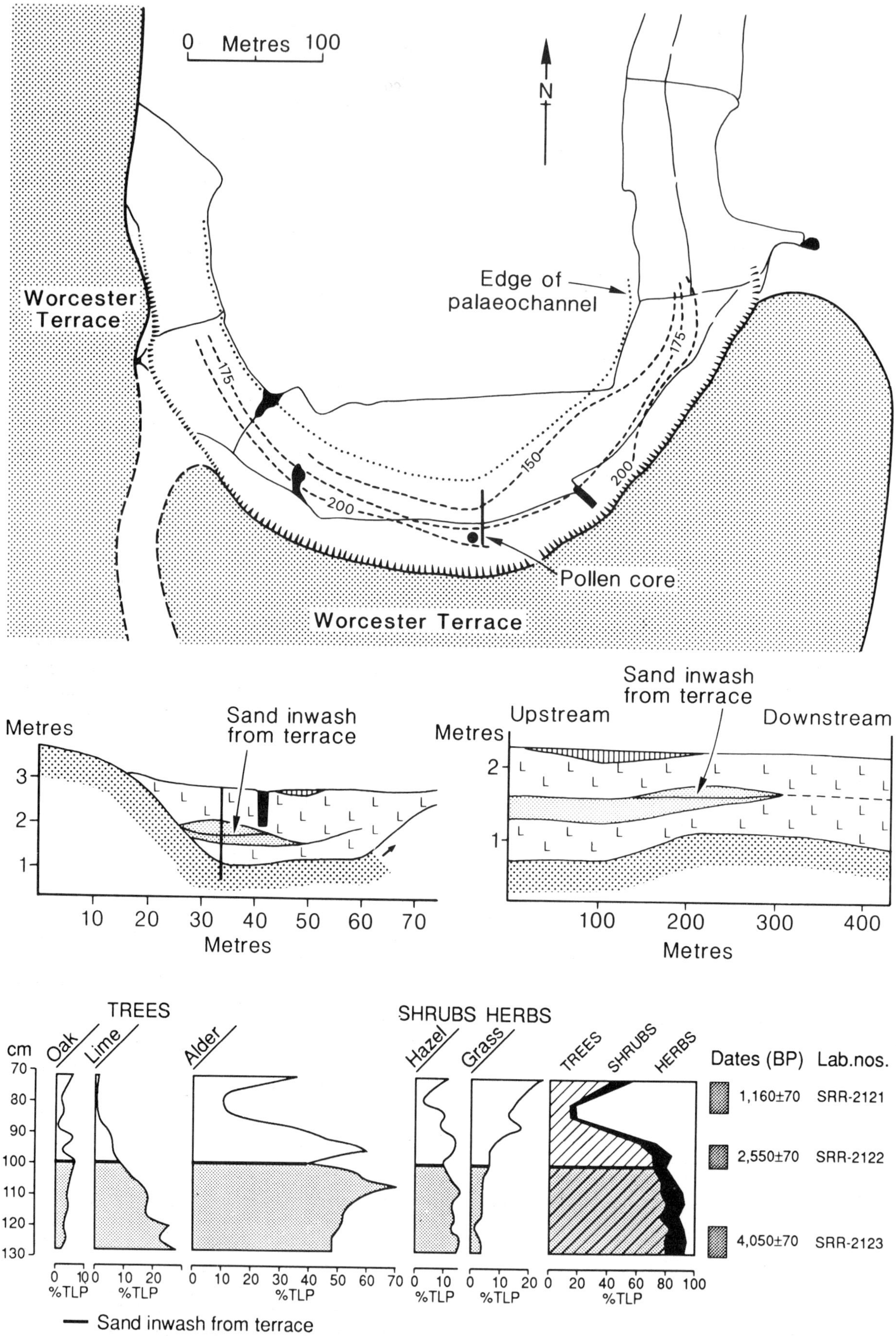


Figure 7.3 Callow End, site plan, sections and simplified pollen diagram. The broken lines are buried channel contours.

but this would be misleading as the depositional response to a storm of a constant magnitude would undoubtedly have increased as vegetation was disturbed and the terrace cultivated.

### Colluvial units in two North German floodplains

Cores from the floodplains of the Ilme and Weser, investigated by Rother (1989) have revealed substantial colluvial inputs. Three cross-sections from the Ilme near Einbeck show Pleistocene gravels covered by older calcareous meadow loam (mixed loamy soil) and younger non-calcareous meadow loam (Figure 7.4). From this valley and others, including the river Werra, it is known that the older meadow loam dates from the eighth to fourteenth centuries A.D. while the younger meadow loam is post-fifteenth century (Rother 1989; Bork 1989). These alluvial units are closely related to two colluvial units; the older and younger alluvial fans, which spread-out over 110 m from the floodplain edge. The author was responsible for the mineral magnetic studies of these cores. Peaks or high values of magnetic susceptibility are caused by soil development and therefore we generally see an enhancement of the magnetic susceptibility in the topsoil or where sediment is derived from topsoil. The magnetic susceptibility profile from B1 shows enhancement towards the base of the older meadow loam (Figure 7.4). The magnetic susceptibility data used in this paper is simple low frequency volumetric susceptibility measured using a Bartington magnetic susceptibility meter either with a well-sensor in the laboratory, as in the case of the German floodplain data, or in the field using a search-loop (for the transect) and a ferrite probe (for some of the Raunds profile data presented later in this paper). These profiles were also measured in the laboratory in order to check the field results. Bulk susceptibility has also been measured in the laboratory, but it shows the same trends as the volumetric data, and, as it is not comparable with the field data, it is not presented here.

A cross-section from the Upper Weser, near Munden, shows a buried terrace and alluvial fan with a colluvial unit spreading out from it upon which soil development has occurred (Figure 7.5). Magnetic susceptibility profile 4 shows some enhancement associated with the surface of both the younger low terrace and younger alluvial fan. Some erosion of the younger alluvial fans is suggested by the peak magnetic susceptibility remaining high right to the top. However, profile 2 does not show enhancement of the meadow loam except towards the base. This may be due to erosion of the younger alluvial fan prior to significant soil development, or possibly the effects of gleying (Dearing *et al.* 1985).

### The Raunds area: slope erosion and cultural features

Magnetic susceptibility may also be a useful tool in tracing the provenance of colluvium which does not reach the floodplain due to a natural or cultural trap such as walls and banks. An example of this is provided by some field boundaries in the Raunds area of the Nene valley in the East Midlands, United Kingdom. The area is the focus of a multi-site, multi-period archaeological project (Foard and Pearson 1985) and it is in connection with this project that the work described was done. In the area convexo-concave valley-side slopes about 30 m in height are underlain by near horizontal strata which include, from top to bottom; boulder clay, limestone, sandstone, ironstone, and Lias clay. The slope profile (Figure 7.6) shows a basal concavity clearly related to the location of a Medieval (or older) track, referred to as the Holloway, and field boundary running on a terrace at the edge of the floodplain. A magnetic susceptibility slope-transect shows increased enhancement over the ironstone and at the slope base. Several pits dug by D. Windell allowed measurement of the magnetic susceptibility profiles. Two pits (A and B) were supplemented by excavations for a pipeline and hand cores. In Pit C the very high enhancement (note the scale) is the result of soil development on ironstone subsequently buried by *c.* 0.3 m of colluvium. Pit B shows over 1 m of enhanced colluvial soil, but Pit A is the most interesting extending 4 m down to the basal sand and gravel. It shows that the morphological terrace sits above the sand and gravel and is a colluvial feature of stratified sand with stone lines and clay drapes. This material which is clearly derived from the adjacent slope, underlies and pre-dates the track. Fractionation of the sediments shows that the silt and clay fraction is enhanced towards the base. Upslope of the track is 2 m of silty sand, which is mixed with no internal structures. Since there was a possibility that the site was affected by sand quarrying a second profile was analysed some 86 m to the south (Figure 7.6b A1). It displays a very similar magnetic susceptibility profile with a slightly higher magnetic susceptibility peak at 50 cm and a fall in values above the sand horizon at about 130 cm. The possibility that the track was partly cut into a natural slope cannot be dismissed, but irrespective of this, the profiles indicate continued erosion of topsoil mixed by ploughing. Archaeological excavations on the slope have revealed a Roman settlement consisting of two buildings with a rectangular enclosure surrounded by small ditches which was in use in the late fourth and fifth centuries AD. Beyond the ditches lay a cemetery. There is evidence of both quarrying of the ironstone, and crop-processing in the form of two corn-driers and quern fragments. Post-Roman evidence on the slope is limited to Saxon pottery concentrated upslope and south of Oak Ditch. The Medieval Holloway has been trenched showing it was about 10 m wide and 1.20 m deep. Medieval pottery found by fieldwalking is assumed to have been deposited by manuring. The evidence is therefore of industry, settlement and agricultural activity on the mid-upper slope during the fourth and fifth centuries AD with probable continued agricultural use into the Medieval period and intensive modern arable cultivation. The linking of archaeological evidence, at the slope scale, to floodplain edge colluvial deposits has the potential to

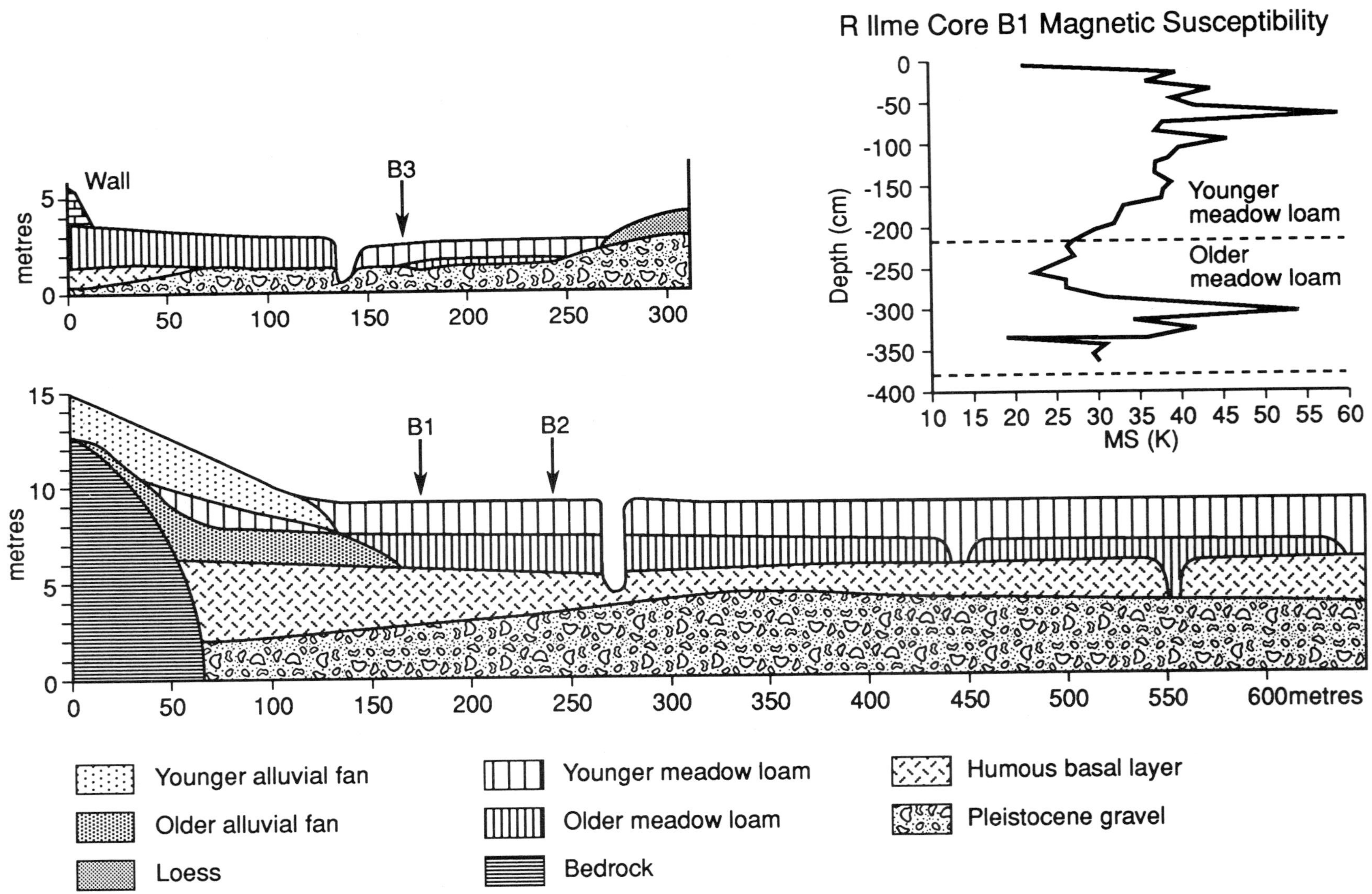


Figure 7.4 River Ilme cross-sections, redrawn from Rother (1989), and the magnetic susceptibility profile from B1. K is volumetric magnetic susceptibility measured in $10^{-6}$GOe$^{-1}$.

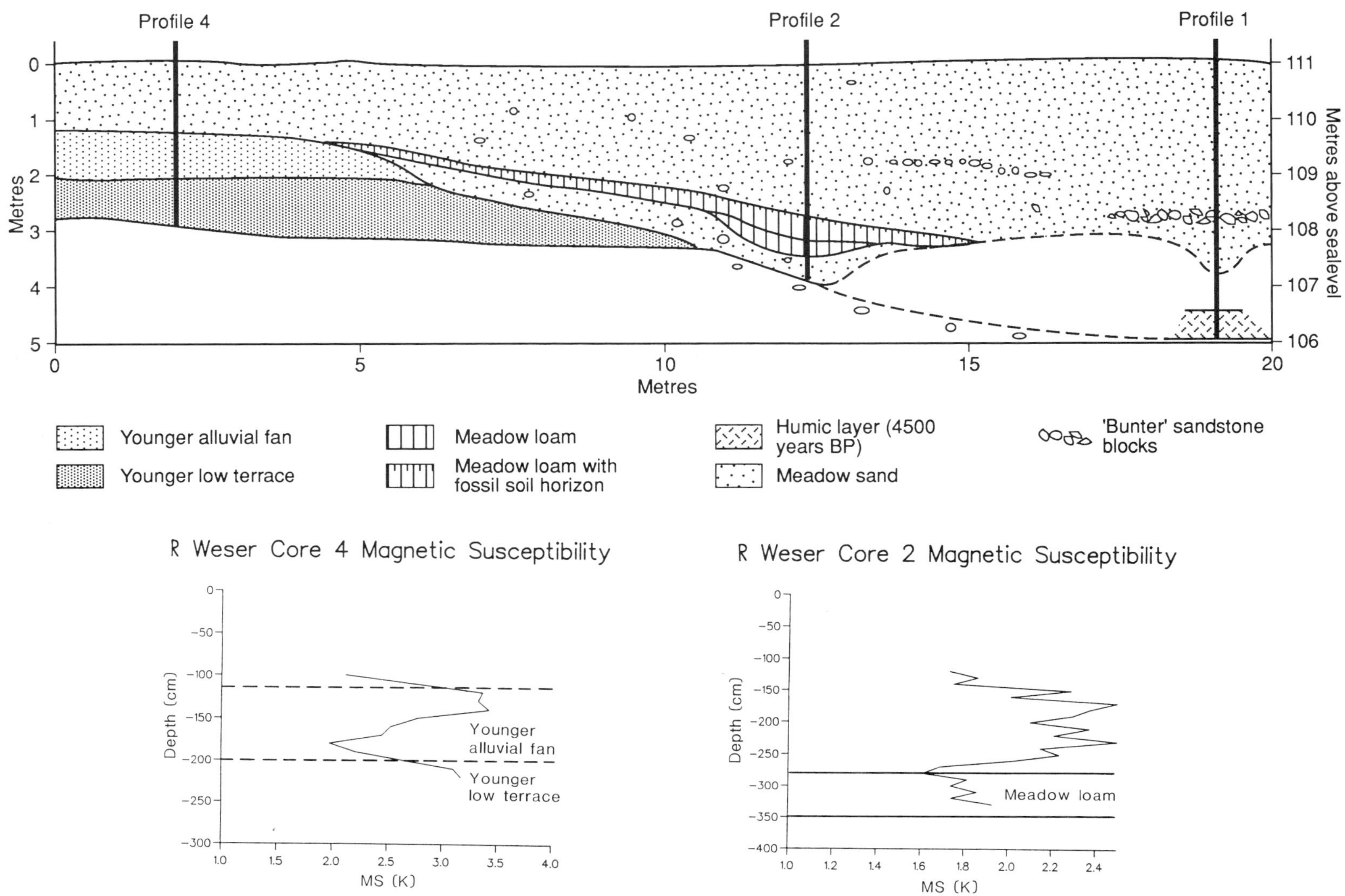


Figure 7.5 River Weser cross-section, redrawn from an original by N. Rother, and magnetic susceptibility profiles from Cores 2 and 4. K is volumetric magnetic susceptibility measured in $10^{-6}GOe^{-1}$.

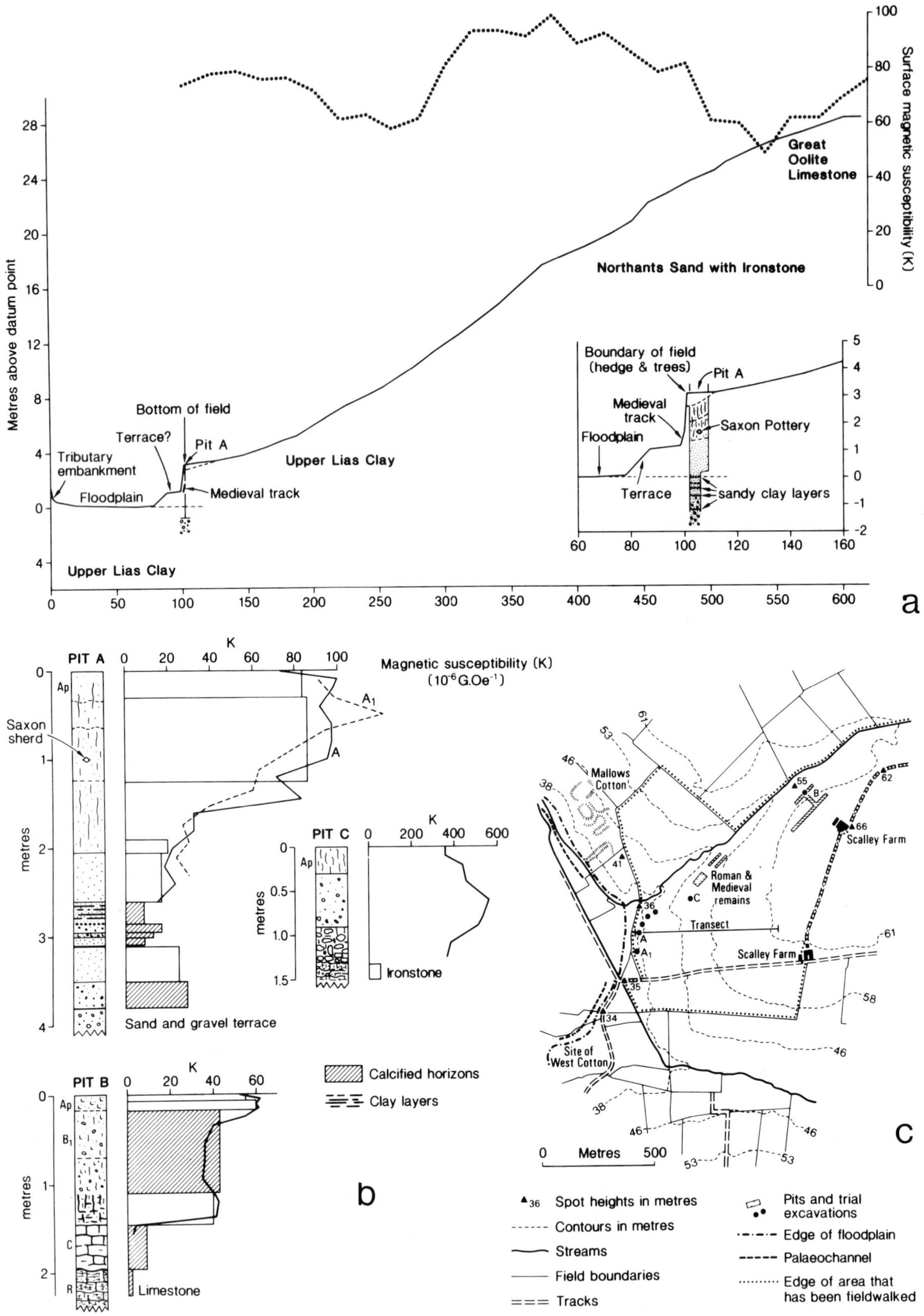


Figure 7.6 Raunds Area data (a) slope profile (b) selected profiles and (c) plan. The broken line on Pit A graph is profile $A_1$. For location see plan.

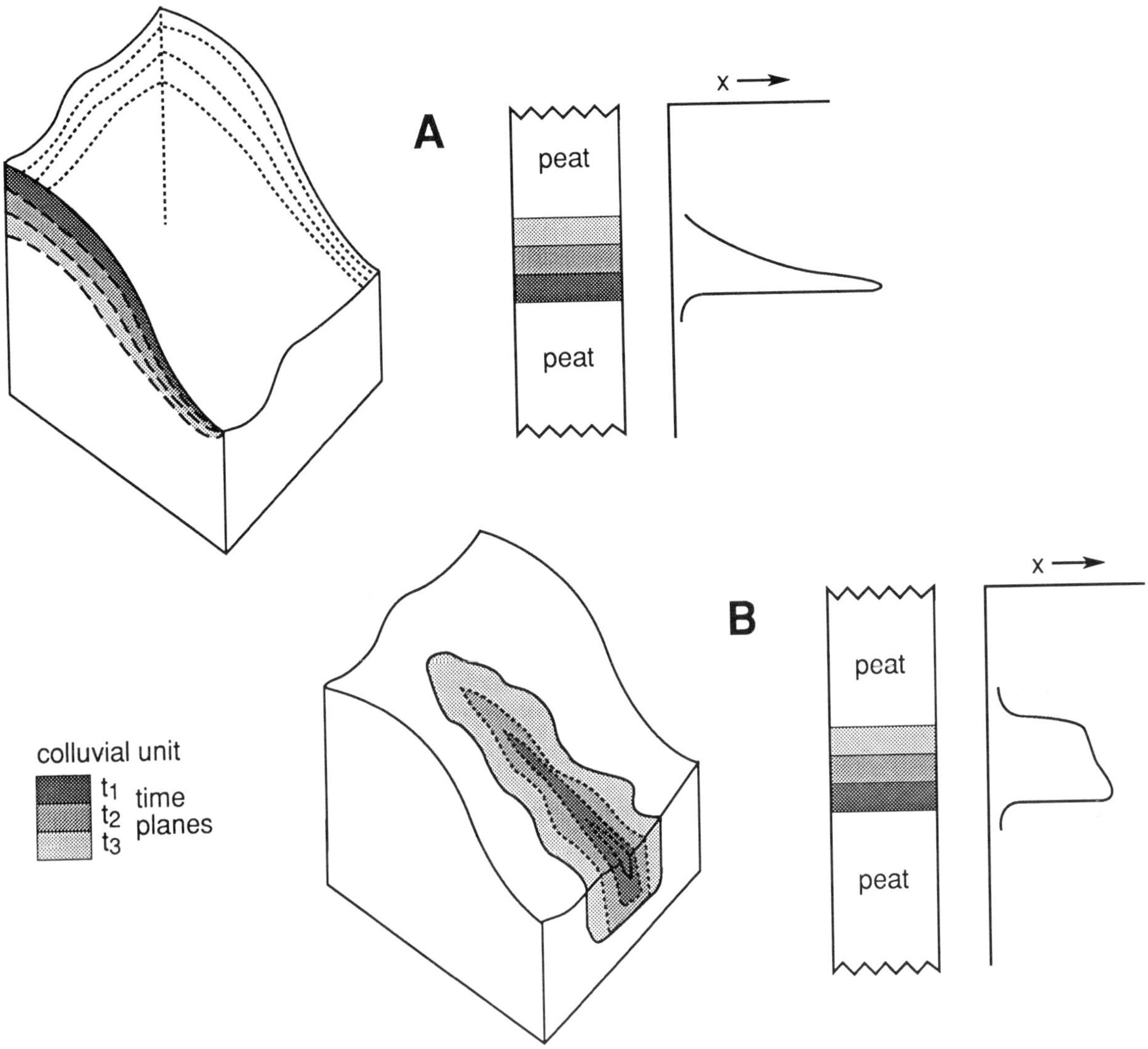


Figure 7.7 A conceptual model of the colluvial magnetic susceptibility profiles likely to result from A truncation and B rilling, where X is magnetic susceptibility.

isolate the impact on soil erosion of specific practices and periods of slope use.

Sustained (i.e. deep) enhancement could be produced by rill erosion with rills moving across the slope, event to event or year to year. A sustained enhanced signal would not result from the sequential truncation of the soil surface or the development of a stable gully network. This is illustrated by the simple model of magnetic susceptibility profiles (Figure 7.7) produced by slope-wash/truncation (A) and rill erosion (B). Truncation would, because of the enhancement of topsoil, result in a magnetic susceptibility peak at the base of the unit whereas rill erosion would produce a lower more prolonged peak depending upon the geometry of the rills and their location year to year. In reality these are idealised poles from a continuum because rills are fed by inter-rill erosion. In the profiles presented in this paper we see both tendencies, truncation/stripping associated with the base of the older meadow loams and continued enhancement associated with Medieval slope deposits at Raunds. However, one problem is that ploughing will mix the profile, and so sites interstratified with wet floodplain sediments such as peats, which are unlikely to have been ploughed, are easier to interpret. In addition it could be expected that earlier colluvial peaks would be more pronounced than later colluvial peaks because of the erosion of virgin woodland soils. These topsoils will have a minimum of 4,000 years of pedogenetically produced topsoil enhancement whereas later soils may be continually disturbed and have a lower organic content tending to reduce topsoil enhancement.

While the aim of this paper is not to document

changes in the rates of colluviation, these German sites and others in Saxony suggest colluviation rates for the eighth to fourteenth century of around 1.7 mm $yr^{-1}$ and lower fifteenth century rates of around 0.3 mm $yr^{-1}$ (Bork 1989). Particularly high rates during the early fourteenth century occurred at the same time as the most devastating European flood of the last 1,500 years which occurred in 1342 (Bork 1989). Colluvium not archaeologically rich and not interstratified with organic-rich floodplain sediments is much more difficult to date but the relation to cultural features and pottery suggests accelerated colluviation on the Raunds slopes is late or Post-Medieval. It was deposited upon colluvial deposits from an older undated period.

## Conclusions

While work on colluvial deposits at floodplain edges is on-going, the data presented here supports four concluding points.

(1) The German examples show that colluvial deposits can form a significant proportion of the alluvial fill of even quite large valleys. In these cross-sections it varies from zero to over 30% of the cross-sectional area. This neglected floodplain sediment type is surprisingly important in areas of low relief and relatively stable channels which have undergone intensive arable cultivation on valley slopes and terraces.

(2) The floodplain edge is an optimal location for colluvial deposition, due to energy loss, *and* the preservation of this colluvium in stratified sequences. If it is undisturbed, it can allow the formative erosional events to be related to conditions in the sediment source area. So if a pollen diagram can be analyzed from the cutoff, or small basin, the pollen source area for most taxa and sediment source area will be largely coincident.

(3) Interdigitation of colluvial units and organic rich alluvial units may provide the best opportunity for dating these events.

(4) Magnetic susceptibility enhancement, although generally the case with colluvial units, will depend upon the age, location, post-depositional history and parent material of the colluvium. But colluvial magnetic susceptibility profiles may be indicative of processes of slope erosion and of relationships between incision by rills and truncation by slope-wash or inter-rill erosion. The frequently encountered basal enhancement of late Holocene colluvial units can be viewed as a transported and inverted soil caused by a period of accelerated soil erosion.

The preferential location of many settlements at the floodplain edge and the differential exploitation of slopes, floodplains and terraces means that environmental reconstruction from sedimentary sequences at these locations is particularly valuable. The study of colluvial deposits at the floodplain edge allows us to study environmental impact from definable areas which can be related to particular settlements and sites. Our understanding of the relationships between processes of hillslope erosion and colluvial characteristics is imperfect and in practice is greatly complicated by post-depositional processes. However, the floodplain edge provides one of the few places where it may be possible to apply geomorphological models which could predict past processes, conditions and rates of accelerated soil erosion in the archaeological record.

## Acknowledgements

The stratigraphic data and samples from the German valleys were kindly provided by N. Rother. Thanks are also due to D. Windell for access to pits and soil data, and to other members of the Northamptonshire County Council, Raunds Area Project Team, especially Steve Parry. Radiocarbon dates for Callow End were provided by the Natural Environmental Research Council and the diagrams were drawn by K. E. Moore and R. E. Pollington. I would also like to thank the three referees for their valuable comments.

## References

Bell, M. 1982 Valley sediments and environmental change. In: Jones, M. and Dimbleby, G. (eds.) *The Environment of Man: the Iron Age to the Anglo-Saxon period*. British Archaeological Reports British Series 87, Oxford, 75–91.

Benda, L. and Dunne, T. 1987 Sediment routing by debris flow. In: Beschta, R. L., Binn, T., Grant, G. E., Swanson, F. J. and Ice, G. G. (eds.) *Erosion and Sedimentation in the Pacific Rim*. International Association of Hydrological Sciences Publication 165, 213–223.

Bork, H-F. 1989 Soil erosion during the past millennium in Central Europe and its significance within the geomorphodynamics of the Holocene. In: Anhert, F. (ed.) *Landforms and Landform Evolution in West Germany*. Catena Supplement 15, 121–132.

Brown, A. G. 1987 Long-term sediment storage in the Severn and Wye catchments. In: Gregory, K. J., Lewin, J. and Thornes, J. B. (eds.) *Palaeohydrology in Practice*. Wiley, Chichester, 307–332.

Brown, A. G. 1988 The palaeoecology of *Alnus* (alder) and the postglacial history of floodplain vegetation: pollen percentage and influx data from the West Midlands, U.K. *New Phytologist* 110, 425–436.

Brown, A. G. and Barber, K. E. 1985 Late Holocene palaeoecology and sedimentary history of a small lowland catchment in Central England. *Quaternary Research* 24, 87–102.

Bull, W. B. 1976 The alluvial-fan environment. *Progress in Physical Geography* 1, 222–270.

Dearing, J. A., Maher, B. A. and Oldfield, F. 1985 Geomorphological linkages between soils and sediments: the role of magnetic measurements. In: Richards, K. S., Arnett, R. R. and Ellis, S. (eds.) *Geomorphology and Soils*. Allen and Unwin, London, 245–266.

De Ploey, J. 1984 Towards modelling colluviation. In: Abstract of *Geomorphology and Soils*, British Geomorphological Research Group Annual Meeting, Hull, September 1984.

De Ploey, J. 1984a Hydraulics of runoff and loess loam deposition. *Earth Surface Processes and Landforms* 9, 533–539.

Dietrich, W. E. and Dorn, R. 1984 Significance of thick deposits of colluvium on hillslopes: a case study involving the use of pollen analysis in the coastal mountains in Northern California. *Journal of Geology* 92, 147–158.

Foard, G. and Pearson, T. 1985 The Raunds Area Project; first interim report. *Northamptonshire Archaeology* 20, 3–21.

Imeson, A. C. and Jungerius, P. D. 1976 Aggregate stability and colluviation in the Luxembourg Ardennes: an experimental and micromorphological study. *Earth Surface Processes* 1, 259–271.

Imeson, A. C. and Kwaad, F. J. P. M. and Mücher, H. J. 1980 Hillslope processes and deposits in forested areas of Luxembourg. In: Cullingford, R. A., Davidson, D. A. and Lewin, J. (eds.) *Timescales in Geomorphology*. Wiley, Chichester, 31–42.

Kwaad, F. J. P. M. and Mücher, H. J. 1977 The evolution of soils and slope deposits in the Luxembourg Ardennes near Wiltz. *Geoderma* 17, 1–37.

Kwaad, F. J. P. M. and Mücher, H. J. 1979 The formation and evolution of colluvium on arable lands in northern Luxembourg. *Geoderma* 22, 173–192.

Lattman, L. H. 1960 Cross section of a floodplain in a moist region of moderate relief. *Journal of Sedimentary Petrology* 30, 275–282.

Mills, H. M. 1987 Variation in sedimentary properties of colluvium as a function of topographic setting, valley and ridge province, Virginia. *Zeitschrift für Geomorphologie Neue Folge* 31, 277–292.

Mücher, H. J. 1973 Micromorphology of slope deposits: the necessity of a classification. In: Rutherford, G. K. (ed.) *Soil Microscopy*. Proceedings of the Fourth International Working Meeting on Soil Micromorphology, Ontario, Canada, 553–566.

Nanson, G. C. 1986 Episodes of vertical accretion and catastrophic stripping: a model of disequilibrium flood-plain development. *Bulletin of the Geological Society of America* 97, 1467–1475.

Reneau, S. L. and Dietrich, W. E. 1991 Erosion rates in the southern Oregon coast range: evidence for an equilibrium between hillslope erosion and sediment yield. *Earth Surface Processes and Landforms* 16, 307–322.

Ritter, D. F. and Miles, C. 1973 Problems of stream terrace correlation and reconstruction of geomorphic history caused by colluvium. *Geografiska Annaler* 55, 85–92.

Rother, N. 1989 Holozane erosion und akkumulation in Ilmetal, Sudeniedersachsen. *Bayreuth Geowissenschaftliche Arbeiten* 14, 87–94.

Thorne, C. R. and Zevenbergen, L. W. 1990 Prediction of ephemeral gully erosion on cropland in the South-Eastern United States. In: Boardman, J., Foster, I. D. L. and Dearing, J. A. (eds.) *Soil Erosion on Agricultural Land*. Wiley, Chichester, 447–459.

# CHAPTER 8

# THE EVIDENCE FOR PAST AND PRESENT EROSION IN THE SLAPTON CATCHMENT, SOUTHWEST DEVON

LOUISE HEATHWAITE AND TIM BURT

*The long-term trend in erosion from a mixed arable and grassland catchment in southwest Devon is examined by combining the palaeolimnological evidence from sediment cores taken from Slapton Ley with a 100 year record of land use change for the catchment. It is thought that the lake arrived at its present hypertrophic state largely as a result of increasing sediment and associated solute inputs from its catchment, particularly since World War II. The current sediment production from different land uses was obtained using plot experiments and is used here with wider spatial analysis of the rainfall-runoff-sediment delivery patterns in the catchment. The current catchment erosion results are used here to assess the relative contribution of different land uses to the lake sediment record and their role in the accelerated eutrophication of Slapton Ley.*

## The lake sediment record for Slapton Ley

Slapton Ley is a coastal freshwater lake divided into the Higher Ley, a 39 ha reedbed and the 77 ha open water Lower Ley (Figure 8.1). The 46 $km^2$ catchment is primarily grassland on the steep valley sides with arable land on the plateau. Topography consists of wide plateau of low gradient dissected by narrow deep valleys (maximum slope 25°). The area is underlain by impermeable Lower Devonian slates and shales, with freely draining acidic and nutrient-poor silty clay loam soils. Mean annual rainfall for the catchment (1961–1988) is 1039 mm; mean annual runoff is 650 mm. The area has a mean annual temperature of 10.5°C.

Radiocarbon dating of the lake sediments suggest that it is unlikely to be more than 1000 years old (Morey 1976). For much of its history, the palaeolimnological evidence suggests that the Ley was a shallow, productive, clear-water lake. Recent evidence suggests that the lake may now be hypertrophic (Heathwaite and O'Sullivan 1991), primarily as a result of increased sediment and solute inputs from its catchment. The current annual load is over 6 t $km^{-2}$ inorganic nitrogen and over 30 t $km^{-2}$ sediment (Heathwaite and Burt 1991). Palaeolimnological evidence is used here to reveal the history of sedimentation in Slapton Ley. The lake sediment record is combined with records of land use in order to establish trends in catchment erosion and the links with land use change. Possible sources of sediment are examined by looking at the current pattern of erosion in the catchment.

The pattern of sedimentation across the bed of the Lower Ley has been studied using multiple coring and whole core magnetic susceptibility (O'Sullivan *et al.* 1991). The objective was to establish whether a consistent pattern of sedimentation occurred within the lake and whether there was evidence for sediment-focussing, particularly around catchment inflows. An undisturbed sediment record is essential if it is to be clearly linked to catchment land use change. Because the lake is shallow and turbulent, the possibility of resorting of the lake sediments once they reached the lake may be high. The lake sedimentation results suggest that most cores show a peak in magnetic susceptibility (K) just below the present sediment-water interface. The susceptibility profile for core SLL3, which was used in the chemical analysis described in this paper, is shown in Figure 8.2. Comparison with the results presented in O'Sullivan *et al.* (1991) indicates that core SLL3 is typical of the pattern of sedimentation in Ireland Bay (Figure 8.1), and is easily correlated with other cores from Slapton Ley. The general pattern of sedimentation in the Lower Ley suggests that there is some sediment focussing along the long-axis of the lake, away from the main inflows of the Gara and Start catchments (Figure 8.1). Sedimentation rates appear to be lower in relatively exposed areas of the lake close to the shingle ridge and higher on the landward side of the Ley (O'Sullivan *et al.* 1991).

The sediment influx record derived from the lead-210 dating method (Olsson 1990) for core SLL3 is shown in Figure 8.3. The method is based on the escape of radon from the Earth, and the subsequent decay of this radioactive gas to the $^{210}Pb$ isotope, which has a half-life of 22.26±0.22 years, making it suitable for dating within the range of 1–150 years.

One of the complications with this dating method is that lake sediments commonly contain an amount of uranium or its daughters, which release a continuous supply of $^{210}Pb$ to the lake sediments at all depths. This $^{210}Pb$ fraction must be subtracted from the $^{210}Pb$ total to give the dating fraction, referred to as unsupported lead. The unsupported $^{210}Pb$ inventory for core SLL3 was 7.73 pCi $cm^{-2}$ $year^{-1}$, a constant influx of 0.24 pCi $cm^{-2}$ $year^{-1}$. In determining a $^{210}Pb$ chronology, it is assumed that the unsupported $^{210}Pb$ (which is mainly derived from atmospheric fallout) once incorporated in the sediment layer, declines with age in accordance with the $^{210}Pb$ radioactive-decay law

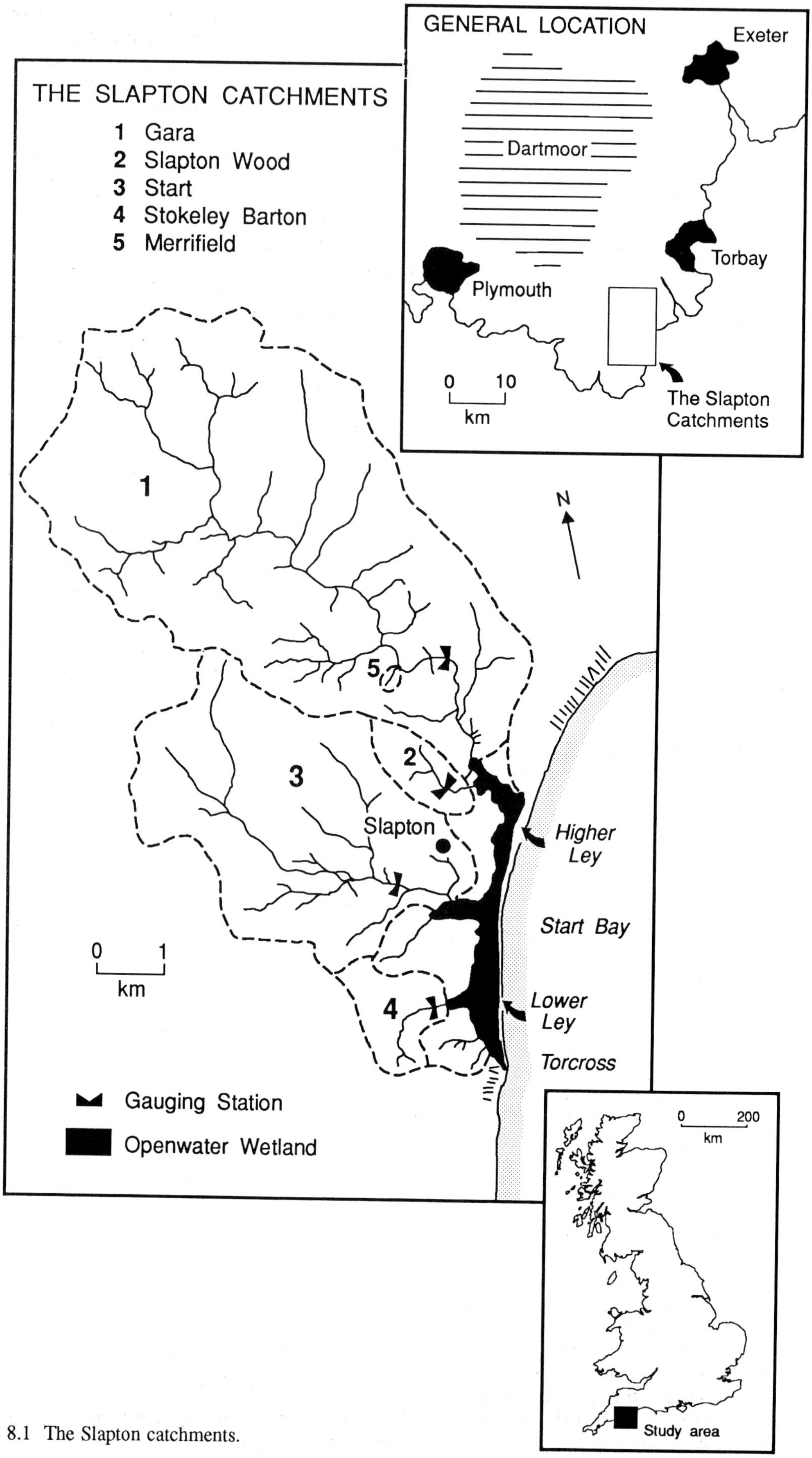


Figure 8.1 The Slapton catchments.

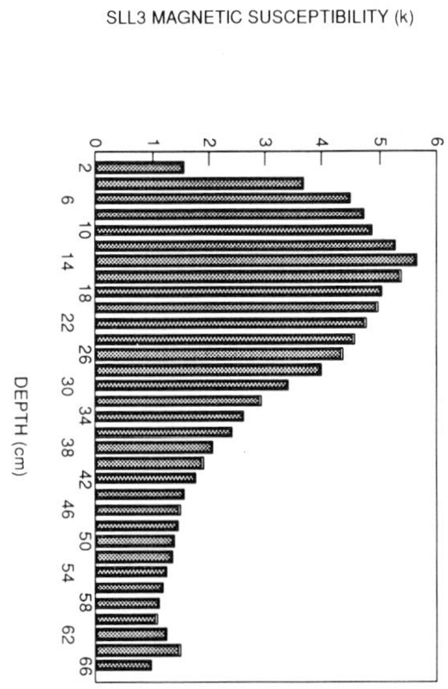


Figure 8.2 Magnetic susceptibility for core SLL3, Slapton Ley

(Oldfield and Appleby 1984) and there is no diffusion of $^{210}$Pb through the pore water of the sediment. Owing to the non-monotonic nature of the $^{210}$Pb profile for core SLL3 (i.e. the sedimentation rate varies but the supply of $^{210}$Pb is constant), the chronology was obtained using the CRS (constant rate of supply) model (Appleby *et al.* 1986) which assumes that the same amount of unsupported lead is supplied for each chronological time unit. Therefore, the model is independent of the net dry mass sedimentation rate of the lake sediments (Oldfield and Appleby 1984).

Using the results for core SLL3 and other dated cores from the Lower Ley, the peak in magnetic susceptibility occurred around 10 years B.P. The results also show that the sediment influx has increased since *c*. 1950, when the sedimentation rate was around 2.4 mm year$^{-1}$ to a peak around 1967 (10.2 mm year$^{-1}$). Sediment influx at the current sediment-water interface is around 8-12 mm year$^{-1}$. The accuracy of the dating method decreases with sediment age. In the upper 15 cm of the core (dated as 1987–1962) the sediment chronology is ± 2 years. Between 15 cm and 25 cm (1962–1915) the accuracy is reduced to ± 3-9 years. Below 25 cm, the dating accuracy is around ± 15 years. The catchment land use records also become less accurate prior to World War I, so the ability to link the lake sediment data to the land use records beyond this time period is reduced.

Calculated on the basis of the dry mass content of core SLL3 together with other cores from the Lower Ley, the current (1987) sediment influx rate is equivalent to approximately 8.7 t ha$^{-1}$ dry matter deposited on the bed of the Ley annually, or a catchment sediment load of 670 t year$^{-1}$ equivalent to an erosion rate of 132 kg ha$^{-1}$ This figure is considerably lower than the 1440 t year$^{-1}$ (315 kg ha$^{-1}$) annual suspended sediment load recorded by Heathwaite and Burt (1991) for the main catchment inputs to the Slapton Ley. The reasons for this difference are discussed later with reference to the sediment sink of the Higher Ley.

Figure 8.4 shows the physical properties of the lake sediments together with the 200-year lake sediment influx record. A major change in the physical properties of the sediment occurs at 24–28 cm depth which correlates with the period 1906–1926. Here the total lake sediment influx declines but the proportion of organic matter increases whilst the carbonate content decreases. This has been linked to changes in the water level of the lake (Heathwaite and O'Sullivan 1991). Above 21 cm depth, dated around 1945, the rate of sediment influx to Slapton Ley increases significantly.

### Catchment land use change and erosion

The long-term land use records for the Slapton Catchment are used here to try to identify those changes in the catchment that might a result in increased erosion and sediment influx to the lake. Figure 8.5 illustrates the land use and livestock numbers in the catchment for the period 1905 to 1990

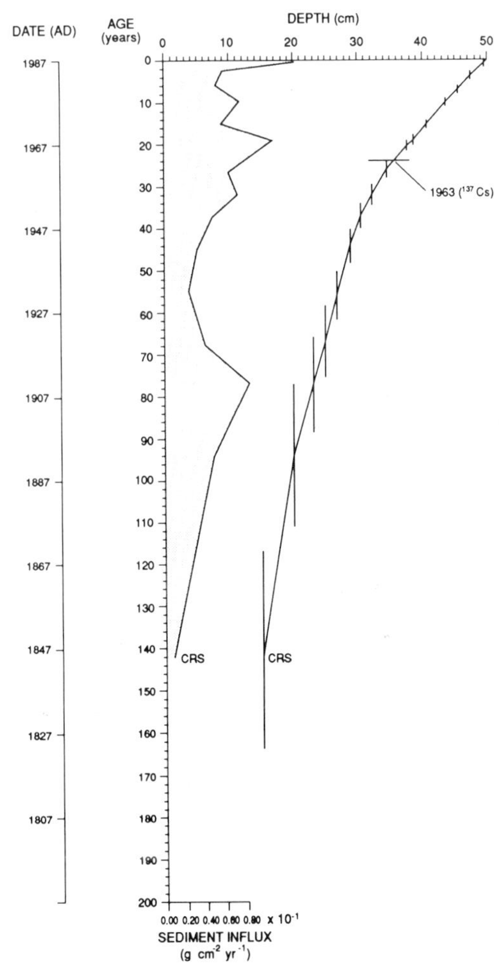


Figure 8.3 $^{210}$Pb chronology for Slapton Ley using CRS $^{210}$Pb dating model

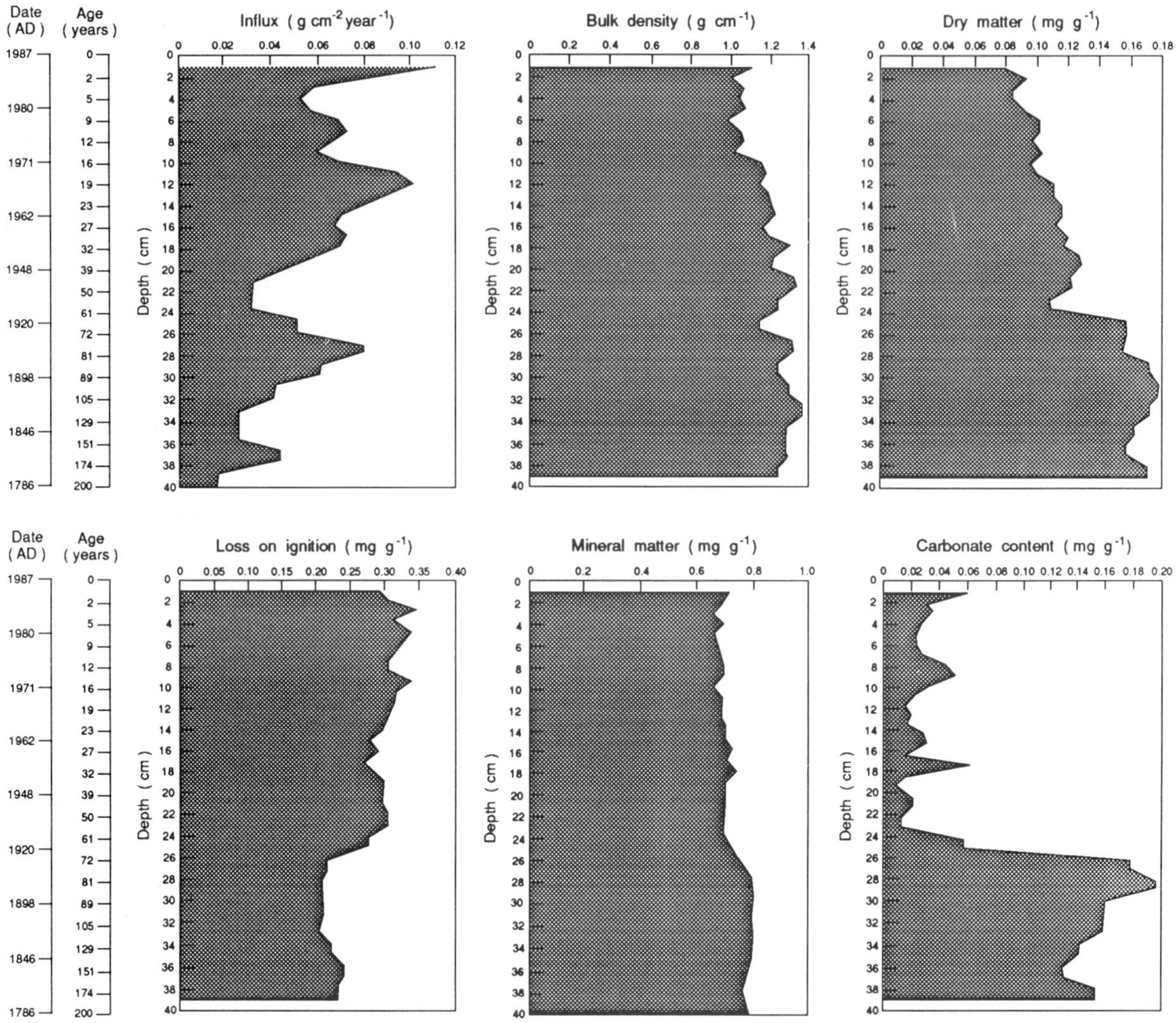


Figure 8.4 Physical properties and rate of sediment influx to Slapton Ley over a 200 year period

(Heathwaite and Burt 1991). Two major trends are shown, both of which would increase the rate of sediment input to the lake. First, the proportion of arable land, and hence the area of ploughed land, increased between 1935 and 1945. Second, the area of temporary grassland increased between 1945 and 1960. This shift in the balance of temporary and permanent grassland again implies an increase in the frequency of ploughing and additionally would increase grazing pressure, especially on the remaining permanent grassland. Runoff from overgrazed grassland has been shown to contribute at least thirty times the sediment load of grassland with intact vegetation cover (Heathwaite *et al.* 1990). Because this land use is commonly found adjacent to the stream channel, a close link with high sediment export, as a result of increased grazing pressure, and increased sediment influx to the Ley, might be anticipated. Sediment delivery to the stream is discussed further below.

**Further evidence for catchment erosion: chemical fractionation of the lake sediments**

The relative contribution, from both allochthonous and autochthonous sources, to the sediments of Slapton Ley were identified using a chemical fractionation technique developed by Engstrom (Engstrom and Wright 1984) and described in Heathwaite and O'Sullivan (1991). The key chemical elements for the interpretation of the erosion record of the lake catchment are K, Al and Si because they are commonly located in the allogenic fraction and are derived from mineral matter embedded in the crystal lattice of the sediments. The allogenic fraction is generally taken to reflect mineral particles washed into the lake from its catchment by streams. Because this fraction is largely unaltered, it is usually assumed to reflect an eroded rather than weathered, catchment sediment source. If weathering predominated in the catchment, runoff waters would be more likely to carry dissolved ions into the lake rather than material which is still embedded in the mineral structure of the sediments. So high influx of the allogenic fraction of these materials is commonly taken to indicate high rates of catchment erosion.

The interpretation of soil development and catchment erosion from sediment geochemistry was pioneered by Mackereth (1966) for the English Lake District. He suggested that high concentrations of Na, K and Mg in the mineral (allogenic) fraction of lake sediments represented periods of active catchment erosion. Alternatively stable soils and deep weathering should reduce the concentration of these elements in the mineral matter of a lake sediment layer because Na, K and Mg would be removed as dissolved ions in catchment runoff *prior* to any subsequent erosion and transport of material to the lake sediments. Sedimentary Na, K and Mg were therefore taken to reflect the intensity of weathering and erosion in the catchment. An eroding catchment would be characterised by a high Na, K and Mg concentration in the lake sediments which is closely related to a high mineral matter concentration (Engstrom and Wright 1984).

There are a number of problems with this hypothesis. Mackereth was dealing with changes in the *concentration* of elements in the sediment cores. He did not have information on the *rate* of influx of material to the lake sediments. It is possible that an increase in the concentration of one element could decrease the relative concentration of another. That is, changes in element concentration in lake sediments are autocorrelated. A second problem was the analytical method used. Mackereth was dealing with *bulk* or *total* element concentration in the lake sediments. This gives no information on the *sources* of the sediment, that is, their authigenic (lake-derived) or allogenic (catchment-derived) origin. It is possible that variation in lake sediment as a result of, for example, land use change would alter the concentration profile of the lake sediments. This was primarily a problem for Mackereth because he measured mineral matter content by loss-on-ignition. Determined in this way, mineral matter would also include authigenic oxides in the analysis which could dilute the allogenic fraction (Engstrom and Wright 1984).

These points can be illustrated with reference to the Slapton Ley sediments. Figure 8.6a shows the relationship between the K and Mg content (by weight percentage) of the allogenic fraction (calculated as common oxides) and the total mineral content of Slapton Ley. It is clear from Figure 8.6a that K and Mg are totally uncorrelated with the inorganic content of the lake sediments. This would suggest that a predictive relationship between the lake sediments and catchment erosion does not exist. However, plotting the results in terms of the total *influx* of K and Mg to the lake sediments reveals a different pattern (Figure 8.6b). Here K influx to the lake sediments clearly increases with increasing mineral matter influx, which suggests that the increase in the rate of sediment deposition to the lake bed is a result of the increased input of eroded material from the catchment. Mg influx is not related to mineral matter influx because it is primarily located in the authigenic fraction of the lake sediments and is mainly associated with carbonate cycling *within* the lake rather than erosion from the catchment (Heathwaite and O'Sullivan 1991).

Figure 8.7 shows the chemical composition of K, Al and Si in the lake sediments. The complete curve represents the total influx to the lake sediments, with the authigenic or biogenic fraction represented by light shading and the allogenic fraction by dark shading. K and Al show a strong upward trend in influx to the lake sediments, with 57-97% of these elements being located in the allogenic fraction. This trend is consistent with increased erosion of these elements from the catchment, particularly since 1940 (21 cm depth), when a high proportion of permanent grassland was ploughed (Figure 8.5). The proportion of biogenic silica, indicative of lake productivity, increases in the upper lake sediments as the lake system becomes more eutrophic (Heathwaite and O'Sullivan 1991).

**The current evidence for erosion in the Slapton catchments**

*Catchment scale studies of sediment delivery*

The current annual suspended sediment load for the Slapton catchments is shown in Table 8.1. The sediment load was computed from continuous discharge observations for each sub-catchment and water samples taken at intervals ranging from 15-minutes during storm events to 24-hours at low flow. The Gara catchment has a particularly high suspended sediment load in comparison with the other sub-catchments. High sediment production might be a function of the steeper slopes and high runoff potential

| Catchment | Area | Suspended* Sediment (Kg $ha^{-1}$) | Land Use Arable % | Land Use grass % | Runoff mm |
|---|---|---|---|---|---|
| Gara | 2362 | 503 | 12 | 81 | 920 |
| Slapton Wood | 93 | 66 | 32 | 36 | 581 |
| Start | 1079 | 225 | 53 | 34 | 950 |
| Stokeley Barton | 153 | 23 | 29 | 66 | 148 |

*Sediment load was computed from continuous discharge observations for each catchment and suspended samples taken at intervals ranging from 15-minutes during storms to 24-hours at low flow.

Table 8.1 The annual suspended sediment load and land use distribution for the Slapton Catchments (1988).

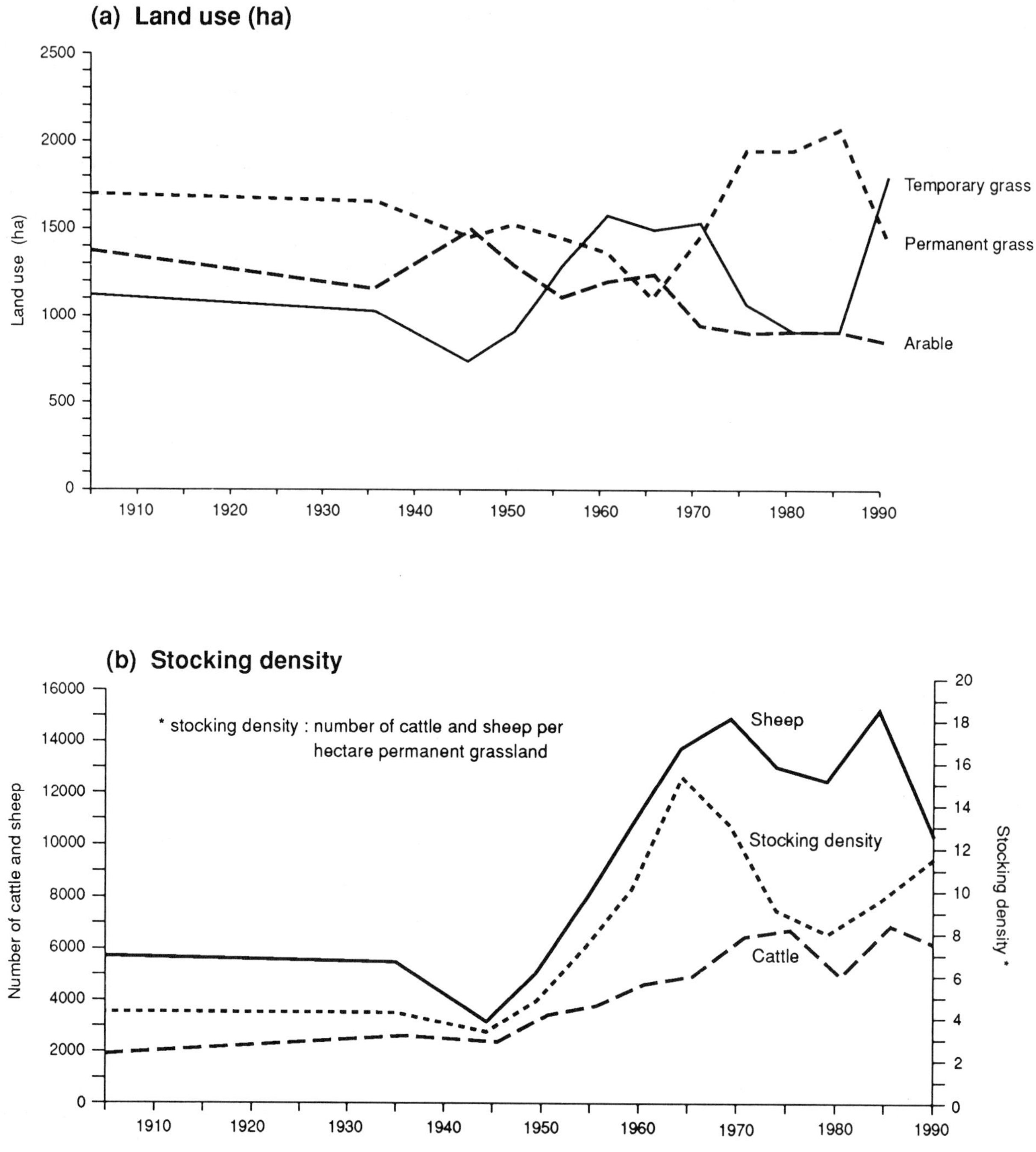


Figure 8.5 Land use and livestock change in the Slapton Ley catchment 1910-1990.

for this catchment which overlays impermeable Dartmouth Slates. The Start sub-catchment has a surprisingly low sediment load considering the high percentage of arable land cover and high surface runoff potential shown in Table 8.1. One factor that might contribute to this lower sediment load is the Meadfoot Beds catchment geology which has a gentler topography. A further difference between the Gara and Start catchments is the number of livestock; over 90% of the total number of cattle in the Slapton Catchments are found in the Gara sub-catchment. The results from the plot experiments, which are discussed in more detail below, suggest that overgrazing of permanent grassland is a major potential source of stream sediment.

It is recognised that not all the catchment sediment load may reach the Lower Ley owing to sedimentation in low-lying valley areas, and in particular in the Higher Ley (Figure 8.1). The balance between monthly sediment inputs to and outputs from the Higher Ley is shown in Table 8.2. With the exception of May and June, when there is little sediment flux, the Higher Ley acts as a net sediment sink throughout the year. It is hardly surprising then that the Higher Ley is becoming increasingly infilled with sediment. Furthermore, the pattern of sediment influx to the Higher Ley is seasonal, with over 40% of the annual sediment input entering the Higher Ley in

| Month | Suspended Sediment load (t) input | output | balance |
|---|---|---|---|
| October | 32.7 | 0.1 | +32.6 |
| November | 198.2 | 1.5 | +196.7 |
| December | 76.0 | 1.8 | +74.2 |
| January | 520.2 | 7.4 | +512.8 |
| February | 198.3 | 56.3 | +142.0 |
| March | 119.9 | 13.9 | +106.0 |
| April | 15.7 | 15.1 | +0.6 |
| May | 7.3 | 14.2 | -6.9 |
| June | 1.4 | 2.3 | -0.9 |
| July | 6.1 | 2.2 | +3.9 |
| August | 6.8 | 1.6 | +5.2 |
| September | 12.4 | 0.7 | +11.7 |
| Total | 1195.0 | 117.7 | +1077.9 |

Table 8.2 The effect of the Higher Ley on sediment delivery to the Lower Ley (1987-1988).

January for the year of study. In effect, the reduction of the proportion of the catchment sediment load reaching the Lower Ley by sedimentation in these low-lying valley areas means that the sediment record in the lake cores must be an underestimate of the absolute sediment load from the catchment over the 100-year period considered here. However, the relative changes in the rate of catchment erosion and sediment input to the lake may remain the same.

*Field scale studies of soil erosion and sediment delivery*

Rill erosion has become a common feature of the south Devon landscape in recent years. The generation of overland flow and erosion seems to be associated with winter cereals, especially where the ground has been rolled after drilling. Poaching in fields containing winter fodder crops has also been a problem in some cases. Since we have not conducted a systematic survey of soil erosion in the Slapton area, the results presented here must remain somewhat preliminary. Even so, given that the area is noted for its production of subsurface stormflow (Burt and Butcher 1985), the recent prevalence of surface runoff, albeit localised, is something of a surprise. In the very wet 1989/90 winter, *all* fields containing winter barley in the Slapton area showed signs of rilling.

*Infiltration capacities of Slapton soils*

The soils of the Slapton area are freely drained acid brown earths overlying impermeable Devonian slates and shales (Trudgill 1983). They are silty loams (median grain size 8-12μm), easily compacted. We have used both a rainfall simulator (Bowyer-Bower and Burt 1989) and ring infiltrometers (Burt 1978) to show the effect of land use on infiltration capacity (Heathwaite *et al.* 1990). Results are shown in Table 8.3. Even light use of soils reduces the infiltration capacity significantly below that of freshly ploughed soil. It is interesting that the infiltration capacity of bare, rolled ground measured using a rainfall simulator was identical to that measured using a ring infiltrometer on a badly rilled field at South Allington: 4 mm $hr^{-1}$. Two episodes of soil erosion are discussed below. Whilst the second event involved somewhat extreme conditions, the first occurred under very modest rainfall intensities.

| Land Use | Rainfall intensity (mm $hr^{-1}$) | Infiltration capacity (mm $hr^{-1}$) | Bulk density (g $cm^{-3}$) |
|---|---|---|---|
| Temporary grass | 12.50 | 12.33 | 0.96 |
| Barley | 12.50 | 11.04 | 1.08 |
| Rolled, bare ground | 12.50 | 4.00 | 0.93 |
| Lightly grazed permanent pasture | 12.50 | 5.85 | 1.12 |
| Heavily grazed permanent pasture | 12.50 | 0.10 | 1.18 |
| Permanent pasture (*) | – | 9 (range 3.6–36) | – |
| Freshly ploughed soil (*) | – | 50 | – |
| Woodland soil (*) | – | 180 | – |

(*) Determined using a ring infiltrometer

Table 8.3 Rainfall simulation of the effect of land use on surface runoff from hillslope plots (after Heathwaite *et al.* 1990, with additional data from Burt *et al.* 1983).

We observed widespread sheetflow at Merrifield (Figure 8.8), a tributary valley of the River Gara, on 23 March 1987 whilst undertaking reconnaissance for a study of subsurface stormflow! Early March was very dry, but Slapton received 13.7 mm of rainfall on 22 March. This rain continued until the evening of 23 March, that day's total being 19.7 mm at intensities between 0.2 and 3.3 mm $hr^{-1}$. We observed extensive sheetflow on two fields - grazed barley stubble and grazed kale - when rainfall was in the range 1 mm $hr^{-1}$. Overland flow was observed in other fields including some with grass, though these always showed signs of heavy tramping by stock.

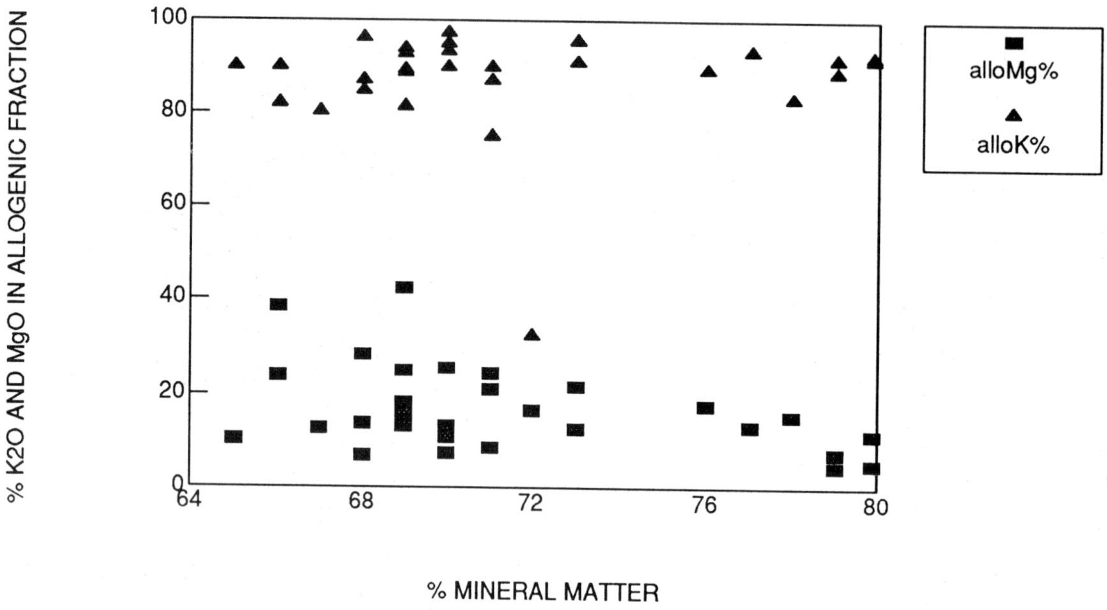


Figure 8.6a Relationship between the Mg and K content of the allogenic fraction (calculated as common oxides) and the total mineral content of the sediments of Slapton Ley, core SLL3.

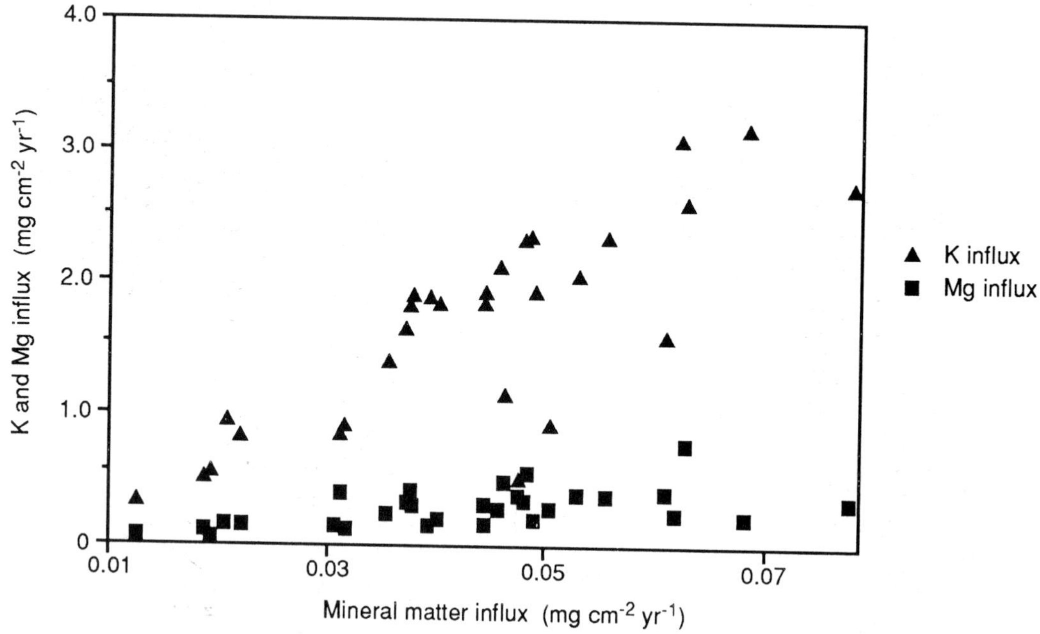


Figure 8.6b Relationship between the Mg and K influx (mg $cm^{-2}$ $year^{-1}$) and the total mineral matter influx to the Slapton Ley, core SLL3.

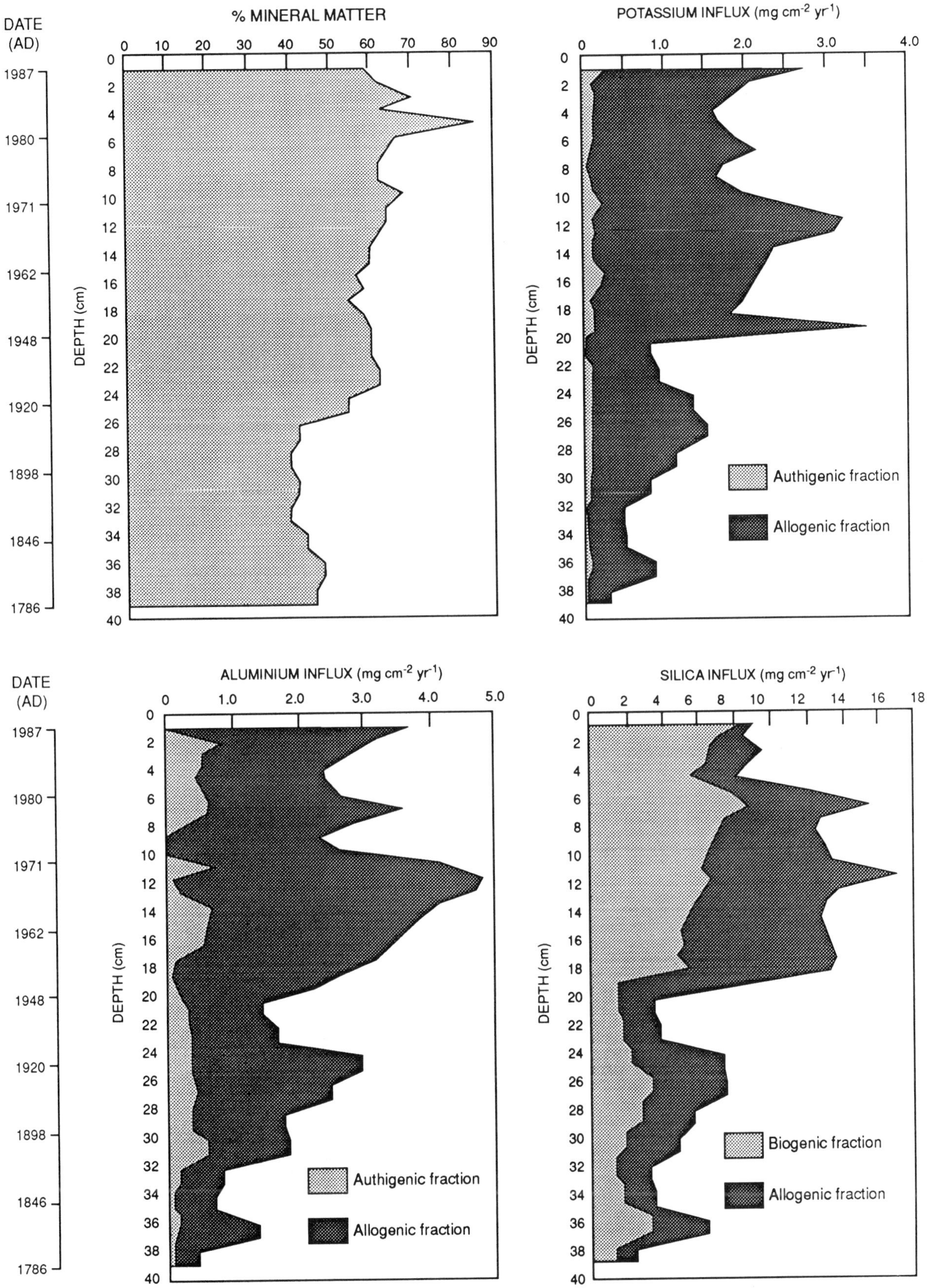


Figure 8.7 K, Al and Si influx to the sediments of Slapton Ley.

Figure 8.8 Sheetflow from overgrazed kale at Merrifield, a tributary valley of the River Gara, 23rd March 1987.

Figure 8.9 Rilling on winter barley at South Allington, maximum slope angle 20°, January 1990.

The worst case of rilling we have seen in this area was a field of winter barley at South Allington, a few kilometres south of Slapton (Figure 8.9). The topography is convexo-concave with a low-angle plateau about 80 m wide above slopes which reach a maximum angle of about 20° Grain size analysis showed a typical result for Slapton soils: 19% gravel, 27% sand, 49% silt and 5% clay, a median grain size of 18.9µm, and an infiltration capacity of 4 mm $hr^{-1}$. The field had been rolled after drilling and was extensively covered by rills, some as deep as 70 cm and between 20 and 50 cm wide. While there was a large amount of deposition at the base of the slope, it was clear that much sediment had entered the local stream, and there was a large amount of flooding and deposition in a lane a few hundred metres down valley. The rilling occurred on 13 December and again on 20 December 1989. The automatic rain-gauge at Slapton showed that rainfall intensities exceeded 4 mm $hr^{-1}$ for four hours on 13 December (6.5, 7.0, 5.0, 11.0) and again for four hours on 20 December (8.0, 4.5, 4.5, 6.0). Figure 8.9 was taken early in January 1990. Further rilling occurred in February 1990, the wettest month at Slapton (259 mm total) since records began in 1959. By the end of February rill erosion was widespread throughout the Slapton area. Every field with autumn drilled cereals which we examined contained rills.

Silty loam soils are particularly prone to surface compaction either by overgrazing or by heavy machinery. Some cultivation techniques such as rolling may cause severe compaction. On the Sussex Downs farmers are strongly advised not to roll fields after autumn drilling, though this advice is usually ignored (Boardman, pers. comm.). It may be that the same policy should be adopted, and enforced, in south Devon. Certainly rilling has become commonplace in south Devon in recent years, suggesting that current land management techniques are detrimental to soil conservation.

As noted above, sediment yields to Slapton Ley are currently about 300 kg $ha^{-1}$. Though this is not an unusual amount for southern England, the source of this sediment is localised to certain fields. Simple calculations of sediment trapped behind walls and hedges suggests that gross erosion rates in these fields may be greatly in excess of the basin-wide figure.

Future research is likely to focus on predicting the effect of land use change on catchment erosion. In the absence of physically based erosion models which can be readily applied at the *catchment* scale, we plan to investigate the use of export coefficient models. These have been used successfully to estimate the nutrient loading into Slapton Ley (Johnes and O'Sullivan 1989) and elsewhere (Johnes 1990). Despite their lack of process realism, export coefficient models have the great advantage of being easy to calibrate. They may well be able to fulfil a satisfactory role in providing interim estimates of soil erosion and sediment delivery in catchments until such times as more complex physically based models can be conveniently used.

## Conclusions

Sequential inorganic chemical analysis of lake sediments, particularly if coupled with sediment dating, can reveal a number of points concerning the history of a lake and its catchment. The sediment influx of allochthonous material to Slapton Ley has increased since 1940, probably as a result of agricultural intensification in the catchment. Large scale conversion of permanent pasture to temporary grass on steep valley sides and in the zone adjacent to the stream network, is likely to be the source of the increased sediment load.

## References

Appleby, P. G., Nolan, P. J., Gifford, D. W., Godfrey, M. J., Oldfield, F., Anderson, N. J. and Battarbee, R. W. 1986 $^{210}$Pb dating by low background gamma counting. *Hydrobiologia* 143, 21–27.

Burt, T. P. 1978 Three simple low-cost instruments for the measurement of soil moisture properties. Huddersfield Polytechnic, Department of Geography Occasional Paper No. 8.

Burt, T. P., Butcher, D. P., Coles, N. and Thomas, A. D. 1983 Hydrological processes in the Slapton Wood catchment. *Field Studies* 5, 731–752.

Burt, T. P. and Butcher, D. P. 1985 Topographic controls of soil moisture distributions. *Journal of Soil Science* 36, 469–486.

Bowyer-Bower, T. and Burt, T. P. 1989 Rainfall simulators for investigating soil response. *Soil Technology* 2, 1–16.

Burt, T. P., Heathwaite, A. L. and Trudgill, S. T. 1992 Catchment sensitivity to land use controls. In: Thomas, D. S. G. and Allinson, R. (eds.) *Landscape Sensitivity*. Chichester: Wiley (in press).

Engstrom, D. R. and Wright, H. E. 1984 Chemical stratigraphy of lake sediments as a record of environmental change. In: Haworth, E. and Lund, J. (eds.) *Lake sediments and environmental history*. Leicester University Press, 11–69.

Heathwaite, A. L., Burt, T. P. and Trudgill, S. T. 1990 Land use controls on sediment delivery in lowland agricultural catchments. In: Boardman, J., Foster, I. D. L. and Dearing, J. H. (eds.) *Soil erosion on agricultural land*. Chichester: Wiley, 69–97.

Heathwaite, A. L. and O'Sullivan, P. E. 1991 Sequential inorganic chemical analysis of a core from Slapton Ley, Devon, U.K. *Hydrobiologia* 214, 125–135.

Heathwaite, A. L. and Burt, T. P. 1991 Predicting the effect of land use on water quality. In: Peters, N. E. and Walling, D. E. (eds.) *Sediment and Water Quality in a Changing Environment: Trends and Explanation*. International Association of Hydrological Sciences No. 203. Wallingford: IAHS Press, 209–219.

Johnes, P. J. 1990 An investigation into the effects of land use upon water quality in the Windrush catchment. Unpublished D.Phil. Thesis, University of Oxford.

Johnes, P. J. and O'Sullivan, P. E. 1989 The natural history of Slapton Ley Nature Reserve XVIII. Nitrogen and phosphorus losses from the catchment - an export coefficient approach. *Field Studies* 7, 285–309.

Mackereth, F. J. J. 1966 Some observations on post-glacial lake sediments. *Philosophical Transactions of the Royal Society of London* B, 250, 165–213.

Morey, C. R. 1976 The natural history of Slapton Ley Nature Reserve IX. Morphology and history of the lake basins. *Field Studies* 4, 353–368.

Oldfield, F. and Appleby, P. G. 1984 Empirical testing of $^{210}$Pb-dating models for lake sediments. In: Haworth, E. and Lund, J. (eds.) *Lake Sediments and Environmental History*. Leicester University Press, 93–124.

Olsson, I. U. 1990 Radiometric Dating. In: Berglund, B. E. (ed.) *Handbook of Holocene Palaeoecology and Palaeohydrology.* John Wiley and Sons, 273–312.

O'Sullivan, P. E., Heathwaite, A. L., Appleby, P. G., Brookfield, D., Crick, M. W., Moscrop, C., Mulder, T. B., Vernon, N. J. and Wilmhurst, J. M. 1991 Paleolimnology of Slapton Ley. *Hydrobiologia* 214, 115–124.

Trudgill, S. T. 1983 The Soils of Slapton Ley. *Field Studies* 5, 835–840.

# CHAPTER 9

# SOIL EROSION IN WESTERN EUROPE: FROM THE LAST INTERGLACIAL TO THE PRESENT

B. VAN VLIET-LANOË, M. HELLUIN, J. PELLERIN
AND B. VALADAS

*During the last interglacial-glacial cycle in Western Europe, soil erosion by rill wash, deflation and solifluction was active at least during the cold and humid phases: the Early Glacial, the beginning and the end of the Upper Pleniglacial and the Late Glacial. Erosion intensity was greatest at the end of the Early Glacial and in the early millennia of the Upper Pleniglacial. The last erosion phase before the impact of humans was in the Younger Dryas, but it was weak because of fairly continuous plant cover. Holocene erosion is rare, usually found in the vicinity of human settlements and related to forest destruction. More extensive erosion began in the Roman period in the loess belt and occurred mainly in the Middle Ages. However, these episodes were much less important than those occurring recently through mechanization and other modifications of agricultural practices since the last war. Present day erosion reaches an intensity similar to that at the beginning of the Upper Pleniglacial.*

## Introduction

Soil erosion in temperate areas of western Europe seems to have been very strong in the last dozen years, even in areas where there is no record of earlier erosion. The record of erosion from the Last Interglacial until today shows that this process was most active during several periods of the Last Glaciation, when it was related to changes in the vegetation cover and climate.

To evaluate the susceptibility of soils to erosion, we must first understand the course of pedogenesis during interglacials or interstadials. Soil horizonation results from interaction of the climate via the drainage and the plant cover and the physico-chemical state of the parent material (texture and pH). Interglacial soil evolution (Eemian or Holocene) follows a path to acidification, starting from the end of the Pleniglacial in loess regions or coversands and/or deglaciation elsewhere. Clay eluviation in many soils is an early phase of pedogenesis, occurring before acidification, mainly from the Bölling-Alleröd Interstadial or in the interstadial just before the Last Interglacial in the loessic belt, but also more recently in calcareous or valley sediments (Van Vliet-Lanoë and Langohr 1981; Van Vliet-Lanoë 1990a). These observations, recently confirmed by dating and archaeology (Buch and Zöller 1990; Fagnart 1991), show that in midwestern Europe, at the onset of an interglacial like the Holocene, pedological development was nearly the same as that observed today. The human impact on the soil climate was rather limited until recently: all modifications accord with the accepted ideas of interglacial soil evolution (leaching, acidification, reduction in organic matter, increasing erodibility and cooling), as developed by Birks (1986), although with some accelerated interaction (acid rain, reduction in soil fauna and organic matter due to agricultural practices).

## The Last Glaciation

### *Processes*

Erosional processes acting during the glacial periods were of three main interacting types, all more or less controlled by frost activities. As frost penetrates into the soil, the most common erosional process seems to be solifluction in the form of frost creep with or without permafrost. This process can be superficial as in present day alpine areas with shallow temporary frost penetration or deeper seasonal freezing of soils. Its efficiency is related to wetness and snow thickness (Van Vliet-Lanoë 1985; Harris 1986). During dry periods, frost creep persists, but with little displacement, as this is controlled by segregated ice in the soil. More effective erosion is related to greater precipitation (>400 mm annual rainfall) and deep frost or permafrost limiting subsurface drainage. In Normandy, the total soil thickness reworked by these processes during the whole Last Glaciation (Lautridou 1985) does not seem to exceed 2 m in littoral conditions.

The formation of ice lenses in the subsoil, or of needle ice in the surface, prepares the soil for deflation by wind or splash and wash erosion by promoting soil disaggregation. As cryogenic structure is poorly resistant to thaw close to the surface, sheets of soil can be easily reworked by both processes. On mountains from the Arctic to the tropics frequent freezing cycles favour these processes which disrupt the plant cover (e.g. Soutadé 1979) or the cryptogamic crust commonly occurring on cold soil surfaces. Disruption is less common but does occur at low altitudes in the present day Arctic: 'liquefaction' of many areas of patterned ground (mud boils, mud flows) is often only splash crusting (Van Vliet-Lanoë 1990) (Figure 9.1). Sealing of the soil by refreezing during the first phases

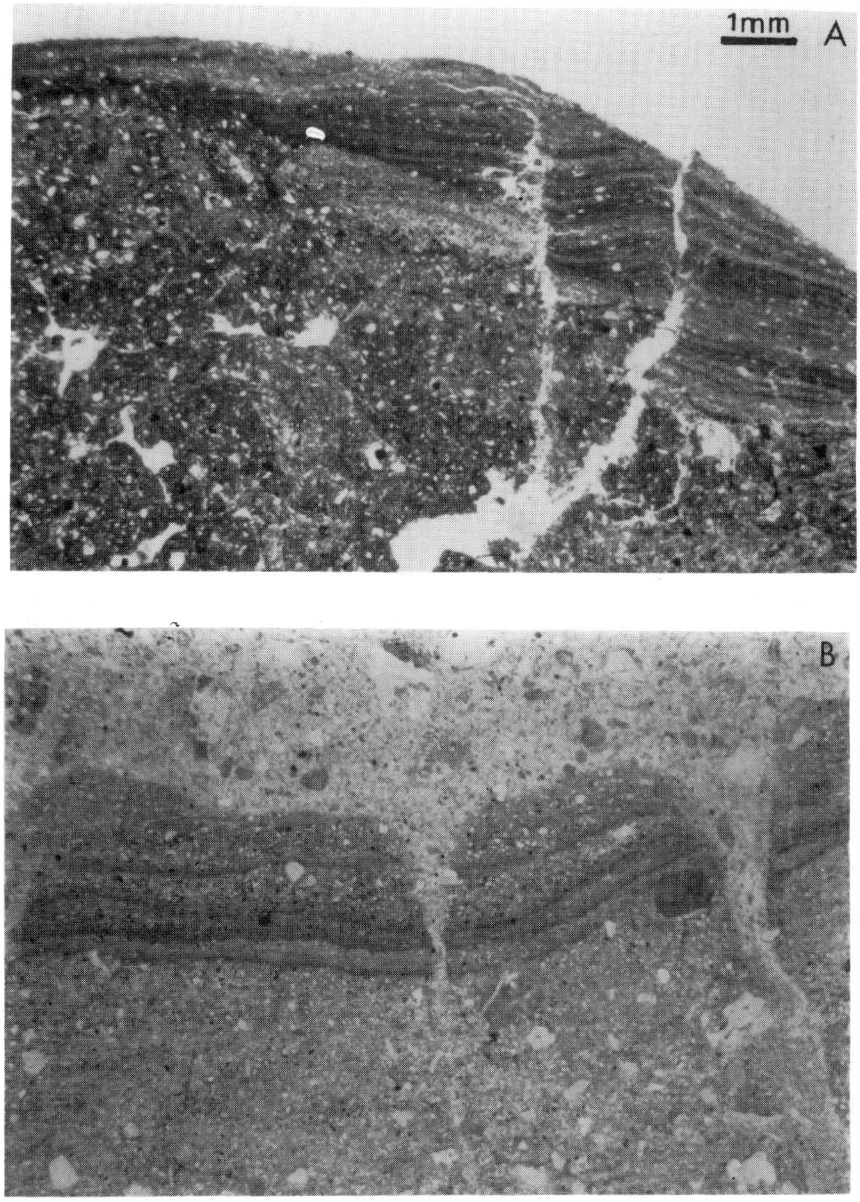


Figure 9.1 (A) Microphotograph of a splash crust, mudboil surface, Ny Alesund, Svalbard, Arctic. (B) Microphotograph of a buried splash crust, Early Glacial, Port Racine, Normandy. Same scale.

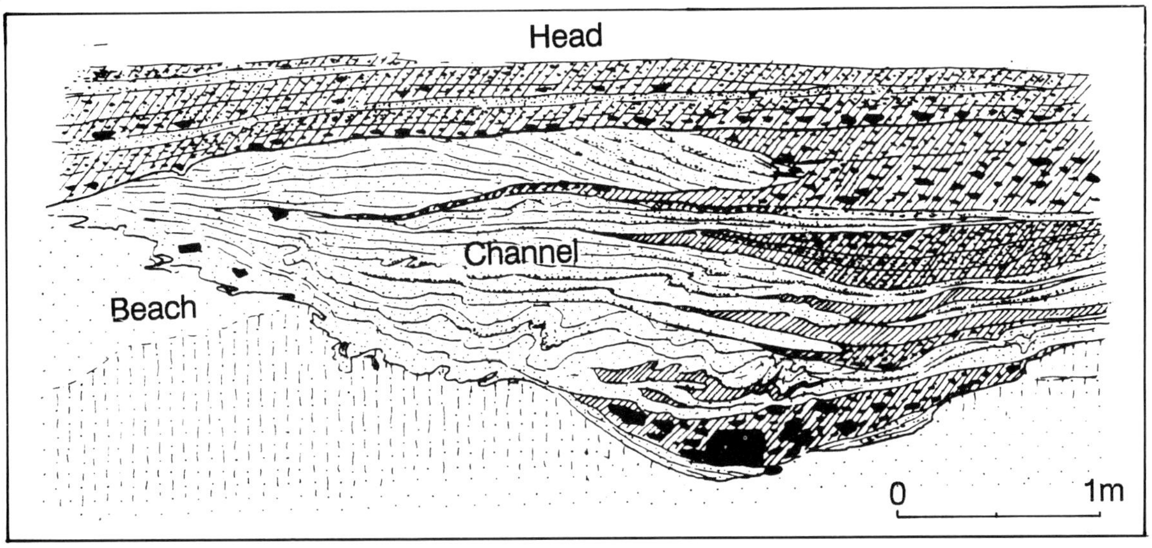


Figure 9.2 Channel cut in an Eemian beach sand, Early Glacial, Le Vougot, Brittany (study in collaboration with B. Hallegouët, Brest)

of the melting season decreases the infiltration capacity and increases first sheet and then rill wash activity (Woo and Steer 1982). As the soil aggregate stability is weak, snowmelt causes abundant rill erosion on bare soils or solifluction lobes (Thorn 1979; Strömquist 1985), but much less where a cryptogamic crust or more vegetation cover exists as this increases the superficial shearing resistance. Greater suspended loads occur in rills draining talus slopes than in rills draining fine soil below a snow patch (Wilkinson and Bunting 1975). Rain increases melting much more than sunny days because of higher energy input (reflectance of the snow and latent heat of fusion furnished by the precipitation): this explains the experimental results of De Ploey (1972) showing that rainfall causes erosion on frozen soil through superficial liquefaction. Rain can cause rapid mass wasting to occur at infrequent intervals in Scandinavian mountains as described by Rapp and Strömquist (1976).

### *Stratigraphical Record*

#### *Early Glacial*

In the Last Interglacial because of a continuous plant cover, there seems to have been little soil erosion except during the final climate deterioration. In cooler phases of the Early Glacial (oceanic oxygen isotope stages 5d and 5b), vegetation cover was modified and specific parts of the landscape were eroded, usually in sites covered late by snow or exposed to the wind. Coastal cliffs, cut in unconsolidated sediments and abandoned by eustatic retreat, were locally affected by large slumps resulting from hydraulic pressure as in Brittany (e.g. Hallegouët and Van Vliet-Lanoë 1986). Gullying resulted from the melting of snow patches (Figure 9.2) and frostcreep resulted in mud infiltrated beach ridges, thinly stratified fans and more or less stratified slope deposits. In this section, strong frost-induced microfabrics (Van Vliet-Lanoë 1985) and afterflow deposits (Figure 9.1) similar to those produced by Mücher and De Ploey (1984) are commonly seen. A grassy cover limited the washing action as today in Svalbard. Inland erosion was also intense, but less spectacular, frequently eroding the interglacial soil and sometimes older deposits. Rill and sheet wash erosion and frost creep were very active, filling shallow depressions and also large ones like the Flemish valley in Belgium (Heyse 1975). Needle ice activity seems to have made an increasing contribution to erosion towards the east and towards the end of the Early Glacial, as the precipitation rate seems to have decreased (Van Vliet-Lanoë 1990b). This period resulted in a continuous erosion surface, often associated with a stone line, somewhat cryoturbated and with traces of aeolian activity (microdunes, ventifacts); valley incision can be observed. The age of this surface was correlated with the end of oceanic oxygen isotopic stage 5 at about 75000 BP (Haesaerts and Van Vliet-Lanoë 1974; Lautridou 1985). Total soil erosion during the Early Glacial was at least 60 cm over the whole surface of northern France and Belgium, with a more drastic phase during stage 5d, particularly in the west (Brittany, Normandy): the interglacial soils (stages 5e to 5a) were partly or totally eroded (Figure 9.3).

#### *The Lower and Middle Pleniglacial*

During this period, wind erosion became more and more prominent, because of a drastic decrease in the

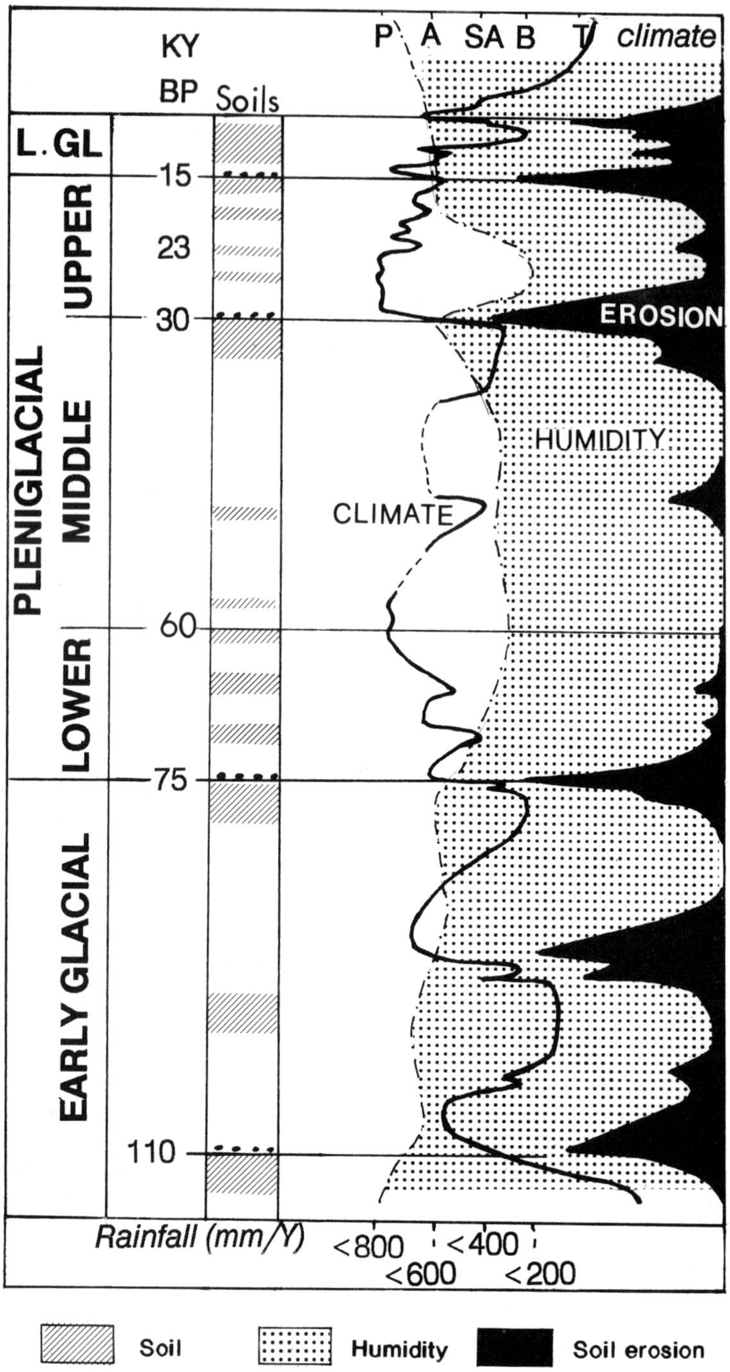


Figure 9.3 Simplified sketch of the erosion phases during the Last Glaciation at 50°N Lat. (from Vliet-Lanoë 1988). P: polar, A: Arctic, SA: Subartic, B: boreal, T: temperate. Notice stone lines *c.* 110, 75, 30 and 15 thousand BP.

vegetation cover. Wash deposits of stage 4 recorded in the Seine Valley (Lautridou 1985) and the Somme Valley (Antoine 1989) are interstratified with subarctic meadow soils (Van Vliet-Lanoë 1986); in Belgium, local aeolian loams are interstratified with arctic meadow soils (Haesaerts and Van Vliet-Lanoë 1974). Rill wash was only sporadic. During stage 3, there was little precipitation but conditions were desert-like in the Netherlands (Vandenberghe 1985) and close to the ice margin in Poland (Maruczszack 1987) as indicated by the absence of periglacial deformations. Vegetation cover was sporadic; spring needle ice activity was very intense (Van Vliet-Lanoë 1990b). Aeolian deposition prevailed in many areas, but on mountains, such as the Massif Central, solifluction (gelifluction and frost creep) continued (Valadas 1984).

At the end of this period, during the Hengelo-Denekamp Interstadial, the climate improved inducing thermokarstic features. Vegetation was almost continuous and precipitation increased to 600 mm/yr in Brittany and Normandy and to 400 mm in Belgium and the Netherlands (Van Vliet-Lanoë 1988). Well developed soils formed south of a climatic boundary located in the south of the Netherlands, crossing Germany to southern Poland (Van Vliet-Lanoë 1988). The climate in the Netherlands remained arctic (Ran 1990). After 28000 BP, an abrupt cooling occurred before the eustatic lowering which resulted in intense wash and solifluction (Figure 9.3). Interstadial soils and also the lower Younger Loesses are commonly eroded in the west, though in eastern countries soils were much better preserved (Demek and Kukla 1969). As loess is very frost susceptible and less stable than soils or stony solifluction deposits, it is more likely to erode. Snow cover and soil ice content probably favoured mudflows and gelifluction. Only after this period did the sea level drop as shown by the Upper Younger Loesses, which are derived from estuarine fans (Lautridou 1985). Polar desert invaded most of western continental Europe, except for western Brittany. This period of erosion was almost as important as that around 75000 BP, but it was effective for a much shorter period (about 2000 years) and the topographical results were somewhat different. The valleys, sometimes located on thermokarst, were asymmetrically enlarged due to a lack of vertical incision (deep frozen soils in spring or permafrost) and microclimate contrasts (Dylik 1969). In mountains, such as the Massif Central, this erosion is also very clear (Valadas 1984).

### *Upper Pleniglacial and Late Glacial*

At about 23000 BP, the climate improved slightly, as indicated by a soliflucted cryoturbated horizon, the Nagelbeek-Kesselt horizon (Haesaerts *et al.* 1981; Van Vliet-Lanoë 1988). This short-lived change from polar desert to arctic climatic conditions is also recorded by an erosion surface associated with local retrogressive thermokarst features. After this period aeolian sedimentation continued interrupted by a milder and more humid type of pedogenesis but the global amount of precipitation rises progressively, leading to the maximum extension of the ice sheets around 20000 BP. The sedimentation budget was very positive, resulting from the low effectiveness of wash and rill erosion.

Close to 15000 BP, the ice sheet readvanced in response to increasing moisture (Van Vliet-Lanoë 1989b). As precipitation reached between 200 and 400 mm in western continental Europe, south of the Netherlands, erosion, again caused by needle ice activity, was very active. Another erosion surface formed, locally marked by a stone line. The volume of soil eroded was less than in the two previous periods, but it was as strong as the 28000 BP episode. However valley shaping began. Reworked loess or drift sands derived from reactivated streams covered the surface. During the Bölling-Alleröd interstadial (Lowe and Gray 1980), surface soil development occurred, with decalcification and clay translocation (Van Vliet-Lanoë and Langohr 1981; Miedema 1986; Van Vliet-Lanoë 1990a). As vegetation cover became continuous, vertical incision and shaping of the present day valleys occurred (Langohr and Sanders 1985; Turkowska 1988; Vandenberghe *et al.* 1991), mostly during the Bölling phase. During the Younger Dryas, drift sands were common in the Netherlands, northern Germany and northern Poland (Vandenberghe *et al.* 1991; Manikowska 1985). In the mountains, frost creep and scree activity were restored, but in most of the loessic lowlands superficial erosion was limited to gentle frost creep deformation.

### The Holocene

During the Holocene, outside the influence of man, soil erosion was inefficient and pedogenesis restricted to biochemical processes. Landscape change was largely the result of incision by rivers because of the extensive forest cover. The loss of forest, considered by Birks (1986) as the oligotrophic phase of the interglacial, began around 5000 BP: the landscape became one of forest, heath, mire and grassland, probably due to clearance by man (Huntley and Birks 1983). Soil development shifted from brunification to podsolization. Today, in many granitic areas, the surface soil is clearly truncated showing a development far too weak for it to have developed in the whole Holocene. This situation prevails in some places in Brittany and in the Massif Central but it is less common in the Vosges. Colluvium, the result of this erosion, accumulated at the foot of slopes or and valleys and is often interstratified with peats or organics (Valadas 1984).

## ***Human Influences***

### *Forest clearance, grazing, and forest exploitation*

In the oligotrophic phase, forest clearance alone was able to modify the hydrological and thermal balance of a soil. It increased runoff and the development of hydromorphic conditions and peats formed in depressions. Climate cooling during the sub-Boreal (e.g. Pearsall 1964; Pennington *et al.* 1972) resulted in reduced evapotranspiration and increased waterlogging of lowlands. Much deforestation occurred in the ninth to twelfth centuries and in the seventeenth and eighteenth centuries (Plaisance 1986; Houzard 1986; Bonnart *et al.* 1986).

Grazing was a factor in the 'steppization' (development of a mull-like A1 horizon) of the profile. Under forests, grazing has a beneficial effect, by importing nitrogen and recycling nutrients, it maintains the soil in a mesotrophic situation (mull humus form) and decreases erosion susceptibility. This practice was frequent on common land up to the time of the French Revolution. The beneficial effect was lost when local meadow or heath were exploited for litter or mattock (imported soil litter - as manuring agent). The subsequent lowering of soil protection by decreased organic content and seedling capability promoted runoff and erosion. In the Massif Central or in the upper Orne valley (Normandy), fields located in the vicinity of farms were progressively enriched (infield) although those close to the commons were abandoned to the heath (outfield).

### *Agricultural practices*

Early agricultural practices consisted of 'slash and burn' clearance (Romans and Robertson 1975), followed by short-term cultivation. After abandonment of the cultivated areas, as fertility declined, shrub forest developed. Certain areas were cultivated continuously from the Neolithic, such as on granitic soils and on coversands: the ploughing technique using a hoe meant that loamy soils could not be worked before the Bronze and Iron Age (Haudricourt and Brunhes-Delamarre 1986; Macphail 1986).

Swing-plough techniques, developed in Western Europe since the Bronze and Iron Ages, are slightly more effective (Haudricourt and Brunhes-Delamarre 1986). Ploughing depth reached about 5 cm with a donkey, and 10 cm with an ox during Roman times. Fields were ploughed in criss-cross furrows (Figure 9.4) (Scheuermeier, in Haudricourt and Brunhes-Delamarre 1986). Ploughs were developed in western continental Europe before the tenth century, and produced straight furrows, generally oriented parallel to the slope. Since the eighteenth century deeper furrows have been common. With the introduction of horse yoking, in the twelfth century, ploughing reached a depth of 18 cm and heavy soils could be cultivated. With the recent mechanisation, tillage depth reached up to 40 cm as for example in the Caen plain. This disrupted the unstable eluvial horizon, promoting the development of ploughpans. Instead of a grassy turning zone ('tournière' or 'fourrière') at the ends of the field, ploughs covered the whole area (Lechevallier 1991) (Figure 9.4). The final smoothing by rolling of the surface after sowing also results in a surface susceptible to erosion.

Changes in manuring techniques were also important. Until the last War, organic manuring and marling, alternating with long periods of fallow, were common for crops, preserving the soil structural stability. Chemical fertilizers do not replace lost organic matter and can lead to a decrease in the structural stability and an increase in sensitivity of the soils to splash (Monnier 1965). Recently, hand weeding has been replaced by the use of herbicides, some of which may also decrease earthworm and other soil populations. For a dozen years, both factors have decreased soil infiltration capacity (Auzet *et al.* 1990).

Changes in crop variety can also modify soil susceptibility to erosion. With winter-sown cereals, the soil is partly protected from winter erosion by young plants and weeds; with the development since the Middle Ages of spring-sown crops (the 'Mars' in Normandy: barley and oats), soils are bare mostly after the winter storms. Some modern crop successions leave the soil exposed to winter rain and frost, sometimes for 5–6 months (Lechevallier 1991).

### *Demographic pressure and change in land-use*

In continental western Europe, demographic pressure seems to have been limited until the Early Middle Ages. Deforestation was localised, around Bronze Age villages in Brittany, but was more extensive around Roman farms, as in Picardie or the Caen plain. During the Roman period, low elevation massifs like Limousin were partly cultivated though the Vosges or Brittany margins were still forested. Brittany was so empty that in the eighth century Saxons sold land in Brittany to Welsh and Cornish immigrants (Chedeville and Guillotel 1984). A second main phase of deforestation began in the eleventh and twelfth centuries. Population pressure resulted in deforestation for domestic and industrial charcoal and ship factories, as in Cotentin (Houzard 1986). In Middle Belgium, charcoal furnace sites are commonly observed in the fields. The extent of deforestation is indicated by comparison of the Cassini maps (AD 1683 to 1744) with more recent documents (Bonnart *et al.* 1986). Maximal deforestation is reached in AD 1815–1820: French forests were reduced to 6.4 million hectares, corresponding only to half of present day cover. In Belgian Hesbaye, the last deforestation occurred between AD 1829 and 1835 (Bolline *et al.* 1980).

The largest populations in the seventeenth century are often responsible for small strip fields, as in Flanders, Alsace or Cotentin. The population decline took place between the French Revolution and the mid nineteenth century ('Chouannerie', Bonaparte wars), but associated with an intensification of agriculture. Nevertheless in the Massif Central it was only effective from the second half of the nineteenth century: densities of 100/km$^2$ are recorded here in AD 1846. This resulted in large changes in the management of the landscape by a shift from open field to meadow bocage (hedges and furrow landscape) in Normandy and on the margins of the Massif Central. In Normandy, 83% of the cultivated soils were transformed to meadows between 1850 and 1914 (Desert 1969; Sion 1909; Hedin 1951), resulting

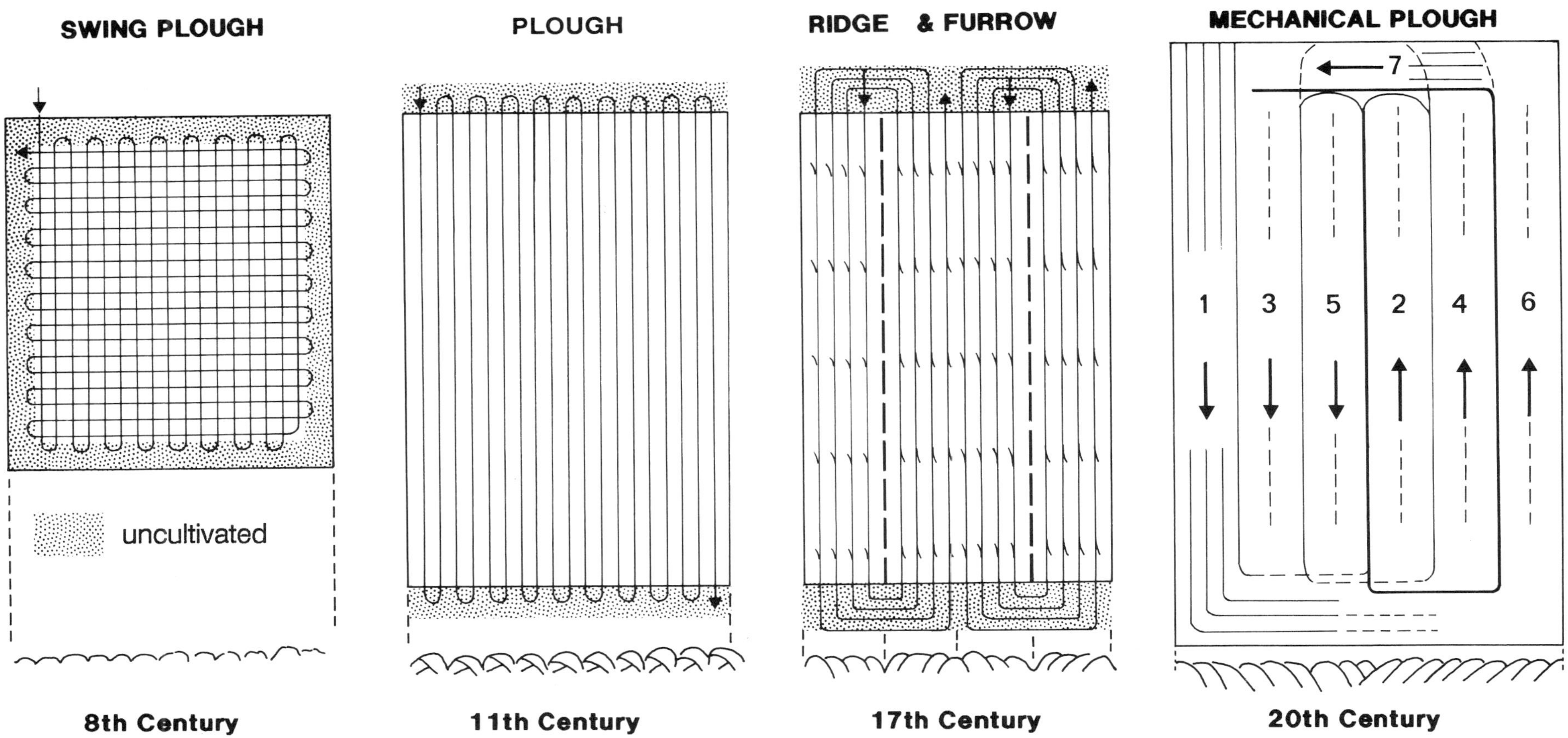


Figure 9.4 Sketch of the different ploughing techniques. Bottom line shows chronological period (modified from Haudricourt and Brunhes-Delamarre 1986 and Lechevallier 1991).

mostly from changes in crop prices and means of transport (Soulet 1977; Vidalenc 1957).

The most recent changes in land-use in France have resulted from economic reorganization after the depletion of the male population, mainly by the First World War, and to the variable decrease of the peasant population. The size of the fields has increased for mechanical and economic reasons (Auzet *et al.* 1990) with destruction of the natural drains and often a doubling of the furrow length (Lechevallier 1991). On loamy soils meadows have been locally ploughed-up. In other places with low economic value or steep slopes, as in the Massif Central, the area of forest has increased (Valadas 1984).

### *Archaeological Record of Erosion*

Loamy soils seem to have been colonized progressively since the Bronze Age but intensive deforestation occurred mainly in the Roman period and reached a maximum at the end of the Middle Ages (Bork 1989; Poeteray *et al.* 1984) as a result of changes in agricultural techniques. When reforested, old cultivated areas are always recognizable by an abnormal thickness of the A1 horizon and the presence of charcoal fragments (Forêt de Brotonne, Normandy; Forêt de Fougères, Brittany: Helluin *et al.* 1991; Massif Central, Valadas 1984 and 1987).

In Brittany, since the prehistoric beginning of field patterns, the land has been continuously cultivated (Flatrès 1986). Briard and Monnier (1976) observed the local preservation of a loess cover below burial mounds, much more extensive than today. In the Forest of Fougères, there was much evidence for many man-made constructions in the still forested landscape. Valley infilling began in the Roman period (second–third century); but most of the sedimentation began in the twelfth–thirteenth centuries. An old Roman road was strongly reworked and displaced at that time (Figure 9.5), attesting to dense traffic and waterlogging problems. Afterwards, only a short period of sedimentation is recorded during the second half of the eighteenth century (Helluin *et al.* 1991) correlated with the opening of a new road. Nevertheless, in Fougères, only very local disturbances are observed though the surface soil is strongly transformed by long-term cultivation.

In Lower Normandy, upland landscapes also evolved very early into cultivated bocage. Erosive events are recorded on uplands, in the Caen plain, and associated with the sites of Roman villages (Bavent, Fleury/Orne). On loessic soils of the Caen plain, the present day deep ploughing only reaches the A2 horizon and the soil profile remains almost complete (Figure 9.6). On old massifs, far from the farms, cultivation was extensive: 3 years of crops alternating with 6 years of fallow sown with broom (*Sarothamus scoparius*). Conversion into grassland occurred from the seventeenth to the end of the nineteenth centuries.

In the Somme valley and in Picardie, as on most of the loessic plateaus, cultivation has continued since the Neolithic but without creation of a bocage. The peaty Holocene infilling, promoted by the growth of travertine dams, ended with the main deforestation of the Roman period, probably because of an increased sediment load in the river. It is now up to 2 m above the river level. Adjacent secondary valleys, as in Coquerel, show up to 2 m of loamy colluvial infilling, burying the surface lessivé soil. Today, on the plateau, ploughing is reworking the Bt horizon.

In the Belgian Hesbaye province, on loess soils, early Neolithic, and later Roman farmers, exploited well to imperfectly drained soils, leaving the soils with fragipan under forest (Van Vliet-Lanoë and Langohr 1981). Today's farmers are ploughing the former upper part of the Bt horizon, whereas during the Bronze Age the soil profile was almost intact. Medieval erosion is also recorded in the form of sunken roads. Similar conclusions to those of Fougères were drawn by R. Souchez (University of Brussels, pers. comm.) in the Zoniën Forest (Belgium): sedimentation on small terraces dated to the eleventh and eighteenth centuries was probably also accentuated by mattock exploitation and by beech monoculture developed for hunting. The last deforestation occurred during the first half of the nineteenth century (Bolline *et al.* 1980).

In the Massif Central, intensive land-use began along the valleys at the end of the Iron Age and was very important in the Roman period. It continued up to today giving rise to 1 or 2 m of accumulated soil, even on very low angled slopes. Large areas were occupied by heath and only weakly eroded. In the Ambazac, Lingas, Aigoual and Lozère massifs, strong valley infilling began during the Merovingian time (eighth century) (Valadas 1984) and continued during the Middle Ages (ninth to eleventh centuries) (Muxart *et al.* 1990; Curt, thesis in preparation), correlated with extensive forest clearance on steep hills with shallow sandy soils on granite (Valadas 1984). This favoured significant erosion in spring (snowmelt) and summer time (Mediterranean thunder storms in the south) (Valadas 1984; 1987). After this period, increasing management of the landscape has also modified the drainage capability and the zones of soil depletion and accretion of colluvium (Figure 9.7). The development of hedgerows and avenues of trees allowed colluvium to accumulate and prevented loss of nutrients on slopes (Baffet 1984): this is a 'recent' Bocage landscape. Since the end of the nineteenth century there has been reforestation in this area especially in the Aigoual and in Ardèche. However, in the southern part, snowmelt and frost at high elevations (Cantal, Monts Dore, Cévennes) associated with Mediterranean storms produce erosion and gullying as in the past (Valadas 1984; Bernard-Allée and Valadas 1990).

### Conclusion

From the evidence of pedostratigraphy, soil erosion was not naturally active during the Holocene, except under human pressure (Neboit 1983). From the beginning of the Neolithic to the middle of the Roman period, geomorphological stability was periodically

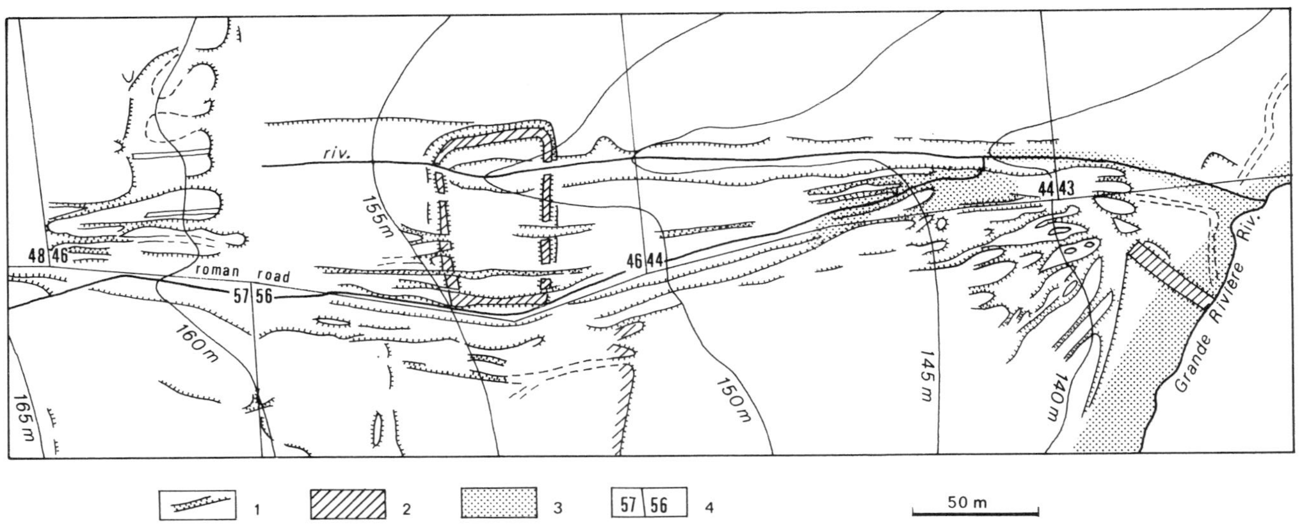


Figure 9.5 Medieval erosion along an old Roman road, related to heavy traffic and drainage problems, Fougères, Brittany (microtopographical mapping: J. Pellerin and M. Helluin).
1. Ditches and sunken roads. 2. Infillings of varied ages. 3. Valley infillings. 4. Actual limits of forestry patches.

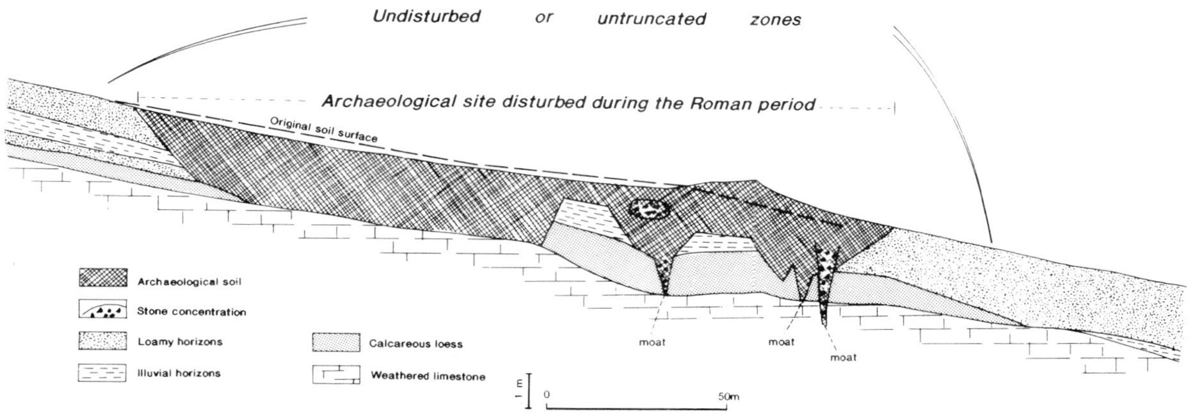


Figure 9.6 Fleury/Orne, Caen plain. After 2000 years of agricultural activities, the soil remains mostly undisturbed outside the main settlement.

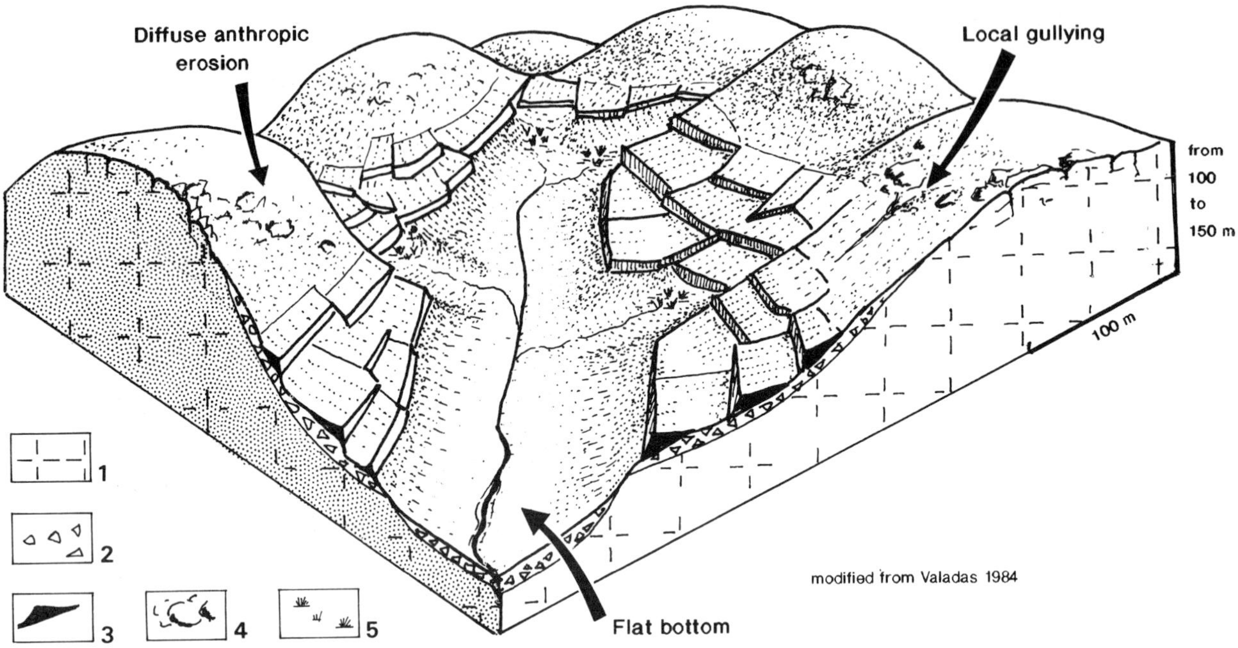


Figure 9.7 Sketch of the landscape management and anthropogenic erosion in the Massif Central. 1) weathered and unweathered granite, 2) periglacial slope deposits, 3) anthropogenic accumulation, 4) rock outcrop, 5) spring zone.

interrupted by human activities causing slight and mainly local soil erosion. But during the Later Middle Ages, very heavy sheet and gully erosion is recorded in continental Europe (Bork 1986; Poeteray *et al.* 1984) resulting from catastrophic rainfall events. Similar situations existed also during cold, wet winters at the beginning of the Little Ice Age (Le Roy Ladurie 1983). Erosion was exacerbated by agricultural modifications. The changes in ploughing techniques, crop rotations and field organisation, all resulting from demographic pressure during the Middle Ages and the eighteenth century, led to longer periods of bare soils in winter and to conditions favouring water concentration and frost activity similar to conditions during certain periods of the Last Glacial (Figure 9.8). The present day modification of cultivation (Auzet *et al.* 1990), mainly under economic pressure, has led to unstable, bare soils, depleted in organic matter, and to long slopes, which are exposed to humid winter conditions, very similar to those which produced erosion around 28000 and 15000 BP (Figure 9.3). The plough pan, together with decreased earthworm activity, behaves like a frozen substratum, favouring sheet erosion. Mild temperatures and dryness in autumn followed by winter rainfall favour the growth of needle ice in winter and lead to detachment of microaggregates (frost mulch). It is therefore not surprising that erosion intensity today is similar to that of the Last Glacial. Changes in climate do not seem to have had a major effect on erosion: at the end of the Little Ice Age, economic and demographic pressures were weak and there was also little erosion. This situation does not seem to be restricted to temperate Europe: in southern Sweden, Dearing *et al.* (1990) have obtained similar results on loamy soils. On inland dunes of Poland (Manikowska 1985) and in Finnish Lapland (Van Vliet-Lanoë *et al.* 1991), erosion by wind is mainly an anthropogenic feature: previous Holocene climatic events led to much less reworking than in the Medieval and present day situations.

## Acknowledgements

We would like to thank in particular Pierre Brunet (Caen) for his help concerning the history of human activities in western France. We express also our gratitude to Michèle Lavollé and Michel Joly for their technical improvement of the figures. Moreover, we would like to express our thanks to an unknown referee and the editors for the improvement of our English.

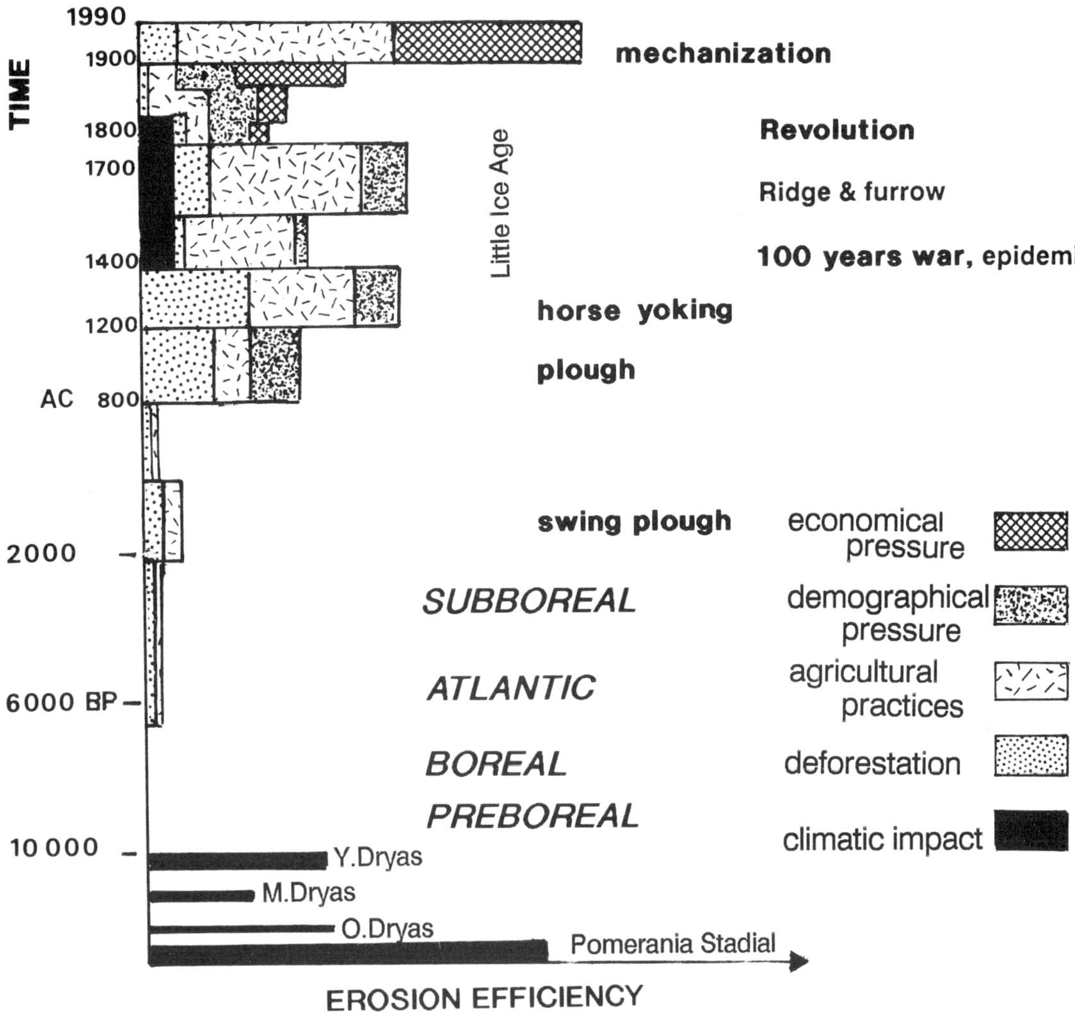


Figure 9.8 Evaluation of the Holocene erosion balance related to the summation of the different climatic and anthropogenic factors. (Note on vertical time, scale AC = AD).

## References


Antoine, P. 1989 *Les terrasses Quaternaires du bassin de la Somme: Etude géologique et géomorphologique. Contribution à la connaissance du paléoenvironnement des gisements paléolithiques.* Doctoral thesis, Université of Lille I.

Auzet, A. V., Boiffin, J., Papy, F., Maucorps, J. and Ouvry, J. F. 1990 An approach to the assessment of erosion forms and erosion risk on agricultural land in the northern Paris Basin, France. In: Boardman, J., Foster, I. D. L. and Dearing, J. A. (eds.) *Soil Erosion on Agricultural Land.* John Wiley, Chichester, 383–400.

Baffet, M. 1984 *Influence de la haie sur l'évolution des caractères physico-chimiques et hydrodynamiques des sols. Exemple des sols lessivés hydromorphes à pseudogley et fragipan du bocage de la Basse Marche (Haute-Vienne).* Doctoral thesis, University of Limoges (Botany).

Bernard-Allée, P. and Valadas, B. 1990 Evolution d'une ravine apparue après une coupe forestière sur le flanc sud du Mont Lozère. *Notes et Comptes-rendus du Groupe Régionalisation du Périglaciaire, Strasbourg* 15, 51–54.

Birks, H. J. 1986 Late Quaternary biotic changes in terrestrial and lacustrine environments, with particular reference to north-west Europe. In: Berglund, B. (ed.) *Handbook of Holocene Palaeoecology and Palaeohydrology.* Wiley, Chichester, 3–65.

Bolline, A., Pissart, A., Bastin, B. and Juvigné, E. 1980 Etude d'une dépression fermée près de Gembloux: vitesse d'érosion des terres cultivées de Hesbaye. *Annales de la Société Géologique de Belgique* 103, 143–152.

Bonnart, N., Duhamel, F. and de Foucault, B. 1986 Corrélation entre l'extension forestière ancienne et des données cartographiques, toponymiques et phytosociologiques dans le Pas de Calais (Nord de la France). *Hommes et Terres du Nord* 2–3, 116–121.

Bork, H. R. 1989 Soil erosion during the past

millennium in Central Europe and its significance within the geomorphodynamics of the Holocene. *Catena Supplement* 15, 121–131.

Briard, J. and Monnier, J. L. 1976 Tumulus armoricains de l'Age du Bronze et couverture loessique Weichsélienne. *Bulletin de la Société Géologique et Minérakiguqye de Bretagne* CB 1/2, 75–88.

Buch, M. and Zöller, L. 1990 Gliederung und Thermolumineszenz-chronologie der Würmlösse im Raüm Regensburg. *Eizeitalter und Gegenwart* 40, 63–84.

Chedeville, A. and Guillotel, H. 1984 *La Bretagne des Saints et des Rois*. Ouest-France, collection Université.

De Ploey, J. 1972 Quelques expériences en rapport avec le rôle éventuel de l'érosion pluviale en milieu périglaciaire. *Bulletin Centre de Géomorphologie du CNRS*, Caen 13–15, 101–115.

Dearing, J. A., Alström, K., Bergman, A., Regnell, J. and Sandgren, P. 1990 Recent and long-term records of soil erosion from southern Sweden. In: Boardman, J., Foster, I. D. L. and Dearing, J. A. (eds.) *Soil Erosion on Agricultural Land*. Wiley, Chichester, 173–191.

Demek, J., Kukla, G. and Collaborators 1969 *Periglazialzone, Löss und Paläolithikum der Tschecoslowakei*. University of Brno.

Desert, G. 1969 *Des transformations de l'agriculture Française*. Faculté des Lettres et Sciences Humaines de Caen.

Dylik, J. 1969 Slope development under periglacial conditions. *Biuletyn Peryglacjalny* 18, 381–410.

Fagnart, J. P. 1991 Nouvelles observations sur le gisement Paléolithique Supérieur de Belloy-sur-Somme. *Gallia Prehistoire* (in press).

Flatres, P. 1986 *La Bretagne*. Presses Universitaires de France.

Haesaerts, P. and Van Vliet-Lanoë, B. 1974 Compte-rendu de l'excursion du 25 Mai 1974 consacrée à la stratigraphie des limons aux environs de Mons. *Annales de la Societé Géologique de Belgique* 97, 547–560.

Haesaerts, P., Juvigné, E., Kuyl, O., Mücher, H. and Roerbroeks, W. 1981 Compte-rendu de l'excursion du 13 Juin 1981 en Hesbaye et au Limbourg Neerlandais, consacrée à la chronostratigraphie des loess du Pléistocène Supérieur. *Annales de la Société Géologique de Belgique* 104, 223–240.

Hallegouët, B. and Van Vliet-Lanoë, B. 1986 Les oscillations climatiques entre 125000 ans et le Maximum Glaciaire d'après l'étude des formations marines, dunaires et périglaciaires de la Côte des Abers (Finistère). *Bulletin de l'Association Française pour l'Étude du Quaternaire* 1–2, 127–138.

Harris, C. 1986 Mechanisms of mass movement in periglacial environments. In: Anderson, M. and Richards, K. (eds.) *Slope Stability*. Wiley, Chichester, 531–559.

Haudricourt, A. G. and Brunhes-Delamarre, M. 1986 *L'homme et la charrue à travers le monde*. La Manufacture, Lyon.

Hédin, L. 1951 Les conditions d'exploitation de la prairie en Normandie depuis le 19e siècle. *Annales de Normandie* 1.

Helluin, M., Pellerin, J., Clet M. and Pilet-Lemière, J. 1991 Anthropisation et remblaiement de vallons de premier ordre dans la région de Fougères (Ille et Vilaine). *AGF*, (in press).

Heyse, I. 1975 *Bijdrage tot de geomorphologische kennis van het noord-westelijk deel van de Vlaamse Vallei en haar randgebieden*. Ph.D. thesis, State University, Gand.

Houzard, G. 1986 Dégradations et restauration en Forêt d'Ecouves de 1666 à nos jours. *Hommes et Terres du Nord* 2–3, 227–230.

Huntley, B. and Birks, H. J. B. 1983 *An Atlas of Past and Present Pollen Maps of Europe; 0–13000 years ago*. Cambridge University Press, Cambridge.

Langohr, R. 1990 The dominant soil types of the Belgian loess belt in the Early Neolithic. In: Cahen, D. and Otte, M. (eds.) *Rubané et Cardial*. ERA, University of Liège, 39, 117–124.

Langohr, R. and Sanders, J. 1985 The Belgian loess belt in the last 20000 years. Evolution of soils and relief in the Zoniën Forest. In: Boardman, J. (ed.) *Soils and Quaternary Landscape Evolution*. Wiley, Chichester, 359–371.

Lautridou, J. P. 1985 *Le cycle périglaciaire Pléistocène en Europe du Nord-Ouest et plus particulièrement en Normandie*. Doctoral thesis, University of Caen.

Lechevallier, C. 1991 L'érosion des terres agricoles en Pays de Caux. *Cahiers Géographiques de Rouen* 1, 97–116.

Le Roy Ladurie, E. 1983 *L'histoire du climat depuis l'An Mille*. Flammarion, Paris, 2 volumes.

Lowe, J. J. and Gray, J. M. 1980 The stratigraphic subdivision of the Late Glacial of NW Europe: a discussion. In: Lowe, J. J., Gray, J. M. and Robinson, J. E. (eds.) *Studies in the Lateglacial of North-West Europe*. Pergamon Press, Oxford, 157–176.

Macphail, R. 1986 Palaeosols in archaeology: their role in understanding Flandrian pedogenesis. In: Wright, V.P. (ed.) *Palaeosols, their Recognition and Interpretation*. Blackwell, Oxford, 263–290.

Manikowska, B. 1985 On the fossil soils stratigraphy and lithology of the dunes in Central Poland. *Acta Geographica Lodz* 52.

Maruczszak, H. 1987 Problems of palaeogeographic interpretation of ice wedge casts in European loesses. In: Pecsi, M. and French, H. M. (eds.) *Loess and Periglacial Phenomena*. Akad. Kiado, Budapest, 285–302.

Miedema, R. 1987 *Soil formation, microstructure and physical behaviour of Late Weichselian and Holocene Rhine deposits in the Netherlands*. Ponsen and Looijen, Wageningen.

Monnier, G. 1965 *Action des Matières organiques sur la stabilité structurale des sols*. Thesis, Paris.

Mücher, H. and De Ploey, J. 1984 Formation of

afterflow silt loam deposits and structural modification due to drying under warm conditions: an experimental and micromorphological approach. *Earth Surface Processes and Landforms* 9, 523–532.

Muxart, T., Cosandey, C. and Billard, A. 1990 *L'érosion sur les hautes terres du Lingas, un processus naturel, une production sociale.* Ed. du CNRS.

Neboit, R. 1983 *L'homme et l'érosion.* Fac. Lettres et Sc. Hum. de l'Université de Clermont Ferrand II, nelle série 17.

Pearsall, W. H. 1964 After the ice retreat. *New Scientist* 383, 757–759.

Pennington, W., Haworth, E. Y., Bonny, A. P. and Lishman, J. P. 1972 Lake sediments in northern Scotland. *Philosophical Transactions of the Royal Society London B* 264, 191–294.

Plaisance, G. 1986 Evolution des paysages forestiers et arborés en Bourgogne et Franche-Comté, de l'An mille à nos jours. *Hommes et Terres du Nord* 2–3, 190–195.

Poeteray, F. A., Riezebos, P. A. and Slotboom, R. T. 1984 Rates of subatlantic surface lowering calculated from mardel-trapped material (Gutland, Luxemburg). *Zeitschrift für Geomorphologie, NF* 28, 467–481.

Ran, E. 1990 Dynamics of the vegetation and environment during the Middle Pleniglacial in the Dinkel valley (The Netherlands). *Mededelingen Rijks Geologische Dienst* 44(3), 139–205.

Rapp, A. and Stromquist, L. 1976 Slope erosion due to extreme rainfall in the Scandinavian mountains. *Geografiska Annaler* 42, 62–206.

Romans, J. C. C. and Robertson, L. 1975 Soils and archaeology in Scotland. In: Evans, J. G., Limbrey, S. and Cleere, H. (eds.) *The Effect of Man on the Landscape: The Highland Zone.* Council for British Archaeology Research Report 11, 37–39.

Sion, J. 1909 *Les paysans de la Normandie orientale.* Paris, A. Colin.

Soutadé, G. 1979 *Modelé et dynamique actuelle des versants surpraforestiers des Pyrénées orientales.* Doctoral thesis, Es Lettres, Bordeaux, Imprimerie Coopérative du Sud Ouest, Albi.

Soulet, M. H. 1977 *Diffusion du progrès technique et formation professionnelle en agriculture, de 1830 à 1944. Analyses à partir de l'exemple bas Normand.* Doctoral thesis (Third cycle), University of Caen.

Stromquist, L. 1985 Geomorphic impact of snowmelt on slope erosion and sediment production. *Zeitschrift für Geomorphologie* 29(2), 129–138.

Thorn, C. E. 1979 Ground temperatures and surficial transport in colluvium during snowpatch meltout; Colorado Front Range. *Arctic and Alpine Research* 11(1), 41–52.

Turkowska, K. 1990 Main fluvial episodes in the Ner valley in the last 22000 years: a detailed study of Lublinek near Lodz, Central Poland. *Quaternary Studies in Poland* 9, 85–99.

Valadas, B. 1984 *Les hautes terres du Massif Central Français: contribution à l'étude des morphodynamiques récenes sur versants cristallins et volcaniques.* Doctoral thesis, University of Paris I, publication EDITEC, Caen.

Valadas, B. 1987 Morphodynamiques récentes dans le Massif Central Français: étude comparée des massifs granitiques du Limousin, de Margeride et du mont Lozère. *Zeitschrift für Geomorphologie, Supplementband* 65, 85–99.

Van Vliet-Lanöe, B. 1985 Frost effects in soils. In: Boardman, J. (ed.) *Soils and Quaternary Landscape Evolution.* Wiley, Chichester, 115–156.

Van Vliet-Lanoë, B. 1986 Le pédocomplexe du Dernier Interglaciaire (de 125000 à 75000 BP). Variations de faciès et signification paléoclimatique, du sud de la Pologne à l'ouest de la Bretagne. *Bulletin de l'Association Française pour l'étude du Quaternaire* 1–2, 139–150.

Van Vliet-Lanoë, B. 1988 *Le rôle de la glace de ségrégation dans les formations superficielles de l'Europe de l'Ouest. Processus et héritages.* Doctoral thesis, University of Paris I – Sorbonne, Editec, Caen (1987).

Van Vliet-Lanoë, B. 1989 Dynamics and extent of the Weichselian Permafrost in Western Europe (Stages 5e to 1). Compte-rendu Session 'The Last Interglacial-Glacial cycle', EUG, Strasbourg, 20–23 mars 1989. *Quaternary International* 3/4, 109–114.

Van Vliet-Lanoë, B. 1990a The genesis of the argillic horizon in Weichselian Loess of Northwestern Europe. *Quaternary International* 5, 49–56.

Van Vliet-Lanoë, B. 1990b Le pédocomplexe de Warneton: où en est-on? Bilan paléopédologique et micromorphologique. Colloque Intern. In: *Methodes et concepts en Stratigraphie du Quaternaire*, Dijon, 5–7, Déc 1988, *Le Quaternaire* 1(1), 65–76.

Van Vliet-Lanoë, B. and Langohr, R. 1981 Correlation between fragipan and permafrost with special reference to Weichsel silty deposits in Belgium and Northern France. *Catena* 8, 137–154.

Van Vliet-Lanoë, B., Seppälä, M. and Käjhkö, J. 1991 Dune dynamics and cryoturbations features controlled by Holocene water level change. Hietatievat, Finnish Lapland. *Geologië en Mijnbouw* (in press).

Vandenberghe, J. 1985 Palaeoenvironment and stratigraphy during the Last Glacial in the Belgian-Dutch Border Region. *Quaternary Research* 24, 23–38.

Vandenberghe, J. and Collaborators 1991 *Excursion guidebook of the Symposium Periglacial environments in relation to climatic change, Maastricht/Amsterdam, 3–6 May 1991*, Vrije Universiteit, Amsterdam.

Vidalenc, J. 1957 L'agriculture dans les Départements Normands à la fin de l'Empire. *Annales de Normandie* 7.

Wilkinson, T. J. and Bunting, B. T. 1975 Overland transport of sediment by rill water in a periglacial environment in the Canadian High Arctic. *Geografiska Annaler* 57A(1–2), 105–116.

Woo, M. K. and Steer, P. 1982 Occurrence of surface flow on Arctic slopes in Southwestern Cornwall Island. *Canadian Journal of Earth Sciences* 19(12), 2368–2377.

# CHAPTER 10

# HOLOCENE COLLUVIAL DEPOSITS ON THE SLOPES OF THE PARIS BASIN

C. KUZUCUOGLU, L. LESPEZ AND J.-F. PASTRE

*Six archaeological sites, dating from the Neolithic to the later Middle Ages, have been used as stratigraphic key beds in an attempt to calculate the amount of sediment removed from the slopes of the northern Paris Basin during protohistoric and historic times. This region has experienced a long period of cultivation (starting about 8000 BP). Field and laboratory work during one year provided information on several parameters controlling the mobility of colluvium. A soil 'accumulation rate' (in t/ha/y) has been calculated for several types of morphodynamical location. It was not possible, however, to observe rhythms in the sedimentary sequence.*

## Introduction

In recent years increasing attention has been given to the measurement and the definition of processes involved in the erosion of farm land in North-western Europe (Auzet 1987; Boardman *et al.* 1990; Bollinne 1982; Chisci and Morgan 1986; Evans and Cook 1986; Morgan 1985). Attention has focused on the role of intensive agricultural practices used on loamy productive soils, since these are thought to be partly responsible for an acceleration of the intensity of the erosion process. Results of measurements collected from the literature (Table 10.2) show that the rich agricultural soils of northern Europe are being steadily lost year after year.

In addition to the data obtained in monitoring contemporary water and soil transfers on cultivated fields, more information is now needed to give a time perspective to the rates obtained today and to consider whether there have been, throughout history, other periods of accelerated erosion on these slopes, more or less linked to agricultural practices (Bell 1987; Bollinne 1982; Evans 1990; Vogt 1953).

In this context, the Environmental Impact Assessment of the construction of the northern branch of the TGV ('High Speed Train') between the cities of Paris and Lille (Figure 10.1) gave birth to a large rescue programme financed by the SNCF (French railways). More than 70 archaeological sites ranging from middle Palaeolithic to later Middle Ages were excavated from 1987 to 1990 (Blanchet and Talon 1989; 1991; Collectif *in press*). The authors surveyed some of these archaeological excavations and measured the volumes of deposits which had accumulated as a result of protohistoric and historic slope erosion. This investigation forms part of a geomorphological and geological research programme concerning Quaternary sediment dynamics in the northern Paris Basin, which was started in 1989 and has been divided into two parts :

(1) October 1989 to September 1990. Attention was focused on the slope deposits, since all excavations were due to end by summer 1990 (Lespez 1990; Kuzucuoglu *et al.*, in press);

(2) Subsequently a large borehole programme was set up in order to understand the stratigraphy of the Weichselian aeolian loams (Antoine in press; Pastre in press, a; Pastre in press, b) covering the plateaux, and that of the Late-glacial and Holocene sediments deposited on valley floors (Pastre *et al.*, in press).

This paper presents the results of the investigations of slope deposits.

## Methodology

The south-north route of the TGV crosses plateaux and slopes formed of Mesozoic (chalk) and Cenozoic (limestones and sands) sediments (Figure 10.1). The land surface is normally covered by a variable thickness, 1-2 m to > 5 m depth of aeolian loam deposited during the last glacial period (Antoine in press; Pastre in press, a; Pastre in press, b). During the period of our work on slope deposits (October 1989 – September 1990), 6 sites of the excavation programme were investigated in detail (Figure 10.1).

A similar investigative methodology was used for all sites. The original archaeological excavation was enlarged and deepened to provide stratigraphic data for a detailed cross section. This was supplemented by a general survey of the depth of colluvium on the total length of the slope and by a series of boreholes to investigate the stratigraphy of the slope colluvium. Sediment samples were analysed for colour, organic matter content, granulometric distribution and carbonate content. Figures 10.2 and 10.3 show the results of these investigations for two sites.

An 'accumulation rate' (in t/ha/y) was calculated for each site. The calculation is based upon a 10 m wide strip from the valley floor to the watershed. Slope materials are assumed to have a density of 1.5. A number of depth measurements in various slope positions are incorporated into the calculation (Figure 10.4). The final figure is based upon the amount of colluvium accumulated above the datable archaeological horizon averaged over both the upper, sediment-contributing areas of the slope and the lower,

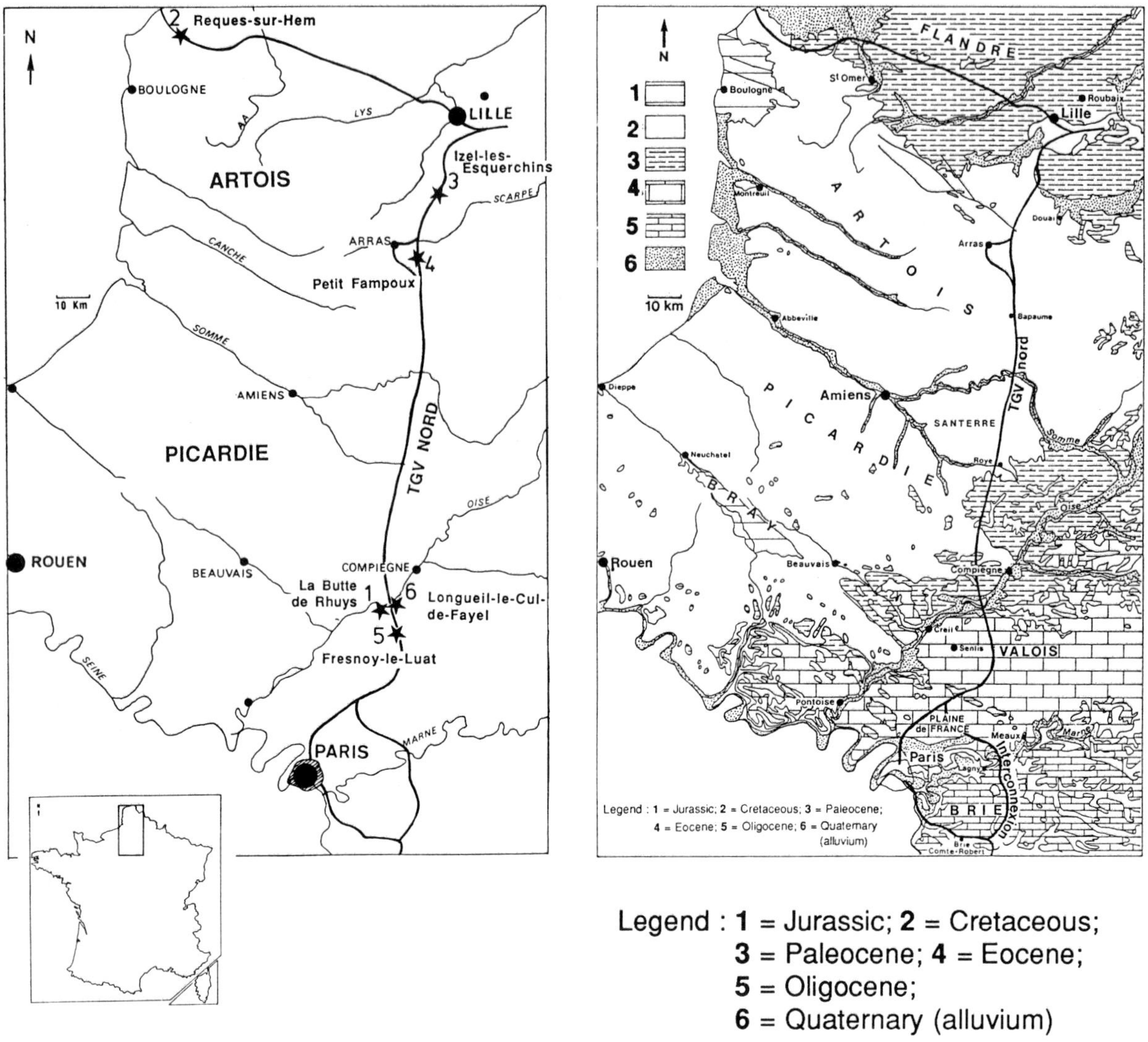


Figure 10.1 Geological map of the northern Paris Basin and location of the sites studied. The numbers have been allocated to the sites according to a chronological scale: 1 is a Neolithic to Bronze Age site; 2 is an Iron Age (Late La Tène) and a Gallo-Roman site; 3 is an Iron Age (La Tène) site; 4 and 5 are Gallo-Roman sites; 6 is a tenth century Medieval site.

| Sites Name (Age of site) | Length (km) | Slope Height (m) | Slope Mean slope value (%) | Orient. | Colluvium Depth of colluvium on slope (mean) (m) | Colluvium Depth of colluvium over the site (m) | Substratum | Accumulation Rate (underestimating the erosion rate) (t/ha/y) |
|---|---|---|---|---|---|---|---|---|
| 1. La Butte de Rhuys (Neolithic to La Tène: 1 site over a hill with 2 trenches) | 0.1 to 0.15 | 2.5<br>2.5 | 1.2<br>2.2 | E<br>S | 0.5 to 1.5 | 0.7 to 1 | alluvial sands (Lateglacial) | 9.0 |
| 2. Reques s/Hem (Gallo-Roman) (La Tène) | 0.6 | 60 | 7.6 | N | 2 | 0.25 to 0.35 m | clay ('Flanders' clay) | 0.4 to 0.8 |
| 3. Izel-lès-Esquerchins (La Tène to Gallo-Roman) | 0.85<br>1.25 | 12<br>22 | 1.6<br>2.1 | S<br>N | 4.5<br>2 | 0.30 m (no site on this slope) | chalk (Senonian) = plateau with a 3 to 7 m thick loess cover | 1.44 |
| 4. Petit-Fampoux (Gallo-Roman) | 1 | 50 | 2.2 | N | 8 to 12.5 | 2.5 m | chalk (Senonian) = plateau with a 3 to 7 m thick loess cover | 2.8 |
| 5. Fresnoy-le-Luat (Gallo-Roman) | 1.1 | 32.5 | 5 | N | >2 | 0.8 to 1.8 m | limestone (5 to 8 m) covering sands (30-40) | 3.1 |
| 6. Longueil-le-Cul-de Fayel (tenth century) | 1 | 33 | 2.5 | N | ~6 (2.5 to 7.5) | 0.7 to 1.10 m | loess (3–7 m) over clay (6–7 m) + sands (15–20 m) over chalk | 2.1 |

Table 10.1 Sites studied: main results, characteristics of the slopes and of the sediment cover.

| Country | Comments | Erosion rate (t/ha/y) |
|---|---|---|
| Belgium | loamy and sandy soils | 15 to 25 |
| | silty soils | >11 |
| France | | |
| – Laonnois | recently cleared field | 40 to 50 |
| – Champagne | | 3 to 5 |
| – N. Paris Bas. | beginning of rill erosion | 8 to 15 |
| – N. Paris Bas. | | 20 |
| Germany | | |
| – Trier | | 7 |
| – N.W. Germany | on soils bare all year | 20 |
| Great Britain | | |
| – N. Norfolk | agricultural soils | 195 |
| – W. Sussex | | 0.5 to 500 |
| – Somerset | | 2.5 to 17.7 |
| – Bedfordshire | | 2.4 to 10.6 |
| – UK | sands and loamy sands | 20 |

Table 10.2 Reference data from other north-western European countries (Belgium, France, Germany, Great Britain)

Sources: mainly from Bollinne, A. (1982), various authors cited in Auzet, A.-V. (1987), various contributions in Boardman, J. *et al.* (1990), Douglas, I. (oral lecture, 1991), Walling, D. E. (1990)

accumulating areas. It assumes a constant annual rate of deposition.

**Presentation of results**

Four types of sites have been recognized on the basis of differences in colluviation during proto-historic and historic times, thus revealing the importance of the recent slope dynamics (a balance between erosion and sedimentation processes) that can be related to land clearing:

*Type 1* This is represented by sites having a thick colluvial cover ( > 1m) and a high accumulation rate (1.4 to 2.8 t/ha/y). They are located on the chalky plateaux covered by deep loess and are dominated by long (about 1.5 km) and gentle slopes (1.5 to 2.8 %).

*Type 2* These sites have shallower colluvial cover (0.8 to 2 m thick) and a rather high accumulation rate (3.1 t/ha/y) due to the erodible nature of the bedrock and the relative steepness of the slope (5 %). These sites are located on the hills shaped in the Cenozoic sandy bedrock but the dominant slopes are steeper than those of Type 1.

*Type 3* These sites have a thinner colluvial cover (0.7 to 1 m thick). The high accumulation rate (9 t/ha/y) is mainly explained by the erodible nature of the bedrock (alluvial sands) although the slopes are gentle and short (1 to 2.2 %). These sites are found on the alluvial sandy hills of the river Oise valley.

*Type 4* These are sites where little erosion has occurred. They are on slopes developed on clay and today are protected by a continuous vegetation cover (forest). Their colluvial cover is thin (0.35 m), even on steep slopes (7.6 %) and the accumulation rate is low (0.6 to 1.0 t/ha/y).

These results must be considered approximate for several reasons :

(i) Only six sites were examined and there is limited evidence of periods before the Late Iron Age La Tène period (i.e. 175-50 BC).

(ii) The sites studied were confined to plateau surfaces or the floors of valleys. There is little evidence from steeper slopes or from mid-slope positions.

(iii) The maximum depth of colluvium above the archaeologically dated horizon was 1.8 m. All sites were therefore affected by soil formation and

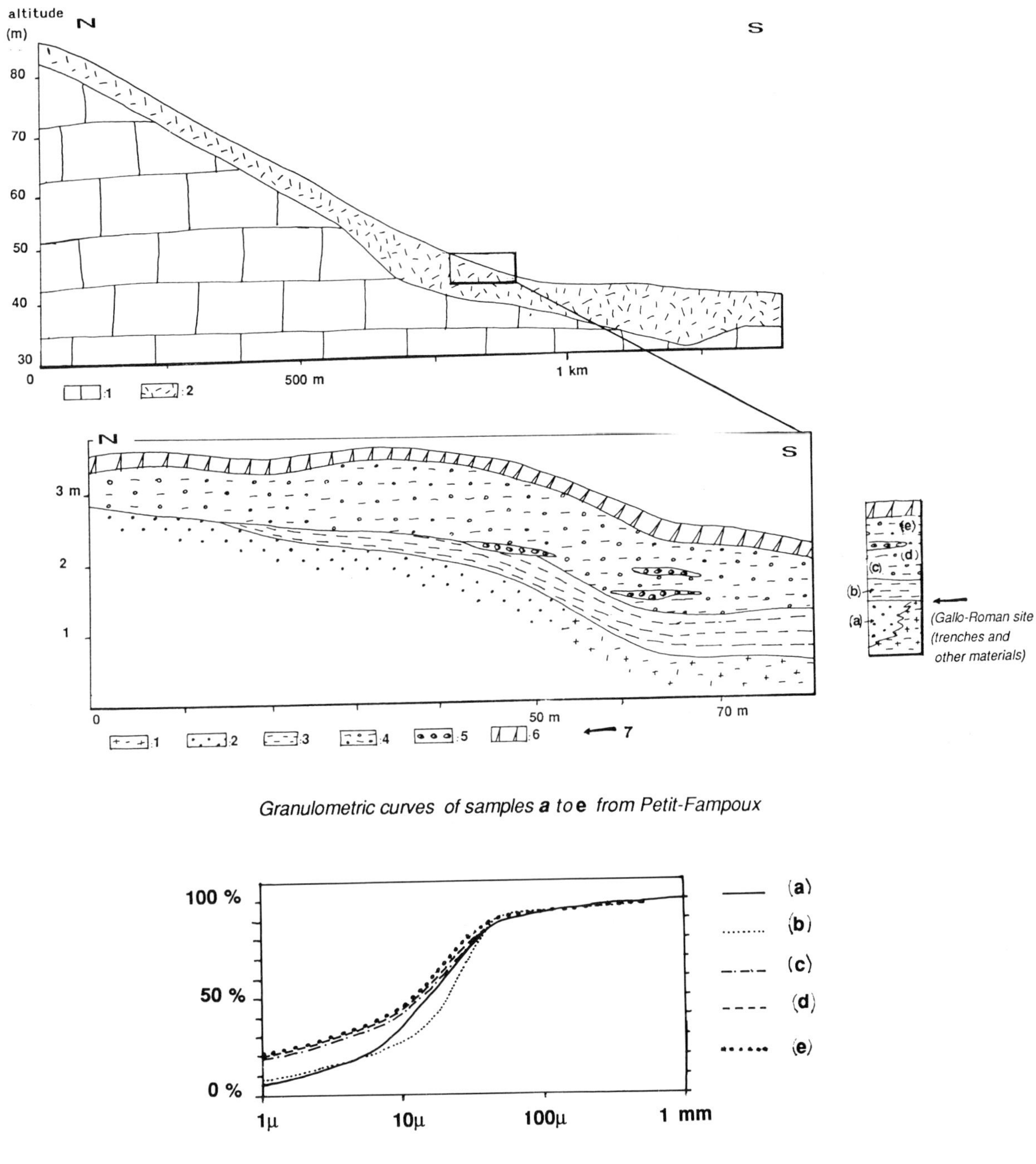


Figure 10.2 Site 4, Petit-Fampoux. Legend of the slope section: 1 = Senonian chalk; 2 = Quaternary formations. Legend of the short section: 1 = argillaceous pseudo-gley; 2 = beige silts with granules of chalk (sample a); 3 = humiferous horizon (boggy palaeosol) (sample b); 4 = brown silt with granules of chalk (samples c, d and e); 5 = frost shattered flints; 6 = plough horizon (Ap); 7 = archaeological site (Gallo-Roman; 2 trenches and other materials). Samples are from the short section; their location is given in the figure and in the legend above.

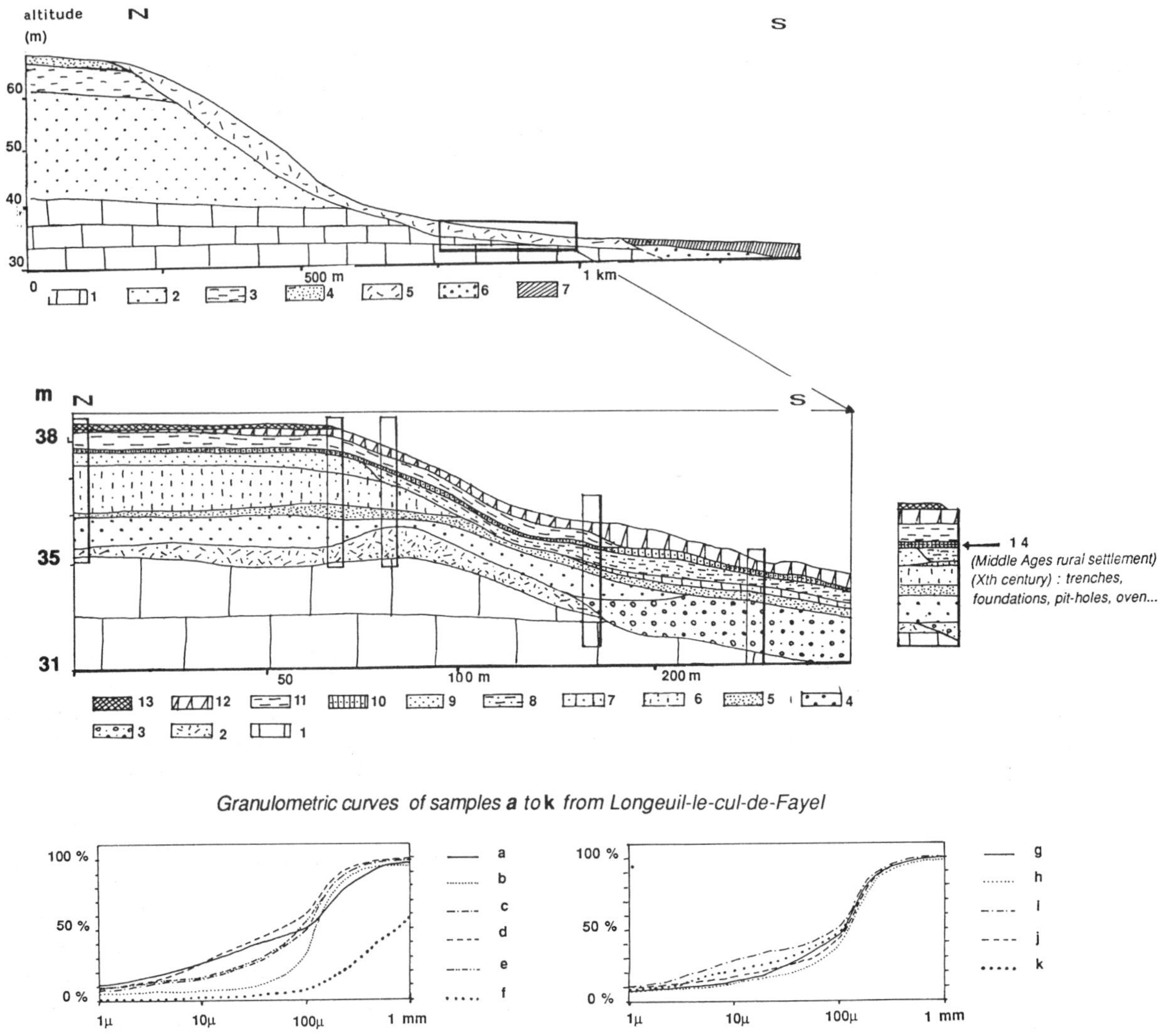


Figure 10.3 Site 6, Longueil-le-Cul-de-Fayel. Legend of the slope section: 1 = Senonian chalk; 2 = Thanetian sands (= 'Bracheux' sands) ; 3 = Ypresian clay (= Sparnacian); 4 = silt over plateaux (Weichselian loess?); 5 = chalky slope deposit; 6 = silts mixed with colluvial and alluvial deposits; 7 = bog.
Legend of the short section and of the 5 bore holes: 1 = Senonian chalk; 2 = Thanetian sands (reworked); 3 = coarse fragments of chalk; 4 = river sands layered in thin or coarse beds (samples a and b); 5 = fine-grained yellow sands (sample c); 6 = calcareous sandy silts (sample d); 7 = calcareous tufa; 8 = sands with evidence of pedogenesis; 9 = yellowish sands (sample e); 10 = clear-coloured sands (sample f); 11 = sandy brown silts (samples g to k); 12 = plough horizon (Ap); 13 = artificial filling; 14 = archaeological site (Middle Ages - tenth century village: trenches; holes; settlement foundations; oven). Samples are from the short section; their situation is given in the figure and in the legend above.

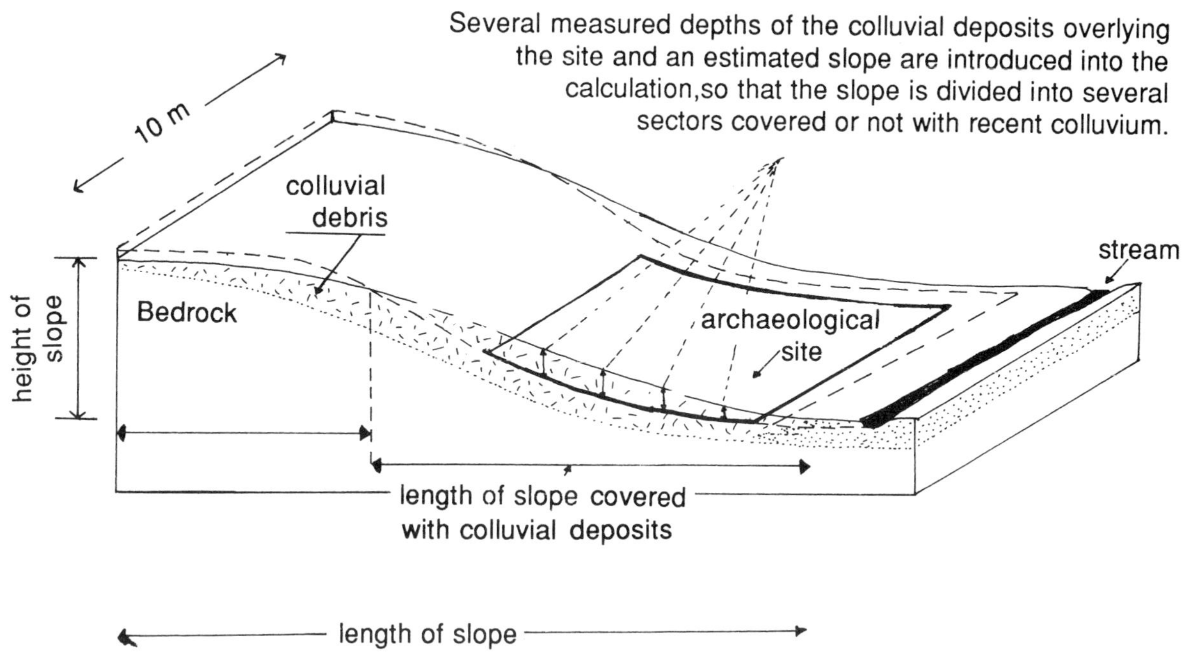


Figure 10.4 Sketch showing the various elements of calculation for the mass of colluvium overlying each site.

cultivation. Vertical migrations as well as horizontal bands of clay may also have occurred. This meant that it was not possible to investigate modes of deposition or identify fine stratigraphic units which may have related to particular episodes of colluviation.

### Discussion

The mean annual accumulation rates presented in Table 10.1 are not erosion rates since the calculation does not take into account loss of colluvium from the slope system. These rates are derived from the measurement of sediment accumulation at the base of slopes and may therefore be underestimating the true erosion rate. Thus the accumulation rate is a minimum rate as far as production of colluvium is concerned. The rates are approximate as explained previously and depend upon local conditions specific to each site, in particular the length and steepness of the slope, the susceptibility to erosion of the soils and sediments and the agricultural land-use history of the site.

The rates obtained do not directly measure soil loss from the slopes as do the erosion rates established by actual monitoring of experimental watersheds (Table 10.2). They take into account soil which has been eroded from the upper parts of the slopes and redeposited at the foot of these slopes over a dated surface. The area dated by the site was extended over the entire slope in order to define an estimated site-aged profile of the slope. Therefore the loss of sediment into the valley, for example the loss that occurred after the abandonment of the site when the human settlements were levelled off by erosion, is lacking in the calculation. The measurements, however, clearly show that impressive volumes of sediment may be moved on relatively gentle slopes (1.5 - 2.5 %), especially considering the fact that these are minimum figures. Comparison with the data obtained from direct monitoring of erosion in England (Boardman 1990; Boardman *et al.* 1990; Evans and Cook 1986; Morgan 1985), France (Auzet 1987) and Belgium (Bollinne 1982), suggests (Table 10.2), that erosion must have occurred over several centuries in these old agricultural plateaux (Bell 1981; Blanchet 1984; Evans 1990; Roblin 1978; Vogt 1953).

We would expect that sedimentation at the sites we have studied occurred in a number of episodes rather than being continuous as our calculations assume. But no archaeological material was found buried in the colluvium so there was no evidence for any chronological sequence. However, beds of frost-shattered flints (*silex gelifracts*) within the colluvium at site 4 (Figure 10.2) suggest vigorous and episodic erosion and deposition but the homogenization of the sediments has prevented us from investigating this further.

The movement of slope sediments may have occurred during periods of severe pressure on agricultural land, for example Neolithic land-clearance; Gallo-Roman intensive exploitation of land for grain and animal husbandry (Bell 1981; 1987); later Medieval population growth and technological innovations (new ploughing and agricultural techniques); renewed population pressure in the eighteenth century (Bollinne 1982; Vogt 1953); and intensification of agriculture since the nineteenth century. At other times, land was abandoned to pasture and woodland or farmed less intensively (the invasions of the fifth to tenth centuries; the plague during the earlier Middle Ages; the epidemics and wars during the fifteenth to seventeenth centuries) for instance, and at these times there may have been a decrease in the movement of sediment on the slopes of the Paris Basin.

## Conclusions

Other workers have shown that soil erosion has been important on the plateaux of north-western Europe since the eighteenth century (Evans 1990; Bollinne 1982; Vogt 1953). Our data suggest that substantial amounts of soil have been eroded from gentle slopes developed on loam-covered sand and chalk landforms of the northern Paris Basin. This erosion may well have occurred in periods associated with agricultural over-exploitation or climatic fluctuation. From our data collected in studies along the TGV route, this acceleration in slope erosion appears to be post Bronze Age (Lespez 1990; Pastre *et al.* 1991).

## Acknowledgements

The authors thank the SNCF (*Société Nationale des Chemins de fer Français*, the French Railways) and the AFAN (*Association pour les Fouilles Archéologiques Nationales*) for their help and contribution to the rescue programme which has enabled them to carry out this study and to perform further research on colluvial and alluvial deposits in this region. We would like to thank Mrs A. Bigonneau who kindly corrected the English. We wish also to express our gratitude to Prof. A. Godard who supported our work, to J.-C. Blanchet, Director of the 'Antiquités d'Ile-de-France et de Picardie', to M. Talon, Co-ordinator of the 'Opération Archéologique du TGV', and to J. Boardman for our fruitful discussions.

## References

Antoine, P. in press Principaux résultats de la campagne de sondages profonds effectuée sur le tracé du futur TGV Nord. *Documents de l'Archéologie Française, Série Spéciale Grands Travaux*. Maison des Sciences de l'Homme Pub., Paris.

Auzet, A.-V. 1987 L'érosion des sols cultivés en France sous l'action du ruissellement. *Annales de Géographie* 537, 530–556.

Bell, M. G. 1981 Valley sediments and environmental change. In: Jones, M. and Dimbleby, G. (eds.) *The environment of man: the Iron Age to the Anglo-Saxon Periods*. British Archaeological Reports, British Series 87,Oxford, 75–91.

Bell, M. G. 1987 Archaeological evidence for the date, cause and extent of soil erosion on the chalk. *Seesoil* 3, 72–83.

Blanchet, J. C. 1984 Les premiers métallurgistes en Picardie et dans le Nord de la France. *Mémoire de la Société Préhistorique Française*,17.

Blanchet, J.-C. and Talon, M. 1989 Etude d'impact archéologique du TGV Nord et de l'Interconnexion. Actes du Colloque 'Archéologie et Aménagement du Territoire'. *Cahiers de l'Urbanisme* 7, Bruxelles, 31–34.

Blanchet, J.-C. and Talon M. 1991 Le programme de sauvetage archéologique du TGV Nord et de l'Interconnexion. *Bulletin de la Société Préhistorique Française* 88, 1, 8–10.

Boardman, J., Foster, I. D. L. and Dearing J. A. (eds.) 1990 *Soil Erosion on Agricultural Land*. John Wiley and Sons, Chichester.

Boardman, J. 1990 Soil Erosion on the South Downs: A Review. In: Boardman, J., Foster, I. D. L. and Dearing, J. A. (eds.) *Soil Erosion on Agricultural Land*. John Wiley and Sons, Chichester, 87–106.

Bollinne, A. 1982 *Etude et prévision de l'érosion des sols limoneux cultivés en moyenne Belgique*. Thèse Liège University , unpub., Liège.

Chisci, G. and Morgan, R. P. C. (eds.) *Soil Erosion in the European Community. Impact of Changing Agriculture*. Balkema, Rotterdam.

Collectif. in press. Les sites archéologiques du TGV Nord. *Documents de l'Archéologie Française, Série Spéciale Grands Travaux*. Maison des Sciences de l'Homme Pub., Paris.

Evans, R. 1990 Soil erosion: its impact on the English and Welsh landscape since woodland clearance. In: Boardman, J., Foster, I. D. L. and Dearing, J. A. (eds.) *Soil Erosion on Agricultural Land*. John Wiley and Sons, Chichester, 231–255.

Evans, R. and Cook, S. 1986 Soil erosion in Britain, *Seesoil* 3, 28–59.

Kuzucuoglu, C., Lespez, L., Pastre, J.-F. in press Can the magnitude and dynamics of colluvial deposits be related, through archaeological data, to land clearing and agriculture? *1st International Regional Conference on Geomorphology (Natural Hazards) Proceedings*. Ankara, May 1991.

Lespez, L. 1990 *L'étude du transport sur les versants aux époques historiques et protohistoriques. Exemples pris le long du tracé du TGV Nord.*. Unpub. Mémoire Maîtrise, Paris I University, Paris.

Morgan, R. P. C. 1985 Soil erosion measurement and soil conservation research in cultivated areas of the UK. *Geographical Journal* 151, 11–20.

Pastre, J.-F. in press Le tracé du TGV Nord et de l'Interconnexion. Présentation géographique et géologique. *Documents de l'Archéologie Française, Série Spéciale Grands Travaux*. Maison des Sciences de l'Homme Pub., Paris.

Pastre, J.-F. in press Le Pléistocène moyen et supérieur de l'Ile de France et de la Picardie. Données récentes sur les tracés du TGV Nord et de l'Interconnexion. *Documents de l'Archéologie*

*Française, Série Spéciale Grands Travaux.* Maison des Sciences de l'Homme Pub., Paris.

Pastre, J.-F., Cecchini, M., Dietrich, A., Fontugne, M., Gauthier, A., Kuzucuoglu, C., Leroyer, C., Limondin, N. 1991 L'évolution Holocène des fonds de vallées au Nord-Est de la région parisienne (France) : premiers résultats. *Physio-Géo*, 22/23, 109–115, Meudon.

Roblin, M. 1978 *Le terroir de l'Oise aux époques gallo-romaine et franque. Peuplement, environnement, défrichement.* A and J. Picard Pub., Amiens.

Vogt, J. 1953 Erosion des sols et techniques de culture en climat tempéré océanique de transition. *Revue de Géomorphologie Dynamique* 3, 4, 157–183.

Walling, D. E. 1990 Linking the field to the river: sediment delivery from agricultural land. In: Boardman, J., Foster, I. D. L. and Dearing, J. A. (eds.) *Soil Erosion on Agricultural Land.* John Wiley and Sons, Chichester, 129–152.

# CHAPTER 11

# EROSION IN THE MEDITERRANEAN LANDS: A RECONSIDERATION OF PATTERN, PROCESS AND METHODOLOGY

JOHN BINTLIFF

*Six phases are suggested for the progress of our understanding of Mediterranean and especially Greek erosional processes during the Holocene, since the pathfinding publication in 1969 of Vita-Finzi's The Mediterranean Valleys. The relative contribution of climatic and anthropogenic factors is reassessed, and particular emphasis given to the role of extreme events in the landscape within sequences of punctuated equilibrium.*

I would like in this paper to discuss the development of our increasing understanding since the 1960s of the interaction between human settlement history and erosional processes in the Mediterranean region, with particular emphasis on Greece.

STAGE ONE commences with the remarkably influential publication in 1969 of Vita-Finzi's *The Mediterranean Valleys*. This began as, and has remained, a controversial book. It simplified late Quaternary fluvial sequences in the circum-Mediterranean into two major phases, widely separated in time, the Older and Younger Fills (essentially last glacial and Medieval in date). This model and chronology have not stood the test of time and the much more detailed local research programmes which in many respects they can be said to have stimulated. The book's second surprise was its rejection of the Uniformitarian model for Mediterranean erosion, whereby landscape degradation is considered to have been continuous and progressive since the first major human clearance of woodlands; instead we might categorize Vita-Finzi's conception of Mediterranean erosional history as one of Punctuated Equilibrium. This second claim has proved to be an enduring insight of immense importance for all of us researching into the impact of erosion processes on the development of Mediterranean societies in prehistory and history. A third element of controversy in Vita-Finzi's book was to suggest that both of the two discrete alluvial fills were best accounted for through natural fluctuations in climate rather than through waves of human landscape disturbance; as regards the Holocene alluvial activity this final component might seem to have been all but conclusively disproved through the detailed research of the early 1980s, yet I hope to suggest that the most recent results in erosional history permit a modest but significant reappearance of climate into the drama of Mediterranean erosional history.

*The Mediterranean Valleys* provided a working model for the 1970s, encouraging rather superficial application by archaeologists involved in the burgeoning field of regional settlement studies. My own Ph.D. research in a series of landscapes in Greece (Bintliff 1977), for example, frequently fitted both reasonably dated and vaguely dated sediments into local landscape evolution schemes based on the Vita-Finzi two-phase model, without pausing to consider whether finding sediment dates that did not *contradict* the model was not a poor substitute for trying to *test* the model independently for each region.

In STAGE TWO a phase of critical examination was inaugurated, both of *The Mediterranean Valleys* and follow-up papers by Vita-Finzi, and of proselyte applications such as my own work in Greece. This critique (cf. Davidson 1980; Wagstaff 1981) explored internal contradictions in the publications concerned, and cited individual case-studies from different parts of the Mediterranean to underline a general view that erosional and depositional sequences were far too variable in time and space to be accommodated into any simple scheme, especially one based on general climatic change. A more plausible case could be made for a correlation with local phases of heightened human impact on the landscape.

In a volume on palaeoclimates in the Eastern Mediterranean which appeared in 1982 I also raised various difficulties with a solely climatic explanation for Mediterranean Holocene alluviation (Bintliff 1982). Previously (cf. Bintliff 1977) I had already been led to suggest on both theoretical and geoarchaeological grounds that Vita-Finzi's periglacial Older Fill should be a multi-phase formation characteristic of each major cold cycle during the last three-quarters of a million years. Now, a closer examination of the palaeoenvironmental and geoarchaeological evidence for Holocene climatic fluctuations in Europe, revealed that well-documented worldwide shifts in secular climatic parameters (the classic example being the Post-Medieval Little Ice Age), are likely to have occurred at regular intervals throughout the Holocene, rather than being merely a very late feature of the Sub-Atlantic phase of the last three millennia. By implication, if purely climatic, the Younger Fill ought likewise to have been a multi-phase formation repeated through the Holocene, which was clearly not the case. I was therefore led to suggest that the prominent late Roman to early Medieval Younger Fill arose through a unique conjunction of a cyclical shift of climate and a vast abandoned landscape preadapted to erosion through the population collapse of the Roman Empire.

However, the basis for STAGE THREE and a revolutionary shift in the quality of the relevant database had already been prepared. In 1980 there had appeared the first of a series of geomorphological studies of small regions of the Mediterranean, in which a quite new level of detailed alluvial history was achieved. This study was by a Belgian team led by Roland Paepe (Paepe *et al.* 1980), and resulted from several years' work in the eastern coastlands of Attica near Athens. That particular landscape showed remarkable stability for the Holocene, typified by well-developed soil formations; yet at infrequent intervals a small number of major erosion events was recorded across the terrain in distinct alluvial units. These events, associated archaeologically with the Early Bronze Age (*c.* 2000 BC), the Classical-Hellenistic period (*c.* 400 BC), Late Roman (*c.* 400 AD) and Post-Medieval times, were still attributed to climatic fluctuations, probably because the settlement history of this region was little researched. But when in 1984 a similar regional geomorphology was published for the South-West Argolid peninsula in Greece by the American team of Pope and van Andel, it could be directly compared with the results for the same landscape of a new kind of equally intensive archaeological field survey, allowing precise correlations of settlements and sediments. Excitingly, the punctuated equilibrium picture of Attica was repeated in the Argolid, and with exactly the same concentrated alluvial events. But now it was abundantly clear that the rare disequilibrium events were tied to disjunction points between phases of unusually dense human population cover and periods of largescale depopulation. In this paper and later discussions of the Argolid evidence van Andel and colleagues (van Andel *et al.* 1986; 1987) provided an incontrovertible case, that the dramatic expansion and contraction cycles of human population and associated land use revealed by this and other field surveys in Greece, were a fundamental cause of such discrete erosion and deposition events. Scope for a climatic factor was essentially dismissed.

In STAGE FOUR the far-reaching implications of these results and insights were absorbed and applied by archaeologists to shed light on settlement history. Meanwhile further regional geomorphologies in Greece and Italy reinforced both the punctuated equilibrium and anthropogenic aspects central to the new model (cf. van Andel and Zangger 1990 for the Argive Plain and the Plain of Thessaly in Greece, and Brückner 1986, 1990 for Lucania in Southern Italy and Southern Attica in Greece).

I shall illustrate the archaeological implications from a regional field survey I have been co-directing since 1978 in Boeotia, Central Greece with Anthony Snodgrass (Bintliff and Snodgrass 1985; 1988a). Like the Argolid survey (van Andel and Runnels 1987), our landscape reveals an unparalleled density of human settlement in climax Classical Greek times (*c.* 400 BC); equally both regions demonstrate a striking depopulation in both town and country in the following late Hellenistic and early Roman period (*c.* 200 BC–300 AD). Although wars and political weakness provide a possible historical causation for depopulation, the prolonged delay in population recovery, of some 500 years (until Late Roman times), requires a deeper explanation. In the light of the Attic and Argolid erosion chronologies we have concluded that the most likely scenario is of overpopulation and overuse of the landscape in Classical times precipitating a catastrophic erosion phase, this being the prime factor in the cultural and demographic collapse of the region after 200 BC. The long timescale for soil recovery provides an appropriate limiting factor on regional recovery. The model is applicable to other cycles at similar wavelengths documented by our own and other Greek surveys.

The resolution of dates available for Greek and other Mediterranean erosion episodes is rarely precise enough to pinpoint the exact stage of cultural florescence and collapse when the erosion event intervenes. Van Andel and colleagues have therefore offered alternative scenarios for interpreting their data, e.g. severe erosion with rapid expansion of a culture, *or* at its maximum extent, *or* after cultural collapse when abandoned terraces and open landscapes offer ideal nurture for erosional forces (cf. van Andel and Runnels 1987).

In the Boeotian case I would presently favour the population climax timing, with the implication that the collapse of Classical Greece was an ecological disaster. Support for this interpretation can be sought in two adjacent regions. In nearby Evvoia, Rust (1978) was able to suggest that the ancient city of Eretria suffered catastrophic erosion and alluviation in its arable hinterland during a very short phase of late Classical times. In South Attica a geomorphological study by Brückner (1990), in association with a detailed field survey by Lohmann (1985), has revealed an elaborate Classical farmstead landscape abandoned after the fourth century BC and never brought back into cultivation, tied to a dramatic inception of soil erosion. Historic sources also suggest a major breakdown in Attic crop production at this time (Garnsey 1988; Bintliff 1991).

In Boeotia we have also exploited the insights of van Andel and co-workers to improve our understanding of archaeological field survey data. With the original intention of clarifying the relationship between surface concentrations of artifacts which were believed to be settlements, and the occurrence of sherds and lithics everywhere else in the landscape (the 'offsite' archaeology), we have been constructing quantitative maps of surface pottery densities for the entire survey zone. Overall, with allowance for localised post-depositional disturbances caused by gullying and alluvial redeposition (processes discussed recently in a British context in an important paper by Allen 1991), we have interpreted the resultant maps as defining settlement sites, infield haloes of concentrated manuring and rubbish disposal around settlements, and wide swathes of agricultural manuring debris. Intriguingly, closely similar maps have been produced from other parts of Europe and the Middle East, yet the artifactual densities show a remarkable range. Indeed, so striking was the surface pottery density variation that I put some of the limited published data for offsite densities on a graph (Figure 11.1), which appeared to show a cline of logarithmically increasing values from North-West Europe to the Arabian peninsula. In a paper in 1988 (Bintliff and Snodgrass 1988b) I suggested that the main variable could be geomorphological: in North-West Europe lower surface artifact densities reflected the interaction of soil erosion and compensatory soil growth; in the Eastern Mediterranean far higher surface densities showed concentrations resulting from cumulative erosion episodes far outweighing soil regrowth; whilst in Arabia wind deflation created stupendous surface artifact accumulations.

Van Andel has estimated (cf. van Andel and Zangger 1990) that the average cumulative depth of soil lost in 4–5 erosive episodes since the early Bronze Age (*c.* 2500 BC) in the Greek Argolid is 40 cm in the mountains and 100 cm in the lowlands. We have recently been able to demonstrate the widespread survival over the Boeotian landscape of the lower horizons of these ancient palaeosols whose loss of A horizon may have been so devastating. Both magnetic susceptibility and trace metal analyses show abnormal concentrations in soils on and around the ancient rural sites, as well as in long transects running towards ancient cities. These soil properties are ascribed by Professor Brian Davies of Bradford University (cf. Bintliff *et al.* 1990) to domestic and farmyard refuse deposited on ancient settlements and through manuring in their hinterlands. Their survival *in situ* requires survival of a contemporary truncated palaeosol.

In STAGE FIVE various problems have emerged with the van Andel model, requiring both fine tuning and perhaps some more radical rethinking. Whereas in the pioneer 1984 publication, Pope and van Andel stressed that the Argolid alluvial sequence was dated without recourse to any *a priori* phases from cultural developments, so that the remarkable fit of settlement fluctuations and erosion cycles was all the more convincing; in more recent papers, especially when later regional studies were included, the much less clear fit to settlement dynamics from the expanding database is visibly overstretching the capacity of the model. Thus the evidence for a long delay in Thessaly between major Neolithic clearance and settlement (6000 BC onwards), and indications of significant erosion (5–4000 BC), is attributed without supportive evidence to changes in farming practices (van Andel and Zangger 1990). The absence in both the Argolid and Thessaly, but not in the Plain of Argos, of a major erosion phase associated with the highly populous Mycenaean civilisation (*c.* 1250 BC) is attributed to rapid reafforestation following collapse, a factor not apparently valid for other civilisational collapse scenarios (*idem*). There is also a recognisable tendency (reminiscent of my own earlier attempts to fit wayward or vaguely-defined sediment dates into a strict Vita-Finzi chronology), to reinterpret broad sediment date ranges to fit more narrowly defined phases of appropriate settlement history (cf. van Andel and Zangger 1990; Brückner 1990; Lewin *et al.* 1991). From squeezing the data to suit climatic determinism, there is a risk that we might be repeating the exercise for a too precise anthropogenic determinism. It appears increasingly likely that erosion events are imperfectly correlated with specific stages in human impact. Fortunately new ideas have been appearing in erosional studies that can shed light upon this problem.

STAGE SIX in our understanding of Mediterranean erosion history commences with the analysis by Professor John Thornes in the early 1980s of contemporary erosion processes in South-East Spain, followed up by joint research with the archaeologist Antonio Gilman on the development of the prehistoric landscape of that region (Thornes and Gilman 1983; Gilman and Thornes 1985). Although bedrock and soil erosional susceptibility, and the degree of vegetative protection, control rates of erosion, Thornes stresses the unique role of extreme rainfall events in creating major episodes of soil stripping. At very much the same time as Thornes' Mediterranean work, the growing interest in soil erosion studies of the British landscape has seen a similar focus, based both on recent empirical observation and on the nature of prehistoric episodic erosion (cf. Boardman and Robinson 1985; Morgan 1985; Boardman 1986; Bell 1987; Allen 1991).

A difficulty already well-recognized in the literature on erosion history in England is that of timescale. Even if the current emphasis in erosion theory in Britain and Spain highlights extreme events, their historical recurrence interval is normally within a human lifetime, in contrast to the recurrence intervals of erosive episodes in the geoarchaeological record, which can be at thousand-year scales. Specific forms of landscape preadaptation might be required, to trigger catastrophic erosion from 'rare' climatic irregularities to the required 'very rare' longer timescale.

The recently-perceived threat to the English Chalkland soils, for example, has been attributed to a conjunction of circumstances operating since 1939: largescale conversion of pastureland to arable, removal of field boundaries and a shift to autumn sowing. The last two decades have also seen a succession of unusual climate conditions rarely paralleled since instrumental records began. On the other hand, similar variations in agricultural conditions are either known or inferred for most periods of history and prehistory by landscape archaeologists, yet preserved major erosional or depositional episodes remain much more infrequent than we would expect from both a climatic and agricultural history viewpoint, or from a combination of the two (cf. Bell 1992; Boardman 1992; Brown 1992). Nonetheless, some of the Greek erosional sequences, with approximately 1000 year recurrences, might suit the empirically-observed wavelength of cultural cycles (cf. Bintliff and Snodgrass 1985).

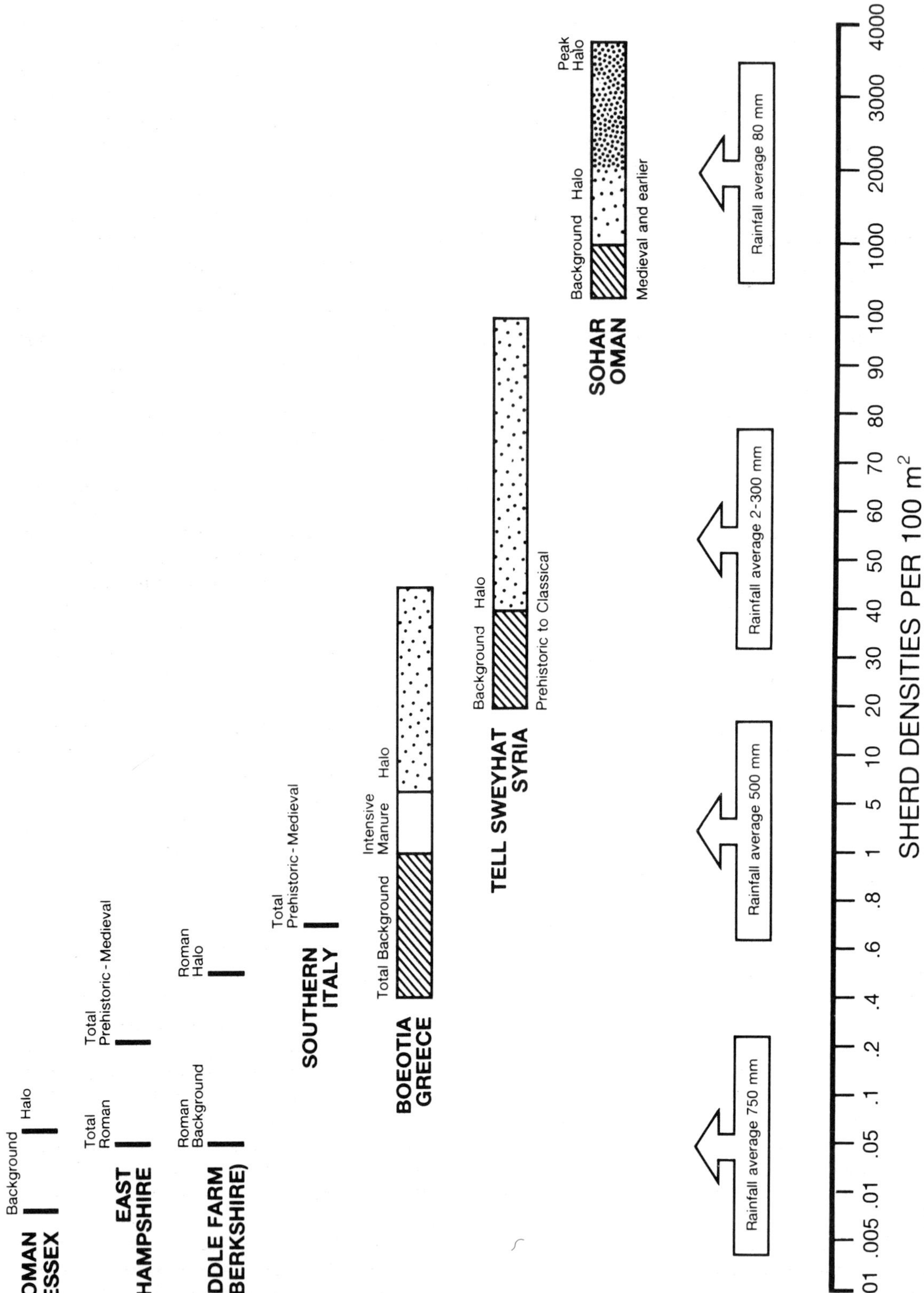


Figure 11.1 Recorded densities of offsite surface pottery in a series of landscapes subjected to intensive archaeological field survey, arranged along a geographical trend from North-West Europe to the Middle East (from Bintliff and Snodgrass 1988b).

Two recently published examples will furtherhighlight the problem. In North Germany a team of archaeologists and a geomorphologist have analysed the development of the landscape in and around a deserted medieval village at Drudewenshusen (Bork 1988). The parish was settled from Neolithic times, but until the high Medieval era there is only limited soil erosion and soil creep, despite clear evidence of major forest clearance and extensive land use. Then around AD 1320 a decisive change occurs with the inception of major gullying activity and rates of surface erosion without parallel for the entire Holocene. The cause is argued to be 'a few scarcely imaginable extreme rainfall events of catastrophic proportions following rapidly one after another'. The village is deserted around AD 1400, but the landscape is recultivated from the 16th century and there is a second, slighter gully phase in the 18th century. Estimates of erosion depth in the major episode include some slopes losing over 2 metres.

My second example is also Medieval, but from West-Central Iran (Brookes 1987). In the Qara Su Basin a geomorphological history running from the late Pleistocene has identified again one single unparalleled event of catastrophic proportions, when the whole basin experienced massive sedimentary deposition. Though short-lived, the exact date is unclear. Brookes suggests however that it could be the result of an extreme cyclonic event with a recurrence interval of a millennium striking a landscape preadaptated to dramatic erosion by the collapse of agriculture following the mid 13th century AD Mongol invasions. (For a related ancient example in Yugoslav Macedonia cf. Folk 1975, with eight, closely-spaced extreme events at Late Roman Stobi.)

If extreme climatic events are so potent, we might expect them to be evidenced in naturally open semi-arid landscapes even without major human impact creating preadaptation. Thornes has indeed suggested that in the almost desert environment of Almeria in South-East Spain (Thornes 1989) colonising farmers of the Copper Age found a landscape already exposed to extensive gully erosion. In the severe southern deserts of Israel Paul Goldberg has drawn attention to two major episodes of valley alluviation (Goldberg 1986) where freak weather fluctuations (perhaps operating for some centuries) had a major impact both on sedimentary history and human farming potential. During the Copper Age of the 4th millennium BC, then again in Late Roman and early Islamic times, large quantities of aeolian material from North Africa, which is normally borne over the Middle East without effect, were rainfall-precipitated throughout the Negev and Sinai, then reworked by water erosion into the lowest relief where it created recognisable and very extensive alluvial fills. This phenomenon had a significant effect on the expansion of settlement in these regions.

Whether extreme events, or short-lived periods of uncharacteristic weather, are more likely to occur at certain phases of the Holocene is a question that may still repay further investigation, even if it must now be clear that strong constraints are imposed by the preadaptation of landscapes through natural susceptibility to erosion or specific types of human land use. Since the full establishment of the semi-arid Mediterranean climate over all of its current region can be dated palynologically only from the 3rd millennium BC (van Zeist and Bottema 1982), then the bunching of erosion events since the early Bronze Age, and the contemporaneous replacement of lowland deciduous by evergreen Mediterranean vegetation (Lewthwaite 1986), may not solely be due to the enhanced human settlement of the Mediterranean area from that period.

An intriguing case-study that raises exactly these questions is that by Jose Peña and colleagues from Zaragoza, of the landscape evolution in later Holocene North-East Spain (Mozota *et al.* 1986; Sancho *et al.* 1988). Analysis of the slopes below prehistoric and historic hilltop settlements in the Ebro Valley, and associated valley fills, shows a cyclical pattern seemingly independent of continued use of the archaeological sites from Bronze Age to recent historic times. Phases of slope pedogenesis and slow soil creep alternate with phases of dramatic gully erosion, scarp retreat, and valley fill, producing a striking pattern of relict triangular slope facets. Periodic shifts in critical climatic parameters are suggested as a key factor alongside human impact.

Equally importantly, this example reminds us that in the Mediterranean there has been a neglect of the potentially complicated relationship between slope processes and valley fills, in contrast to the important work now being done in Britain (Bell 1987; Allen 1991). We need greater understanding of the trigger factors that convert a stable slope with long-term accumulations of soil creep into a severely gullied eroding soil reservoir feeding aggrading valley systems (cf. for Spain, Sala 1983; Thornes 1989). This interface could contain vital clues to the long recurrence intervals for major erosion events.

## References

Allen, M. J. 1991 Analysing the landscape: a geographical approach to archaeological problems. In, Schofield A. J. (ed.) *Interpreting Artefact Scatters*. Oxbow, Oxford, 39–57.

Bell, M. G. 1987 Archaeological evidence for the date, cause and extent of soil erosion on the chalk. In: Burnham C. P. and Pitman J. I. (eds.) *Soil Erosion. SEESOIL* 3, 72–83.

Bintliff, J. L. 1977 *Natural Environment and Human Settlement in Prehistoric Southern Greece*. British Archaeological Reports, International Series 28, Oxford.

Bintliff, J. L. 1982 Palaeoclimatic modelling of environmental changes in the East Mediterranean region since the last glaciation. In, Bintliff J. L. and van Zeist W. (eds.) *Palaeoclimates, Palaeoenvironments and Human Communities in the Eastern Mediterranean Region in Later Prehistory*. British Archaeological Reports,

International Series 133, Oxford, 485–527.
Bintliff, J. L. 1991 Review of Garnsey P., *'Famine and Food Supply in the Graeco-Roman World'. American Historical Review* 96, 145.
Bintliff, J. L. and Snodgrass, A. M. 1985 The Cambridge/Bradford Boeotian Expedition. The first four years. *Journal of Field Archaeology* 12, 125–161.
Bintliff, J. L. and Snodgrass, A. M. 1988a Mediterranean survey and the city. *Antiquity* 62, 57–71.
Bintliff, J. L. and Snodgrass, A. M. 1988b Off-site pottery distributions: a regional and interregional perspective. *Current Anthropology* 29, 505–513.
Bintliff, J. L., Davies, B., Gaffney, C., Snodgrass, A. and Waters, A. 1990 Trace metal accumulation in soils on and around ancient settlements in Greece. In, Bottema S., Entjes-Nieborg G. and van Zeist W. (eds.) *Man's Role in the Shaping of the Eastern Mediterranean Landscape*. Balkema, Rotterdam, 159–172.
Boardman, J. 1986 The context of soil erosion. In: Burnham C. P. and Pitman, J. I. (eds.) *Soil Erosion. SEESOIL* 3, 2–13.
Boardman, J. and Robinson, D. A. 1985 Soil erosion, climatic vagary and agricultural change on the Downs. *Applied Geography* 5, 243–258.
Bork, H.-R. 1988 Mittelalterliche Relief-, Sediment- und Bodenentwicklung im Bereich der Wüstung Drudewenshusen. *Archäologisches Korrespondenzblatt* 18, 89–95.
Bottema, S., Entjes-Nieborg, G. and van Zeist, W. (eds.) 1990 *Man's Role in the Shaping of the Eastern Mediterranean Landscape*. Balkema, Rotterdam.
Brookes, I. A. 1987 A medieval catastrophic flood in central west Iran. In, Mayer L. and Nash D. M. (eds.) *Catastrophic Flooding*. Allen and Unwin, Boston, 225–246.
Brückner, H. 1986 Man's impact on the evolution of the physical environment in the Mediterranean region in historical times. *GeoJournal* 13, 7–17.
Brückner, H. 1990 Changes in the Mediterranean ecosystem during antiquity - a geomorphological approach as seen in two examples. In, Bottema S., Entjes-Nieborg G. and van Zeist W. (eds.) *Man's Role in the Shaping of the Eastern Mediterranean Landscape*. Balkema, Rotterdam, 127–137.
Davidson, D. A. 1980 Erosion in Greece during the first and second millennia BC. In, Cullingford R. A., Davidson, D. A. and Lewin, J. (eds.) *Timescales in Geomorphology*. John Wiley, Chichester, 143–158.
Folk, R. L. 1975 Geologic urban hindplanning, *Environmental Geology* 1, 5–22.
Garnsey, P. 1988 *Famine and Food Supply in the Graeco-Roman World*. Cambridge University Press, Cambridge.
Gilman, A. and Thornes, J. B. 1985 *Land Use and Prehistory in South-East Spain*. Allen and Unwin, London.
Goldberg, P. 1986 Late Quaternary environmental history of the Southern Levant. *Geoarchaeology* 1, 225–244.
Lewin, J., Macklin, M. G. and Woodward, J. C. 1991 Late Quaternary fluvial sedimentation in the Voidomatis Basin, Epirus, Northwest Greece. *Quaternary Research* 35, 103–115.
Lewthwaite, J. 1986 The transition to food production: a Mediterranean perspective. In, Zvelebil M. (ed.) *Hunters in Transition*. Cambridge University Press, Cambridge, 53–66.
Lohmann, H. 1985 Landleben im klassischen Attika. *Jahrbuch Ruhr-Universität Bochum*, 71–96.
Morgan, R. P. C. 1985 Soil erosion measurement and soil conservation research in cultivated areas of the UK. *The Geographical Journal* 151, 11–20.
Mozota, F. B., Elorza, M. G. Peña Monne, J. L. and Marcen, C. S. 1986 Geomorphological processes as indicators of climatic changes during the Holocene in the North-East Spain. In, *Quaternary Climate in Western Mediterranean. Proceedings of Symposium on Climatic Fluctuations during the Quaternary in the Western Mediterranean Regions*, 31–44.
Paepe, R., Hatziotis, M. E. and Thorez, J. 1980 *Geomorphological Evolution in the Eastern Mediterranean Belt and Mesopotamian Plain*, Report for the International Geological Correlation Programme Project 146: River Flood and Lake Level Changes.
Pope, K. O. and van Andel, T. H. 1984 Late Quaternary alluviation and soil formation in the southern Argolid: its history, causes, and archaeological implications. *Journal of Archaeological Science* 11, 281–306.
Rust, U. 1978 Die Reaktion der fluvialen Morphodynamik auf anthropogene Entwaldung östlich Chalkis. *Zeitschrift für Geomorphologie N.F. Suppl. Bd.* 30, 183–203.
Sala, M. 1983 Fluvial and slope processes in the Fuirosos basin, Catalan Ranges, north east Iberian coast. *Zeitschrift für Geomorphologie, N.F.* 27, 393–411.
Sancho, C., Gutierrez, M., Peña, J. L., Burillo, F. 1988 A quantitative approach to scarp retreat starting from triangular slope facets, Central Ebro basin, Spain. *Catena Supplement* 13, 139–146.
Thornes, J. B. 1989 Solutions to soil erosion. *New Scientist* 3 June, 1989, 45–49.
Thornes, J. B. and Gilman, A. 1983 Potential and actual erosion around archaeological sites in South-East Spain. *Catena Supplement* 4, 91–113.
van Andel, T. H., Runnels, C. N. and Pope, K. O. 1986 5000 years of land use and abuse in the southern Argolid, Greece. *Hesperia* 55, 103–128.
van Andel, T. H. and Runnels, C. 1987 *Beyond the Acropolis*. Stanford University Press.
van Andel, T. H. and Zangger, E. 1990 Landscape stability and destabilization in the prehistory of Greece. In: Bottema S., Entjes-Nieborg, G. and van Zeist, W. (eds.) *Man's Role in the Shaping of the Eastern Mediterranean Landscape*. Balkema, Rotterdam, 139–157.
van Zeist, W. and Bottema, S. 1982 Vegetational

history of the Eastern Mediterranean and the Near East during the last 20,000 years. In: Bintliff J. L. and van Zeist, W. (eds.) *Palaeoclimates, Palaeoenvironments and Human Communities in the Eastern Mediterranean Region in Later Prehistory*. British Archaeological Reports, International Series 133, Oxford, 277–323.

Vita-Finzi, C. 1969 *The Mediterranean Valleys. Geological Changes in Historical Times*. Cambridge University Press, Cambridge.

Wagstaff, J. M. 1981 Buried assumptions: some problems in the interpretation of the 'Younger Fill' raised by recent data from Greece. *Journal of Archaeological Science* 8, 247–264.

# CHAPTER 12

# NEOLITHIC TO PRESENT SOIL EROSION IN GREECE

EBERHARD ZANGGER

*The first Post-glacial events of soil erosion occurred time-transgressively in different parts of Greece. Regardless of the area of observation, the earliest Holocene phase of soil instability has been most significant with respect to the quantity of moved material. Furthermore, human clearance of the natural vegetation on slopes, accompanying the introduction of widespread agriculture, appears as the most likely cause of the first landscape instability. Later soil erosion occurred with increasing frequency and decreasing magnitude, since most fragile soil had already been washed away. New data from the Berbati-Limnes area provide striking evidence in support of a human cause of prehistoric deforestation and soil erosion at the Neolithic/Early Bronze Age transition. The first seeds of agricultural communities arose in the Argive hinterland during the Middle Neolithic. By the Final Neolithic/Early Bronze Age transition their number had increased by a factor of five; subsequently, during Early Helladic II, it soared even more dramatically. A pollen diagram from the coastal swamps near Lerna records a drastic decline of the formerly dense oak forests in favour of Hornbeam (Carpinus) and evergreen shrubs and trees which coincided with the population increase at 4000–3000 BC. The palynological evidence by itself argues unequivocally for a strong human impact on the natural vegetation ultimately resulting in large-scale soil erosion.*

## Introduction

Imagine a survey team of six people searching in a sparsely vegetated terrain for artifacts. The first team member (Figure 12.1A:1) finds exposed white marl with big, square limestone boulders and many well-rounded pebbles but no sherds. The second person (Figure 12.1A:2) reports a dark brown soil with many artifacts, while her neighbour to the right finds only sterile, exposed marl. The fourth team member walks on red soil, and the fifth on white marl and finally the sixth on red soil again. All individuals notice big, square limestone boulders and much smaller, well-rounded pebbles. What is going on?

The answer is quite simple: they are surveying along the boundary between exposed marl and an alluvial fan (Figure 12.1B). The large unrounded limestone boulders have fallen down from a nearby bedrock slope while the smaller rounded pebbles were originally interbedded with the marl and have weathered out. All surfaces, limestone, marl and alluvial fan have been stripped of their soils. The two bedrocks, limestone and marl, lie barren and exposed, while the alluvial fan displays a truncated Pleistocene palaeosol (Table 12.2). With the Holocene soil, most evidence for human occupation has been washed away. Only in one patch, under the tree in the second transect, do remnants of the previous soil cover remain and these do indeed contain artifacts indicating past human land use. Pottery fragments and grinding stones indicate an agricultural community (Figure 12.1C). How extensive this site was and how thick the soil is everybody's guess. The evidence is gone. From this place we cannot even determine when it disappeared. At an earlier stage, before human occupation, the whole area may have had a considerably thicker soil which probably supported a dense vegetation (Figure 12.1D).

This introductory example was drawn from observations made during fieldwork conducted from 1987–90 in the Berbati-Limnes area just east of Mycenae on the Peloponnese. Today, this landscape represents an erosional environment dominated by barren limestone hills. Numerous terraces have been erected to prevent the last vestiges of soil from further destruction. The erosion which occurred here and elsewhere in Greece before such terraces were built is the subject of this study. The questions to be answered are: when did the erosion occur, what caused it and what are its consequences for the archaeological record?

## A status report

These questions are not new. Only ten years ago it seemed as if a straightforward answer had been found for both the date and mechanism of soil erosion in Greece (Bintliff 1977; Vita-Finzi 1978). During the last decade, however, a number of geoarchaeological projects have been carried out in Greece (Figure 12.2), for instance in Volos (Zangger 1991), the Argive Plain (Finke 1988), the Berbati-Limnes area (Wells *et al.* 1990), Asine, the Skourta Plain, in the Thessalian basin (Demitrack 1986; van Andel *et al.* 1990), the plain of Marathon (Baeteman 1985), the southern Argolid (Pope and van Andel 1984), Nemea (Wright *et al.* 1990) and in Kammena Vourla (Niemi 1990). Two recent synoptic articles summarized the most important results of these projects with respect to soil erosion and redeposition in Greece (van Andel and Zangger 1990; Zangger 1992a), concluding that the first Post-glacial soil instability occurred time-

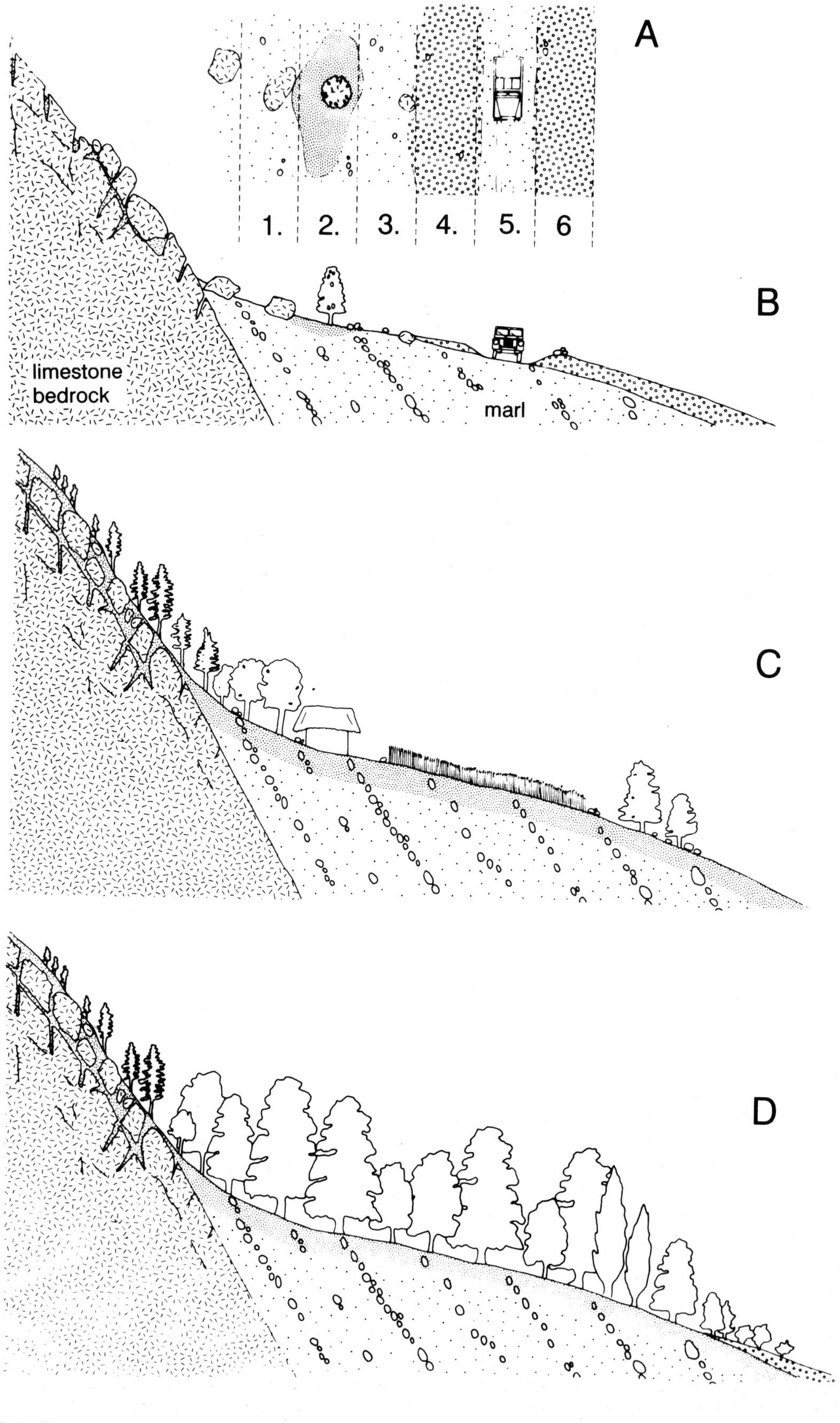


Figure 12.1 This diagram explains how an erosional landscape can be reconstructed by archaeological survey. See text for details.

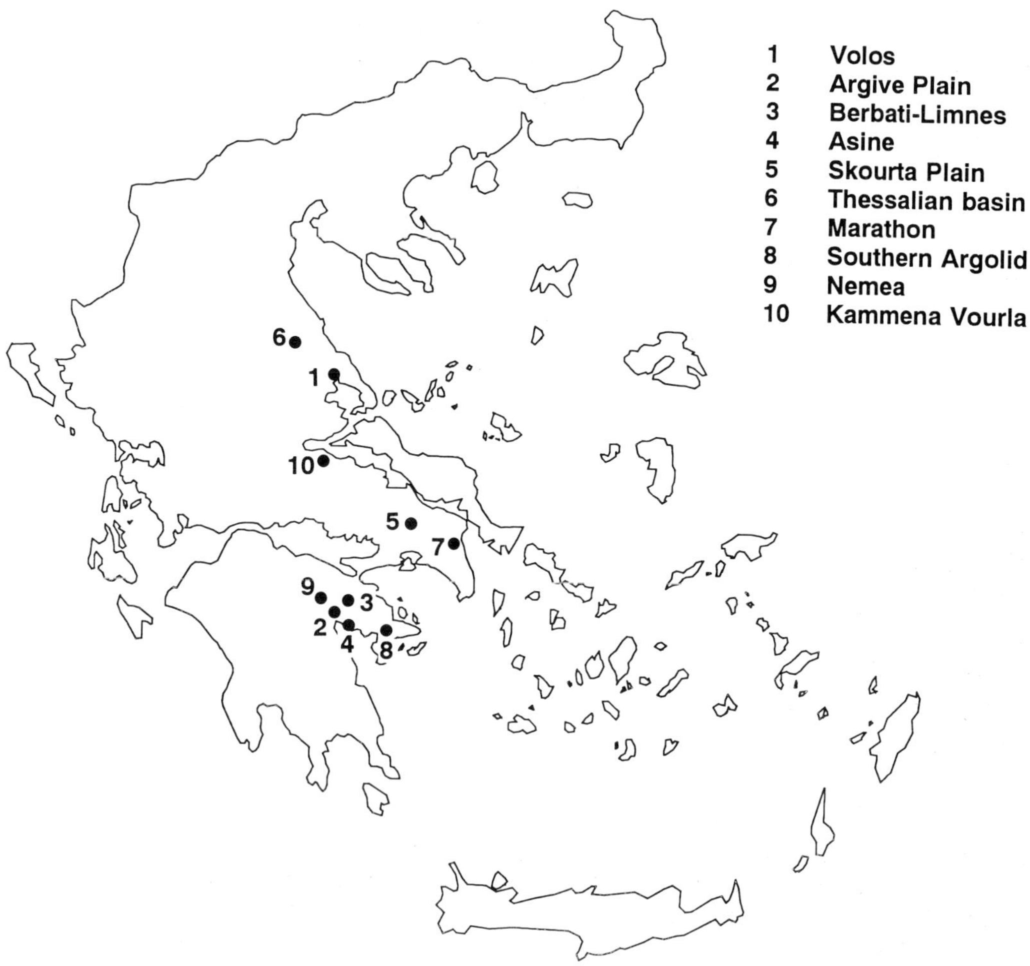


Figure 12.2 Locations of recent geoarchaeological projects in Greece.

transgressively in different parts of Greece. For instance, the shallow bay of Volos and parts of the Thessalian basin saw enhanced sedimentation during the Late Neolithic, while the Argive Plain and the Southern Argolid experienced soil erosion and redeposition mainly during the Early Bronze Age. The time transgressive occurrence of this initial phase of landscape instability follows the introduction of widespread agriculture in the respective areas with a few centuries delay. Thus it was concluded that human-induced clearance of the natural vegetation on meta-stable slopes represents the most likely trigger of the soil erosion. Regardless of the area of observation, this first Holocene phase of instability was also the most significant one in terms of quantity of eroded material. After the equilibrium had been lost, erosion occurred more frequently but with decreasing magnitude, since most fragile soil had already been washed away. The aim of the present publication is to illustrate these conclusions and to verify their validity in the light of new results obtained from an area that is characterized by soil erosion, the Berbati Valley.

Agriculture was introduced to Greece during the seventh millennium BC. For three to four thousand years, however, the farmers utilized only the most fertile and tillable soils in the plains leaving the hill slopes untouched. In remote areas of Greece, far away from human settlements, one can still today find examples of such largely undestroyed environments as they must have prevailed in the Neolithic. In these areas mature soil reminiscent of *Terra Rossa* and usually 20–100 cm thick covers the limestone upland providing the substrate for a healthy vegetation. In other places, where the equilibrium collapsed at some point in the past the mountains were stripped of their soil cover. Masses of eroded material were transported downhill and redeposited in topographically lower areas. Today, most Greek mountains are barren – with little or no soil cover left, they can only support a limited vegetation cover. The plains, on the other hand, are now filled with fertile alluvium often 5–10 m thick.

Soil erosion consequently caused the destruction of archaeological sites too, almost completely eroding settlements at higher elevations. Today, we can only discover vestiges of these sites without being able to tell how extensive they may have been and how many others may have disappeared completely (Cherry *et al.* 1988). At the same time the redeposition of the eroded material buried other sites in lower areas under several metres of alluvium, also concealing them from archaeological investigation. A Middle Neolithic site, marked by deposits rich in organic material, pottery fragments, charcoal and shells all dating to the middle of the sixth millennium BC, was found at 5–6 m depth in one auger core from the Argive Plain (Zangger 1992b). Such deeply buried or almost entirely eroded sites are chance finds and little can be said about the site pattern prior to the erosion.

Late Bronze Age residences in the plain of Argos, however, show thresholds level with today's surface. Thus, at least in this particular area, soil erosion and redeposition must have ceased by the end of the Bronze Age. While the above mentioned Middle Neolithic site at 5–6 m depth provides a maximum date for the alluviation in the Argolid, the Mycenaean houses built on top of the alluvium give a minimum date. Yet another settlement, dating to the Early Helladic II or Early Bronze Age II period at roughly 3000 BC, was found within the alluvium, thereby yielding a date for the landscape collapse itself.

Numerous indications supporting an Early Bronze Age date for the first widespread soil erosion were found during the work in the Argolid. The majority of sites and sherds dated to Early Helladic II, the time of the House of Tiles at Lerna and an impressive circular structure in Tiryns (Kilian 1986). Both buildings symbolize the existence of a thriving community. The Berbati and Nemea surveys in adjacent areas have shown a dramatic increase in population density during the Early Bronze Age (Wright *et al.* 1990; Wells *et al.* 1990). Furthermore, Professor Pullen from the Nemea project discovered the figurine of a draft animal dating to Early Helladic II pointing towards the importance of farming at that time (Pullen 1992). Although these indications of enhanced human exploitation are helpful, they do not provide solid evidence for man-induced land abuse. This evidence, however, comes from a pollen core that was taken from the swamps near Lerna during the course of the Argive Plain Project in 1987 and which has now been examined in Göttingen (Jahns 1990).

The core extended to the surface of the Pleistocene at 7 m depth, but the lowest 1.5 m did not contain pollen. The diagram starts at the time of the Post-glacial climatic optimum in the sixth millennium BC at 5.5 m depth. A characteristic of the lowest part of the pollen diagram is the prevalence of deciduous oak comprising an average 40% of the pollen material. Jahns presumes that oak forest covered higher elevations in the sixth and fifth millennium BC, while shrubs dominated the plain. Towards the end of the fifth and throughout the fourth millennium the climate became wetter or cooler enabling the oak forests to expand to lower elevations. At the same time non-tree pollen diminished pointing towards an unusually dense vegetation with few glades. The pollen diagram provides no indications for human interference during Neolithic times. Towards the middle and end of the fourth millennium BC, however, the vegetation changed rapidly and drastically. The percentage of deciduous oak dropped dramatically. Other species which thrive on cleared land (*Pinus, Erica* and *Carpinus orientalis*) increased pointing to human tree felling. Similar observations were made from pollen cores taken from Lake Kopais in central Greece (Turner and Greig 1975; Rackham 1983; Allen 1990). The beginning of the Bronze Age also saw an increase in olives but whether these were domesticated or wild olives which could have grown in the *maquis* following the land clearance cannot be established (Renfrew 1972; Runnels and Hansen 1986).

Deforestation by itself would probably not have sufficed to cause the extensive soil erosion which occurred in prehistoric Greece. Only the combination of tree-felling and expanding pastoralism prevented the vegetation from recovering. To illustrate the effects of pastoralism on individual plants Figure 12.3 shows a prickly oak which is one of the main *maquis* bushes in Greece usually about 1 m high. This particular plant, however, grew in the courtyard of a monastery where it was undisturbed for many centuries and where it could grow unrestrictedly to impressive size. Figures 12.5 and 12.6 show photographs taken (in Tunisia) on the same day at places only 10 km apart. The landscape in Figure 12.5 has been used for grazing, but the one in Figure 12.6 was abused by grazing.

If, however, soil erosion was triggered by local human land abuse, then one should be able to find places where no such extensive exploitation took place and where the landscape is still intact. Such a landscape can be found in the area around Skourta on the northern slopes of the Parness mountains between Boeotia and Attica (Figure 12.4). This area has few inhabitants and fields; hence, plenty of soil is left on the limestone supporting a healthy forest of deciduous and coniferous trees. For much of Greece's history the Skourta area was a borderland between opposing kingdoms (Munn and Munn 1989). It belonged to Attica although it is more accessible from Boeotia. Since the Boeotians were not allowed to utilize the land and the Athenians could not get to it, because it lies beyond the Parness mountains, the landscape was saved from human abuse. The extent of these forests can be traced on satellite images, showing that they roughly coincide with the boundary between the two states in Classical times. In Boeotian territory the

Figure 12.3 A prickly oak (*Quercus ilex*) in the yard of a monastery could grow unimpeded until it had reached twice the height of the nearby church

Figure 12.4 The slopes on the southern side of the Skourta Plain

plains were then, and are today, used for intensive agriculture. The Boeotian mountains are completely bare of vegetation and the same is true for Attica, with the exception of a protected no man's land in between.

In summary, the archaeological, geological and vegetational records in Greece point towards human induced soil erosion in prehistoric times. The onset of landscape instability varied from area to area depending on the cultural history. Places of intensive Neolithic cultivation often show extensive soil erosion accompanying human land use (Zangger 1991). Other areas experienced soil erosion at later stages; the Argive Plain, for instance, first became a major cultural centre in the Early Helladic II period and suffered soil loss at the same time. Again other areas in deserted mountainous regions, such as the Skourta Plain, may never have been over-exploited by man and therefore still possess intact soils and vegetation. Thus, there is no general pattern for environmental reconstruction. Each area must be examined individually.

### The Berbati Valley: a case study

To validate the above conclusions a new project was conducted in the hinterland of Mycenae, the Berbati Valley (Figure 12.7), which constitutes an exclusively erosional environment lacking recent deposits that could conceal prehistoric sites. Although the low preservational potential of the area reduced the chance of discovering prehistoric monuments or undisturbed settlements that would help date Holocene processes, the lack of extensive alluvium did provide an opportunity to examine a representative cross-section of the artifactual evidence from all periods of human occupation. In this respect the area contrasts markedly with the adjacent Argive Plain, where Holocene deposits are up to 8 m thick, often concealing prehistoric sites completely. Yet, the two areas are related, although inversely, because much of the redeposited material in the Argive Plain's coastal zone may have had its provenance on the slopes of the Berbati-Limnes hills. Such a reciprocal relation between two adjoining landscapes confirms the above conclusion that there is no standard answer to Holocene landscape reconstructions.

Geologically speaking the Berbati Valley is relatively stable; there are no indications of recent tectonic movements, no significant Holocene deposits and although the landscape is characterized by soil erosion, this process is by no means excessive. Because this landscape was sheltered from dramatic natural changes in the recent past it has a high potential for recording human disturbance. Places with little or no disturbance of surface deposits are extremely rare and incipient soils on Holocene sediments (Table 12.1) only occur in the immediate vicinity of the ephemeral stream at the Mastos, the only Mycenaean citadel in the valley, while mature but truncated Pleistocene soils are found on alluvial fans (Table 12.2).

At first sight erosional environments may seem of lesser appeal to geoarchaeological investigation than depositional ones. In reality, however, the interrelations in eroded landscapes between bedrock, topography, soils, soil erosion and human land-use are of foremost importance to the understanding of the historic man/landscape interdependence. In addition to reconfirming the picture of man-induced environmental destruction in prehistoric Greece, the study of the Berbati Valley illuminates the multifarious links between landscape evolution and its history of utilization. Site location, site preservation and land-use have been recognized as dependent on the geological setting (Wells *et al.* 1990). After the completion of the fieldwork the list of detected interrelations can be extended to include for instance (Figure 12.8):

- topography and bedrock
- bedrock and soil distribution
- soil distribution and land-use
- land-use and site location
- soil erosion and site preservation

Many other links exist, for instance, between water and building stone availability and site locations. The one parameter in this ecosystem that is independent of the others is bedrock – it determines topography, soils and human land-use. The overall character of the survey area is dominated by two markedly different geological units: (1) metamorphic limestone in the Limnes upland terrain in the eastern half reaching almost 800 m above sea level, (2) erodible flysch and marl in the depression of the Berbati Valley itself which has its lowest point at about 140 m.

1. The so-called Pantokrator limestone dominates much of eastern Greece including the Argolid and the Limnes uplands. It formed during the Lower Triassic to Middle Liassic when carbonate-rich sediments accumulated in a shallow-water basin. These terrains are exceptionally homogenous and monotonous with respect to topography, soils and vegetation.

2. The flysch, on the other hand, originated during a later phase of tectonic activity in the Upper Jurassic to Lower Cretaceous. It is rather diverse and intercalated with marls, sandstones and conglomerates. The flysch and marl terrains are infinitely more suited to agriculture. In addition, their interlayered conglomerate and sandstone bodies offer valuable building stone resources and effective foundations for construction. Thus, the twofold division of the bedrock in the survey area is likely to be reflected in differences in land-use too.

Some of the interrelations detected in the study area include:

#### *1. Bedrock/Topography*

The most obvious effect of the distinct bedrock geology is a twofold division of the topography of the

Figure 12.5 Vegetation in the Zaghouan Valley, northeast Tunisia, used for husbandry without massive damage.

Figure 12.6 Vegetation ten kilometres from the area shown in Figure 12.5, this time hampered by overgrazing.

| Depth (m) | Horizon | Munsell Colour (wet) | | Munsell Colour (dry) | | Texture | Structure | Consistency | Miscellaneous |
|---|---|---|---|---|---|---|---|---|---|
| 0.0–0.1 | A | 10YR 4/2 | dark greyish brown | 10YR 4/2 | dark greyish brown | silt loam | weak, very fine angular blocky | hard, sticky | incipient soil, very homogeneous deposit, no stratigraphy, no sherds, no human impact |
| 0.1–0.3 | B1 | 10YR 5/4 | yellowish brown | 10YR 6/4 | light yellowish brown | silt loam | weak, fine subangular blocky | hard, very sticky | |
| 0.3–1.9 | C | 2.5YR 5/4 | light olive brown | 2.5YR 7/2 | light grey | silt | none | hard, slightly sticky | violently effervescent |

Table 12.1 Soil profile (BV–SP–09) in a scarp near the dry stream a few hundred metres west of the Mastos in the Berbati Valley exposing almost 2 m of unstratified recent alluvium. One of the few places where material has been deposited, not eroded.

| Depth (m) | Horizon | Munsell Colour (wet) | | Munsell Colour (dry) | | Texture | Structure | Consistency | Miscellaneous |
|---|---|---|---|---|---|---|---|---|---|
| 0.0–0.2 | Ap | 7.5YR 3/4 | dark brown | 10YR 4/4 | dark yellowish brown | clay loam | weak, medium granular | weakly coherent, slightly sticky | violently effervescent |
| 0.2–0.3 | B1 | 7.5YR 3/4 | dark brown | 7.5YR 4/4 | dark brown | silty clay loam | weak, coarse granular | weakly coherent, slightly sticky | strongly effervescent |
| 0.3–0.7 | B2k | 5YR 4/4 | reddish brown | 7.5YR 4/6 | strong brown | silty clay loam | moderate, fine-coarse angular blocky | very hard, slightly sticky | strongly effervescent, lime in filaments |
| 0.7–1.0 | B2t | 5YR 4/4 | reddish brown | 5YR 3/4 | dark reddish brown | silty clay loam | moderate, subangular blocky | very hard, sticky | strongly effervescent some thin clay films |

Table 12.2 Soil profile (BV–SP–07) in a road scarp a few hundred metres south of the Roman bath in the Berbati Valley showing soil profile on a Pleistocene alluvial fan.

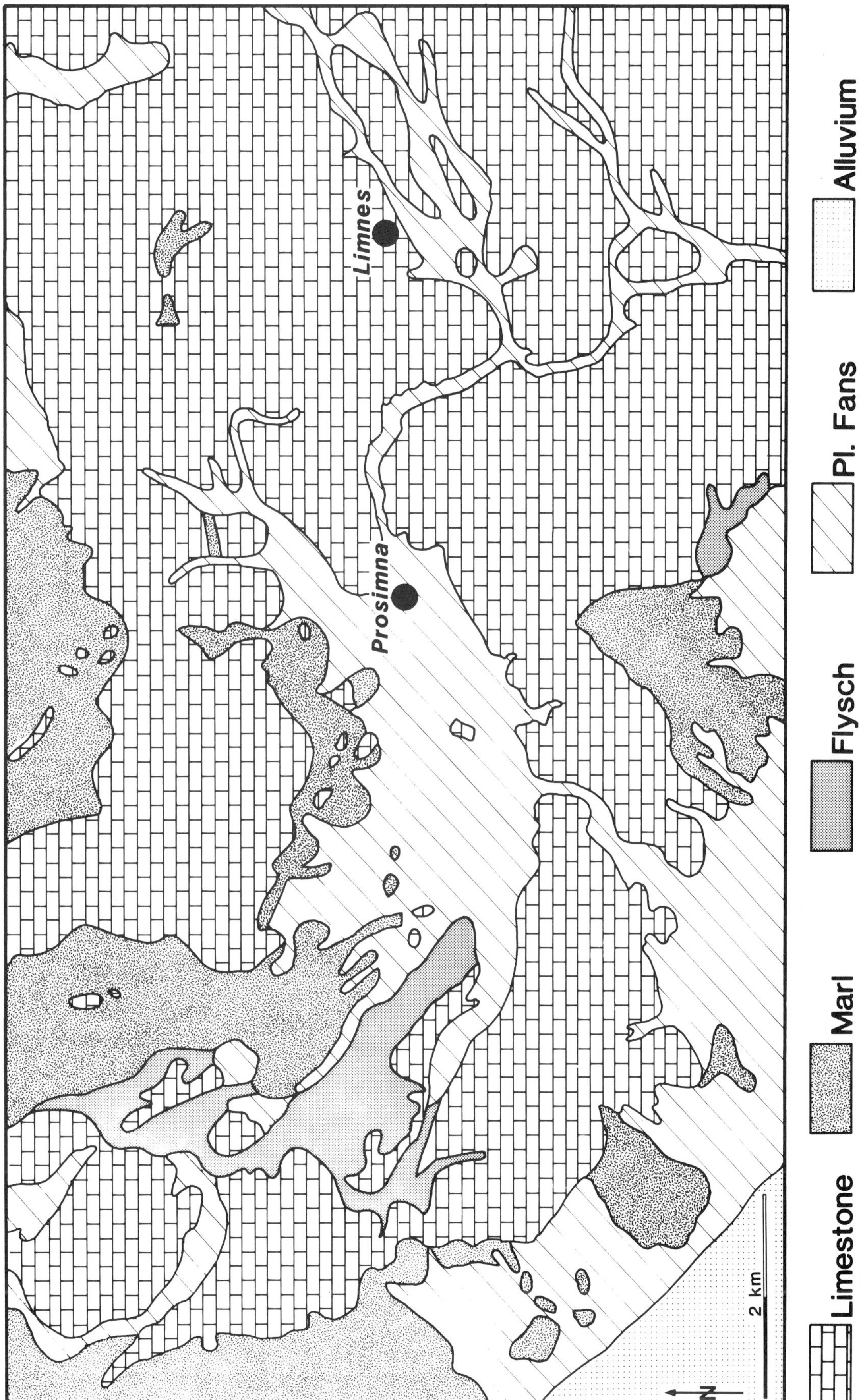


Figure 12.7 Bedrock distribution in the Berbati Valley according to the geologic map of Nauplion. Mesosoic limestone forms the Limnes uplands in the east and deeply eroded marl and flysch around Prosimna form the Berbati Valley. Prosimna is the modern name for the village which used to be called Berbati.

survey area. Because limestone withstands erosion effectively, it tends to form ridges often protruding several hundred metres above the floor of the valley, whereas the softer marl and flysch units form topographic depressions. The most sizable of these is the Berbati Valley itself. These physiographic differences – even without considering their impact on soil and soil erosion – have consequences for the human land-use, for instance, because olives do not grow at high elevations.

*2. Bedrock/Soil*

Even more important for agriculture are the quality and quantity of soil which again depend on the bedrock, considering that the climate is identical throughout the study area. The limestone terrains generally develop a clay-rich, dark red soil reminiscent of *Terra Rossa.* This soil forms exceedingly slowly; it is thin, highly fragile, but unusually fertile. Numerous limestone boulders within the soil often render it not tillable. Furthermore, limestone terrains usually include steep slopes, reducing the value of the land for agriculture. Soils on marl and flysch, on the other hand, form more quickly and are deeper than limestone soils, because precipitation and temperature fluctuations penetrate deeper into these less consolidated rocks. The tillage properties of the marls in the study area, though exceeding those of limestone soils, are still not excellent, because of intercalated layers of boulders and pebbles. Finally, marl and flysch terrains usually form gentle slopes which are quite suitable for agriculture.

*3. Soil erosion/Topography/Soil*

The level surface of softer bedrock areas is a sign of ongoing erosion. Wherever erosion takes place soils will not form, unless soil genesis exceeds the rate of erosion which is certainly not the case in the Greek dry climate where a mature soil probably requires a few thousand years to form even on soft rock like marl. Limestone soils are eroded because they are thin, fragile, often on steep surfaces and generally underlain by a flat interface to the unweathered bedrock. Marl and flysch soils, on the other hand, erode mainly because they rest on unstable substrates. Thus, soil distribution and erosion are attributes of a highly fragile ecosystem whose chief determining components are bedrock and climate. An undisturbed natural vegetation, however, might be able to provide a metastable equilibrium within this system; although it will not prevent erosion in general, for example through stream incision, it will reduce the potential of *soil* erosion.

*4. Settlement and land use*

Wherever people utilize a natural environment some particular characteristics of this ecosystem will be more important for them than others. The presence of fertile, arable soil and water, shelter from wind, protection from the sun or exposure to it, availability of building stones, strategic control from elevated heights, remoteness and inaccessibility – all these attributes of the natural landscape have influenced people's site choice and land-use at one time or another. These interrelations between human land-use and natural environment and their fluctuations with time are evident from the archaeological record of the Berbati Valley.

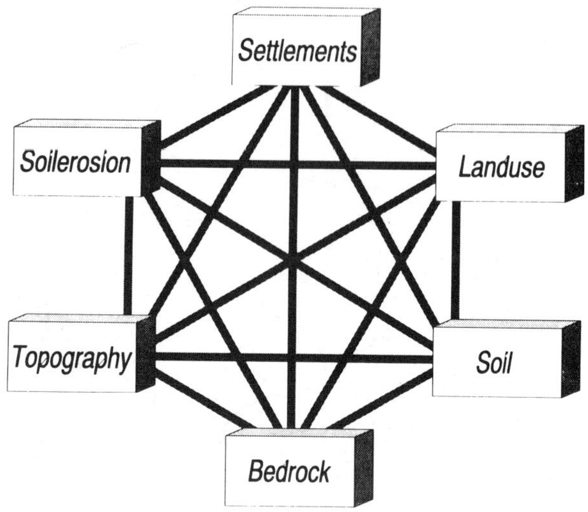


Figure 12.8 Interrelations between different components of the environment as recognized during the study of the Berbati Valley.

The recently concluded Berbati-Limnes Archaeological Survey (Wells *et al.* 1990) has shown that the first human occupation of the Berbati Valley occurred during the Middle Palaeolithic when temporary visitors entered the area. A contemporary cave site at Kefalari in the Argive Plain may well have served as the permanent domicile of these people. For their unprotected open-air sites they only required access to perennial rivers and springs. Similar open-air camps of occasional visitors existed during the Mesolithic in the Kleissoura gorge, the entrance to the Berbati Valley, while the valley itself was apparently left undisturbed and only used as a hunting ground until the Middle Neolithic (*c.* 5500–5000 BC), when agriculture was introduced to the valley. The first farmers utilized different resources of the natural environment compared to those used by Palaeolithic and Mesolithic hunters and although they were again looking for access to water, among their prime concerns was deep, arable soil. Thus, agricultural communities arose near the centre of the valley at the Mastos and at Findspot 400 near Ayios Nikolaos.

Figure 12.9 First generation terraces on abandoned slopes between the Berbati Valley and the Argive Plain (east of Monastiraki). The terraces themselves are not visible from a distance, only the lines of shrubs whose roots provide additional support for the soil.

Figure 12.10 These terraces in the Limnes uplands belong to the second generation of terrace construction. They have been erected during the last 200 years and are still maintained today.

Figure 12.11 Detail of the terraces in Figure 12.9 showing the size of the boulders. Some of the rocks are up to 60 cm in diameter and because of their weight cannot have been brought there by individual farmers.

During the Late and Final Neolithic a prosperous village farming economy had been established on grain agriculture while the thriving settlements expanded into the higher plateaus. The increasing size of population and settlements, however, required clearing of so far undisturbed land for farming, including the bedrock slopes surrounding the valleys. The deciduous oak forests which initially covered these limestone hills were cleared to produce more grazing and farming land; the soil lost the protection from splash erosion and the support of roots and as a consequence it was washed away quickly.

During the Late Bronze Age when the Argive nobles established their residences on strategically elevated heights, such commanding positions began to play a role for site locations in the Berbati-Limnes area too. Nevertheless, habitation on lower ground continued in the form of small settlements, possibly farmsteads, on fertile land. The Mycenaean demise, however, brought a total abandonment of the area until the later Geometric period when it was resettled again. Within a few centuries, the population expanded dramatically to reach a high density in Classical times when even the remotest arable land was being exploited. At that time, however, terraces appear to have been in place to stabilize the soil and to prevent the kind of erosion which the land has suffered during the initial cultivation.

One of the most puzzling geoarchaeological problems in the Berbati and Limnes area are the agricultural terraces. They can be divided into two groups. Firstly, those still maintained today which tend to be 1–2 m high, 1–2 m wide and 10–20 m apart (Wells *et al.* 1990). They are typically found on limestone surfaces around Limnes and consist of relatively small boulders which can be carried by one person (Figure 12.10). The other generation of terraces dates back to an earlier time when every surface but limestone was artificially stabilized. Many of these former fields are found on the slopes surrounding the Berbati Valley which are no longer used for agriculture (Figure 12.9). These terrace walls, made of large boulders up to 60 cm in diameter (Figure 12.11), are often buried or overgrown by dense brush.

While the younger group of terrace walls dates to the last 200 years the age of the earlier one is unknown. It appears logical that they must have been erected during a time of dense occupation when even the remotest soils had to be utilized for agriculture. Such periods of high population density existed in the Berbati-Limnes area only during the Final Neolithic/ Early Helladic, the Mycenaean and the Classical periods. The first of these possibilities can be

Figure 12.12 The only place where first generation terraces are still maintained in modern (second generation) style. A whole sequence of walls has been erected in this area to stabilize a steep slope along the boundary of limestone and flysch.

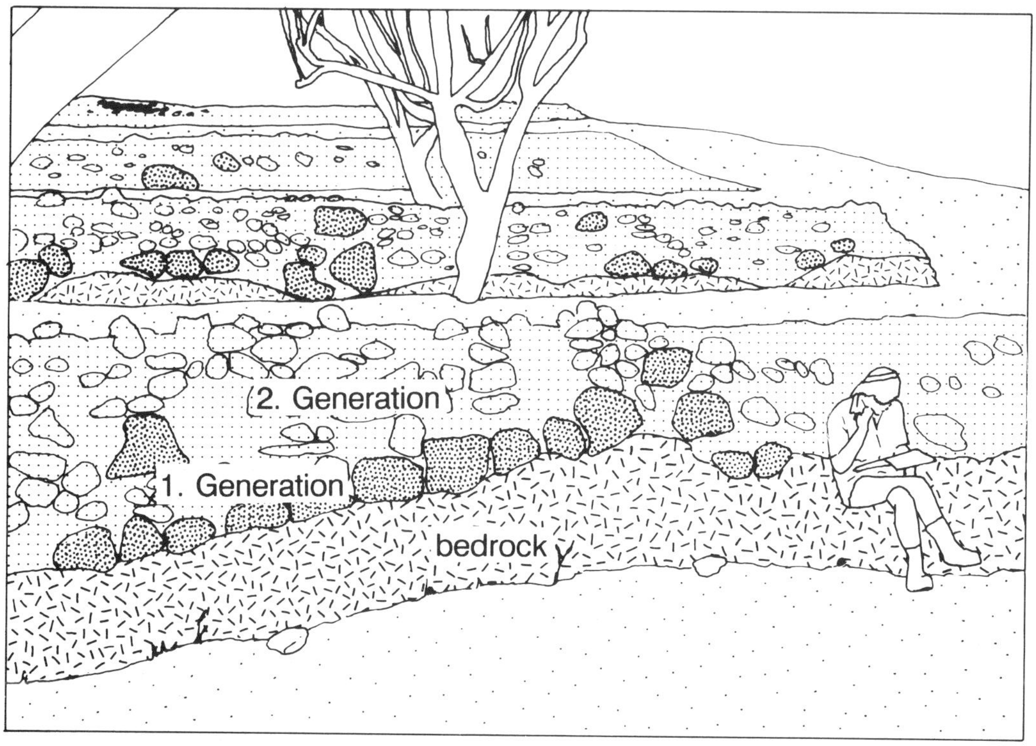


Figure 12.13 This sketch of the photograph in Figure 12.12 highlights the difference between first generation terraces consisting of large boulders and second generation terraces made up of smaller pebbles.

disregarded as a time of terrace construction, because the soil erosion in the Argive Plain post-dates this period and such large scale destabilization could not have occurred if the fields had already been protected by terraces. Thus, the Mycenaean and Classical periods are left as the most likely periods for terrace construction. The uniform character of the walls gives the impression that they had been erected during a relatively short time, possibly in some kind of community effort, since the boulders are too large to be handled by individual farmers. Furthermore, some ancient terraces had fertile top soil dumped behind them, which again points to a major and well-planned construction effort (Wells *et al.* 1990). Unfortunately, we are still unable to determine the date of this construction from the terrace walls themselves but one might be able to date them in the archaeological context.

Only in one remote area relatively high up in the mountains between Berbati and Limnes are terraces of the older generation combined with the later generation (Figures 12.12 and 12.13). In this area, the high relief between a limestone plateau and a pass of flysch has been very effectively stabilized by a sequence of up to 2–3 m high terraces. Some vertical bedrock faces appear at the bottom of the walls, which begin with a layer of large boulders, characteristic of the first generation terraces. The upper half of these walls, however, consists of modern-style terraces made up of small boulders. The whole system is evidently still maintained today. Sherds found in the area only date to the Classical period which therefore may well have been the date of the terrace construction. In fact, similar large-scale terracing projects from the Classical period have recently been described from southern Attica, where no artifacts from any other period are present (Lohmann 1985), thus at least the terraces in Attica are almost certainly of Classical age.

### Conclusions

Population density in the Berbati-Limnes landscape has been low during the last few centuries and it was probably never much higher in earlier times. Because of the relative landscape fragility, however, the continuous disturbance of the natural environment by people has resulted in some profound changes, causing a decrease of soil quality, quantity and natural vegetation. As a consequence surface water has disappeared from an area where water mills were part of the scenery until the end of Turkish occupation. The groundwater level is sinking at an exponential speed today, due to the over-exploitation by farmers mainly from the eastern Argive Plain where groundwater has virtually disappeared. Soon, the human abuse of the natural resources in the Berbati Valley may once again, after the disastrous effects of deforestation in the Early Bronze Age, surpass nature's tolerance, causing unforeseen and potentially irreversible damage to the environment.

The study of the Berbati Valley has shown how different aspects of the natural environment and its utilization by man depend on each other and how these can be usefully determined employing an holistic approach. Like all other current palaeo-environmental studies in Greece it has demonstrated that landscape evolution cannot be reconstructed with a search for over-simplified model-oriented processes in mind. Although similarities and interrelations between different areas do exist, most land-forming processes are extremely complex and sometimes even inversely related. Hence, environmental evolution must be viewed as an intricate, multi-dimensional system with many interdependent parameters – one of which is the history of human land use.

### References

Allen, H. 1990 A Postglacial record from the Kopais Basin, Greece. In: Bottema, S., Entjes-Nieborg G. and van Zeist W. (eds.) *Man's Role in the Shaping of the Eastern Mediterranean Landscape.* Balkema, Rotterdam, 173–182.

Baeteman, C. 1985 Late Holocene Geology of the Marathon Plain (Greece). *Journal of Coastal Research* 1, 173–185.

Bintliff, J. 1977 Natural environment and human settlement in prehistoric Greece. *British Archaeological Reports*, Supplementary Series 28, Oxford.

Cherry, J. F., Davis, J. L., Demitrack, A., Mantzourani, E., Strasser, T. and Talalay, L. 1988 Archaeological Survey in an Artifact-Rich Landscape: A Middle Neolithic Example from Nemea, Greece. *American Journal of Archaeology* 92, 159–176.

Demitrack, A. 1986 *The Late Quaternary Geologic History of the Larissa Plain (Thessaly, Greece) – Tectonic, Climatic and Human Impact on the Landscape.* Ph.D. dissertation at Stanford University, University Microfilm International, Ann Arbor, Michigan.

Finke, E. 1988 *Landscape Evolution of the Argive Plain (Greece): Paleoecology, Holocene Depositional History and Coastline Changes.* Ph.D. dissertation at Stanford University, University Microfilm International, Ann Arbor, Michigan.

Jahns, S. 1990 Preliminary notes on human influence and the history of vegetation in S. Dalmatia and S. Greece. In: Bottema, S., Entjes-Nieborg, G. and van Zeist, W. (eds.) *Man's Role in the Shaping of the Eastern Mediterranean Landscape.* Balkema, Rotterdam, 333–340.

Kilian, K. 1986 The Circular Building at Tiryns. *Studies in Mediterranean Archaeology* 76, 65–71.

Lohmann, H.-G. 1985 Landleben im klassischen Attika – Ergebnisse einer archäologischen Landesaufnahme des Demos Atene. *Jahrbuch der Rur-Universität Bochum*, 71–96.

Munn, M. H. and Zimmermann Munn, M. L. 1989 Studies on the Attic Boiotian Frontier: The Stanford Skourta Plain Project, 1985. *Boeotia Antiqua* 1, 73–127.

Niemi, T. M. 1990 Paleoenvironmental History of

Submerged Ruins on the Northern Euboean Gulf Coastal Plain, Central Greece. *Geoarchaeology* 5, 323–347.

Pope, K. O. and van Andel, T. J. 1984 Late Quaternary alluviation and soil formation in the southern Argolid: its history, causes and archaeological implications. *Journal of Archaeological Science* 11, 281–306.

Pullen, D. J. 1992 Ox and Plow in the Early Bronze Age Aegean. *American Journal of Archaeology* 96, 45–54.

Rackham, O. 1983 Observations on the Historical Ecology of Boeotia. *The Annual of the British School in Athens* 78, 291–351.

Renfrew, C. 1972 *The Emergence of Civilization – The Cyclades and the Aegean in the Third Millennium BC.* Methuen, London.

Runnels, C. and Hansen, J. 1986 The Olive in the Prehistoric Aegean: the evidence for domestication in the Early Bronze Age. *Oxford Journal of Archaeology* 5, 299–308.

Turner, J. and Greig, J. R. A. 1975 Some Holocene pollen diagrams from Greece. *Review of Paleobotany and Palynology* 20, 171–204.

van Andel, T. H., Zangger, E. and Demitrack, S. 1990 Land use and soil erosion in prehistoric and historic Greece. *Journal Field Archaeology* 17, 379–396.

van Andel, T. J. and Zangger, E. 1990 Landscape stability and destabilisation in the prehistory of Greece. In: Bottema, S., Entjes-Nieborg G. and van Zeist, W. (eds.) *Man's Role in the Shaping of the Eastern Mediterranean Landscape.* Balkema, Rotterdam, 139–157.

Vita-Finzi, C. 1978 *Archaeological Sites in their Setting*. Thames and Hudson, London.

Wells, B., Runnels, C. and Zangger, E. 1990 The Berbati-Limnes Archaeological Survey – the 1988 season. *Opuscula Atheniensa* 18, 207–238.

Wright, J. C., Cherry, J. F., Davis, J. L., Mantzourani, E., Sutton, S. and Sutton, R. F. 1990 The Nemea Valley Archaeological Project – a preliminary report. *Hesperia* 59, 579–659.

Zangger, E. 1991 Prehistoric coastal environments in Greece: the vanished landscapes of Dimini Bay and Lake Lerna. *Journal of Field Archaeology* 18, 1–15.

Zangger, E. 1992a Prehistoric and historic soils in Greece: assessing the natural resources for agriculture. *Symposium Acta*, Swedish Institute in Athens (in press).

Zangger, E. 1992b *Geoarchaeology of the Argolid.* Deutsches Archäologisches Institut, Athens (in press).

# CHAPTER 13

# A SURVEY OF SOIL EROSION IN VIOTIA, GREECE

D.A. DAVIDSON AND S.P. THEOCHAROPOULOS

*This paper reports the preliminary results of a land evaluation project in Viotia, Greece which makes considerable use of a GIS (Geographical Information System). Data on soil erosion were collected as part of a 1:100,000 land resource survey for the agricultural area. Assessment of erosion classes on agricultural land was based on the extent to which subsurface horizons have become exposed as a result of past and present erosional processes. For the mapped part of Viotia, soil erosion is not evident on 52.5% of the area. The incidence of erosion is found to correlate closely with slope angle; in addition entisols and inceptisols suffer from greater erosion. The practical importance of soil erosion and land suitability maps in assisting with the formulation of sustainable land use systems is highlighted.*

Although there is frequent reference in the literature to the magnitude and extent of soil erosion in Greece, both in the past and in the present, there is a dearth of survey data on the incidence of such erosion. In archaeological studies in Greece, the importance of investigating the nature of landscape change has long been recognised; as an example Wagstaff and Cherry (1982) incorporate physical landscape change into their analysis of settlement evolution from prehistory to the present on the island of Melos. The essential changes are hillslope erosion, basin and valley alluviation and coastal sedimentation though it must be stressed that there has been much controversy regarding chronology, spatial patterns and causal factors (Davidson 1980; Wagstaff 1981). The importance of land resource data for Greece is now being realised given the land use objectives within the European Community of agricultural diversification and selection of optimum land use systems to ensure sustainability and environmental quality. Of particular concern in a Mediterranean environment is the susceptibility of soils to degradation, especially to erosion.

Yassoglou (1987) provides ratings for sensitivity to degradation for the main soil types of southern Europe. He argues that it is important to evaluate the impact of land degradation processes, not only on single soil properties, but also on the performance of soil systems. His plea is thus for more data on such systems as well as on rainfall erosivity, soil loss/yield relationships and the impact of irrigation on structural change and erosion. A project also dealing with the prediction of soil erosion at the Mediterranean scale is described by Giordano *et al.* (1991). This investigation is within the CORINE programme of the European Community and has the objective of providing first estimates for soil erosion risk and land quality at the European level. The method is based on the principles of the Universal Soil Loss Equation whereby scores for erodibility, erosivity, slope angle and soil cover are multiplied to provide an assessment of actual soil erosion risk. The scale of analysis is 1:1,000,000 since the soil data comes from the 1:1,000,000 Soil Map of the European Communities. The results when available will be of value for strategic planning at the EC level. There are substantial problems in translating the results of land resource analysis from one scale to another and thus there is no substitute for field survey at more detailed scales (e.g. 1:50,000 or 1:100,000) if regional assessments are required and there is a lack of primary data. Such a strategy had to be adopted in Viotia, Greece.

A land resource survey by the Forestry Research Institute has been underway in Greece for some years (Nakos 1983). This project was initiated in the mountainous and hilly areas and using a landscape ecological approach, land mapping units (associations) at 1:50,000 are characterised according to climate, vegetation, landforms, soil depth, soil erosion, slope and aspect. The approach is based on aerial photo interpretation followed by fieldwork. Land resource maps and land capability maps for forestry have been produced though they are not available for purchase. Erosion is assessed in three classes (none, moderate and severe) though mapping units are categorised according to the dominant and subdominant erosion classes. In the Eurytania project area of central Greece, Nakos (1983) found that the dominant erosion description was none and moderate covering 47.5% of the area.

Soil survey of agricultural areas is one responsibility of the Soil Science Institute of Athens. The aim of this paper is to report some preliminary results on the incidence of erosion in one region of Greece (Viotia) on the basis of a collaborative research project between the Soil Science Institute of Athens and the University of Stirling. Overall, the objective is to develop a land resource information system for a Mediterranean environment using GIS technology (Davidson *et al.* 1990). In detail the aims are to produce single factor and land suitability maps for specific crops. In this paper the results of soil erosion mapping are presented and spatial correlations with

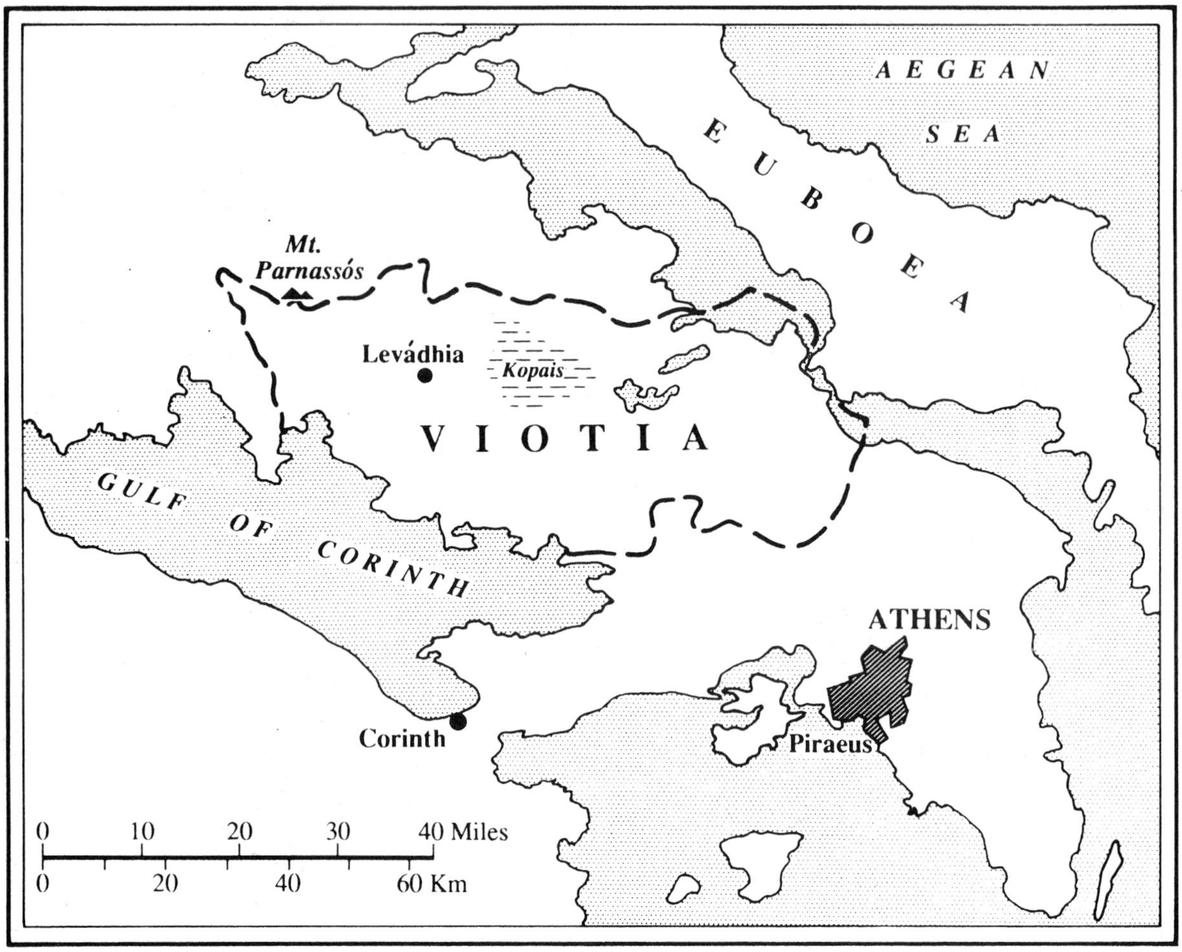


Figure 13.1 Location of Viotia to the north west of Athens

selected environmental attributes are investigated. The Soil Science Institute of Athens has been responsible for the field survey and the subsequent laboratory analysis; GIS processing is undertaken by the Department of Environmental Science, University of Stirling.

The selected study area is Viotia to the immediate north-west of Athens (Figure 13.1). Viotia is typical of many basins in Greece - an extensive plain mantled with Pleistocene and Holocene sediments and surrounded by uplands and mountains formed of Tertiary and Cretaceous limestones and flysch (sediments ranging from clays and marls to sandstones and conglomerates). An outstanding feature of the plain is the reclaimed Lake Kopias. Before reclamation early this century, Kopias was a vast reedy lake in winter whilst in summer it became a marshy plain. Former areas of peat have disappeared through cultivation, fire and shrinkage; an overall lowering of 3.50 m has been estimated on the former lake bed between 1882 and the present. Entisols and gleys occur on the former lake sediments whilst inceptisols and vertisols are associated with the fringing terraces and fans. Alfisols are found at higher levels, either on the upper stable terraces or on the upland plateau areas whilst entisols are dominant on the colluvial areas and steeper slopes.

The soils of Viotia were mapped by the free soil survey approach with emphasis being given to aerial photo interpretation and subsequent field checks and sampling. Such an approach proved particularly suitable in Viotia given the clarity of soil-landform assemblages; for example the former lake area and the sequence of coalescing alluvial fans round the perimeter of the plain are very distinct. Mapping was also restricted to agricultural areas. A database on soil mapping units and on soil sampling points has been established and integrated with the GIS. Within the main mapping units, at least one full soil profile description was obtained as well as subsequent analytical data. Sites of brief inspection using an auger were also recorded. In summary, the data produced by the survey for each mapping unit can be listed as follows:

Polygon type (e.g. mapped, not mapped, sea)
Drainage (classes)
Texture (classes and for 3 depths)
Gravel
Slope (classes)
Erosion (classes)
Calcium carbonate (classes)
Soil order, suborder and great group
Irrigibility
Variability class and limitations
Mean annual rainfall

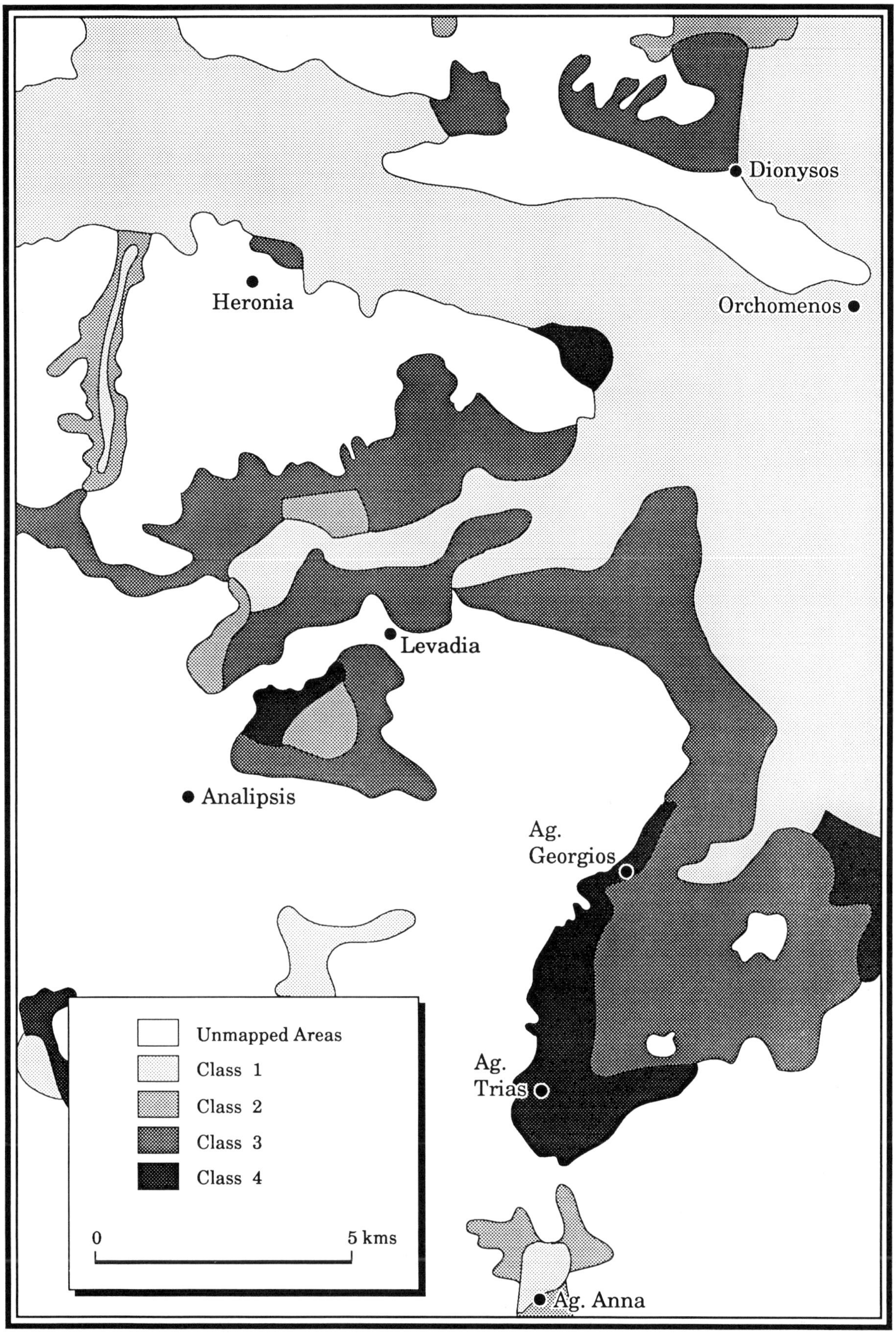


Figure 13.2 Distribution of soil erosion classes in the vicinity of Levadhia.

For the mapping units a code was entered for each of these properties and thus a decision had to be made during field survey on the dominant category. In effect each mapping unit had an associated range of labels for the various attributes. An inevitable problem was variability within mapping units; this was tackled by assigning a variability class (from 1 to 4) to each mapping unit with the most variable properties being indicated by the use of letters (for example, 3t to indicate class 3 variability with variation in topography being dominant). A summary description of erosion classes is given in Table 13.1. Surface wash and rill erosion are the dominant processes on agricultural land with gullying very evident in particular upland areas. Assessment of the erosion classes on agricultural land was based on the extent to which subsurface horizons had become exposed as a result of past and present erosion. Irrigibility was assessed according to the availability of water for irrigation whilst mean annual rainfall was predicted on the basis of elevation-rainfall regression results. In addition to digitizing the land mapping units and associated attribute codes, geological and land use maps were also input.

| Erosion Class | Extent in Lowland Area % |
|---|---|
| 1 | 52.5 |
| 2 | 12.7 |
| 3 | 26.5 |
| 4 | 8.3 |
| 5 | 0.0 |

Explanation of Erosion Class

1. None
2. Subsurface horizon exposed in <30% area
3. Subsurface horizon exposed in >30% area
4. All of A horizon removed
5. Most of profile removed

Table 13.1 Definition and extent of soil erosion classes in Viotia

The land resource survey also generated site data from individual sampling points. Soil profile descriptions have not been coded, but the following analytical results have been input:

Sand, silt and clay
Electrical conductivity
Salt
pH
Calcium carbonate
Organic matter
Cation exchange capacity

For input and processing of the Viotia data, use was made of the Laser-Scan software running on a microVAX II with a Sigmax 6000 graphics terminal. In the digitizing module, line segments were digitized for mapping units, contours and coastline; attribute codes for the variables listed above were added to each seed point within each mapping unit. A similar approach was also adopted for the input of attribute codes for sampling points (soil analytical data). Following digitizing of mapping unit segments, the vector 'spaghetti' had to be put through geometry idealisation and structure formation to create polygons which had labels taken from seed point data.

As already mentioned, field survey included the assessment of soil erosion for each mapping unit. Five categories of erosion were defined though only four occurred within the mapped area of Viotia (Table 13.1). With the use of a GIS it is a simple matter to select one attribute code such as erosion and to plot the results; a sample extract is shown in Figure 13.2 for the area round Levadia. Mapping was limited to agricultural areas. The publication of a soil erosion map for Viotia at 1:100,000, the first of its kind for Greece, is one important output from the project. It is no surprise to find that much of the Kopias basin is distinguished by the absence of soil erosion; the same applies to central and very gently sloping land in smaller enclosed basins as at Kallithea, Ellopia and Thisvai. Areas of moderate and severe erosion (classes 3 and 4) predominate on the southern foothills of the Kopias basin extending from Levadia in the west (Figure 13.2) to Thivai in the east. Such erosion is also dominant on the more rolling landscapes of south-eastern Viotia round Kallithea. Moderate and severe erosion is also very evident in enclosed upland basins as at Antikira and Arachova. The spatial extents of the erosion classes are given in Table 13.1. Half the mapped area or about a quarter of the whole of Viotia displays no erosion. About a quarter of the mapped area is in erosion class 3 (moderate: subsurface horizon exposed in >30%). Such figures provide a very useful summary statement on the magnitude and extent of soil erosion in this rather typical area of mainland Greece.

The use of a GIS facilitates analysis to determine if the incidence of soil erosion in the mapped area of Viotia can be related to slope angle, soil order and surface texture. As is to be expected, there is a close association between slope angle and the incidence of soil erosion (Table 13.2). Land with slopes of 3% or less is distinguished by the lack of soil erosion; almost half the land between 3 and 6% suffers from erosion class 2 with about three quarters in erosion class 3 for land between 6 and 12%. For the slope range between 12 and 18%, soil erosion is equally divided between erosion classes 3 and 4; land steeper than 18% is dominated by erosion class 4.

The occurrences of soil erosion class 1 are associated with long established and stable soils - vertisols and alfisols (Table 13.3). Such soils also occur on areas of very gentle slopes. In contrast, the younger soils, inceptisols and entisols, display much greater variability in incidence of erosion. However, it

| Slope Classes (%) | Erosion Classes 1 (%) | 2 (%) | 3 (%) | 4 (%) |
|---|---|---|---|---|
| 0–3 | 99.5 | 0.5 | 0.0 | 0.0 |
| 3.1–6 | 15.8 | 47.6 | 30.6 | 6.1 |
| 6.1–12 | 1.5 | 6.2 | 78.6 | 13.7 |
| 12.1–18 | 0.0 | 0.8 | 48.7 | 50.4 |
| 18.1–25 | 0.0 | 0.0 | 16.7 | 83.3 |
| 25.1–35 | 0.0 | 0.0 | 0.0 | 100.0 |
| Totals | 52.5 | 12.7 | 26.5 | 8.3 |

Table 13.2 Incidence of soil erosion according to slope

| Soil Orders | Erosion Classes 1 (%) | 2 (%) | 3 (%) | 4 (%) |
|---|---|---|---|---|
| Vertisols | 98.2 | 1.8 | 0.0 | 0.0 |
| Alfisols | 90.6 | 8.9 | 0.5 | 0.0 |
| Entisols | 39.5 | 10.7 | 33.7 | 16.1 |
| Inceptisols | 35.6 | 25.4 | 38.6 | 0.4 |
| Totals | 52.5 | 12.7 | 26.5 | 8.3 |

Table 13.3 Incidence of soil erosion according to soil orders

| Texture Groups | Erosion Classes 1 (%) | 2 (%) | 3 (%) | 4 (%) |
|---|---|---|---|---|
| SCL, CL, SiCL | 35.3 | 11.4 | 42.4 | 11.2 |
| SL | 32.1 | 27.0 | 40.9 | 0.0 |
| SC, SiC, C | 68.4 | 13.9 | 11.4 | 6.3 |
| SiL, Si, L, vfSL | 53.6 | 3.9 | 33.3 | 9.2 |
| Totals | 52.5 | 12.7 | 26.5 | 8.3 |

SCL – sandy clay loam
CL – clay loam
SiCL – silty clay loam
SL – sandy loam
SC – sandy clay
SiC – silty clay
C – clay
SiL – silty loam
Si – silt
L – loam
vfSL – very fine sandy loam

Table 13.4 Incidence of soil erosion according to texture of topsoil

should be noted that 40% of entisols and 36% of inceptisols exhibit no erosion. The pattern between soil erosion and texture of topsoil is similarly varied (Table 13.4). As is to be expected, the heavier textured clay soils are dominated by erosion class 1, but no other correlations are evident from Table 13.4. It is likely that the variations in incidence of erosion within areas mapped as entisols and inceptisols can be attributed primarily to slope differences; the same explanation can also be applied to variations with the different types of loams. The influence of calcium carbonate content on erosion was also investigated, but no correlations were obtained.

It is unlikely that land resource surveys as carried out in Viotia will be done for many other regions of Greece in the immediate future. Thus there is the need to be able to predict soil erosion hazard from readily available data. The simplest approach is to base such predictions on slope measures; the validity of such an approach is demonstrated in the results in Table 13.2 for Viotia, and slope values can be determined from digital terrain models. An alternative approach is to determine soil erosion risk using the methodology described by Giordano *et al.* (1991) though such a model requires data on soil texture, depth, slope, stoniness and mean monthly precipitation and temperature. For Viotia, this model was applied to data for selected mapping units which varied markedly in erosion as determined by field survey. The results indicated that the model was not sufficiently sensitive to variations within Viota with, at most, only two classes of erosion risk being predicted. An avenue for further research is to link the soil erosion hazard and land suitability assessments with the results from the Viotia archaeological site survey directed by Dr. John Bintliff. Such an analysis offers the possibility of investigating the changing relationships between sites and land resources.

The Viotia project is at the stage of having produced a wide range of single factor and land evaluation maps at a scale of 1:100,000. The intention is that the erosion map will be used for targetting areas which require the adoption of soil conservation strategies. The results from land suitability analysis indicate that areas of higher erosion also coincide with land of lower suitability for many crops. The implication is that farmers in such areas should be encouraged to adopt more extensive forms of land use which provide better soil protection. The use of a GIS permits the production of high quality maps which are essential requirements for planning of sustainable land use systems.

## Acknowledgements

The Viotia project is supported by the Soil Science Institute of Athens, the British Council and NATO. Thanks are expressed to members of the Pedology Section of the Soil Science Institute of Athens for the field survey and to Mr. John McArthur in the University of Stirling for processing the data in the GIS.

## References

Davidson, D. A. 1980 Erosion in Greece during the 1st and 2nd millennia BC. In: Cullingford, R. A., Davidson,. D. A. and Lewin, J. (eds.) *Timescales in Geomorphology*. Wiley, Chichester, 143–158.

Davidson, D. A., Papadopoulos, N. and Theocharopoulos, S. P. 1990 Using a GIS as an aid to land resource assessment in Viotia, Greece. In: *Egis '90: Proceedings First European Conference on Geographical Information Systems* (Amsterdam) 236–244.

Giordano, A., Bonfils, P., Briggs, D. J., Menezes de Sequeira, E., Âoquero de Laburu, C. and Yassoglou, N. 1991 The methodological approach to soil erosion and important land resources evaluation of the European Community. *Soil Technology* 4, 65–77.

Nakos, G. 1983 The land resource survey of Greece. *Journal of Environmental Management* 17, 153–169.

Wagstaff, J. M. 1981 Buried assumptions: some problems in the interpretation of the 'Younger Fill' raised by recent data from Greece. *Journal of Archaeological Science* 8, 247–264.

Wagstaff, M. and Cherry, J. F. 1982 Settlement and resources. In: Renfrew, C. and Wagstaff, M. (eds.) *An Island Polity: the Archaeology of Exploitation in Melos*. Cambridge University Press, Cambridge, 246–263.

Yassoglou, N. J. 1987 The production potential of soils: Part II - sensitivity of the soil systems in Southern Europe to degrading influxes. In: Barth, H. and Hermite, P. L. (eds.) *Scientific Basis for Soil Protection in the European Community*. Elsevier Applied Science, London, 87–122.

# CHAPTER 14

# AGRICULTURAL TERRACES: THE VASILIKOS VALLEY, CYPRUS

MALCOLM WAGSTAFF

*The paper reports the results of a morphological examination of two small systems of man-made terraces in the Vasilikos valley area of south-central Cyprus. It suggests ways in which this type of study can produce hypotheses about the relationships between the form of terraces and their function, and it outlines some methods for producing both relative and absolute chronologies for terrace systems.*

## Introduction

As human artifacts, agricultural terraces are widespread on a world scale and have attracted a certain amount of rather sporadic scholarly attention over the years (e.g. Perry 1916; Spencer and Hale 1961). Particularly striking examples of agricultural terrace systems are found at various locations in the Mediterranean region, but the focus here is on Cyprus and, in particular, on agricultural terraces in the Vasilikos Valley. The Vasilikos is one of a series of perennial rivers which drain the south-eastern slopes of the Troodos massif (1954 m) (Figure 14.1). The upper half of its catchment (total area 151.4 $km^2$) is composed of igneous rocks of Upper Cretacaeous age (the Troodos Ophidite complex) at elevations between 1200 m and 200 m above sea level. The lower section of the catchment consists of sandstones (Athalassa formation), marls and sandy limestones with gypsum (Dhali Group), and chalk (Lefkara Group), all again of Upper Cretacaeous date (Bagnall 1960; Pantazis 1967). In this area, the floor of the Vasilikos Valley is about 250 m wide down to the coast and bordered by a set of alluvial terraces (Gomez 1982). The whole is set in rolling hill country, with summits ranging between about 100 and 180 m in height above sea level. Considerable archaeological research has been carried out in the lower catchment under the overall direction of Professor Ian Todd of Brandeis University.

Whilst much of the attention has been archaeological, including excavations at the aceramic Neolithic site of Kalavasos Tenta (Todd 1977; 1986; 1987), the fluvial geomorphology of the lower valley has also been investigated. Gomez (1987) found that the Younger Fill in the valley consisted of two distinct units. The first is a coarse channel zone deposit. The other is a finer overbank sediment. Whilst the unstratified lower third of the overbank deposit contains igneous material which can only have originated in the upper reaches of the Vasilikos catchment, the upper two-thirds lacks this material but contains angular chalk gravel which probably came from slopes on either side of the lower valley. Radiocarbon dates from a firepit suggest that the overbank deposit began to form at about the time of the first aceramic Neolithic occupation of the valley (*c*. 5800–5250 B.C.), whilst dates from near the top of the deposit suggest that it was still being formed in the fifteenth century A.D. It is in this context that the agricultural terraces of the surrounding area become important. They might be seen, for example, as a response to erosion on the sides of the main valley and the consequent loss of soil. Dating their formation might indicate when the problem was first tackled by local people and, at the same time, allow terracing to be related to general questions of land use in the area.

Agricultural terraces in the area of the Vasilikos appear to be grouped into discrete but often spatially related systems. Three types have been recognised on the basis of form and these are similar to the types identified elsewhere (e.g Donkin 1979; Spencer and Hale 1961; Stager 1976).

(1) In the bottom of the Vasilikos valley itself, where the river channel still exists, the terraces are often wide (20 m or more) and their retaining walls are comparatively low (up to 1.0 m).
(2) Terraces in the main side valleys are much smaller, rhomboid in shape, though with their long axes either roughly parallel or roughly transverse to the axis of the valley floor, and enclosed by steep side slopes. They are supported by transverse walls which are higher than those in Type 1 (mean of 1.5 m in the two sequences examined below; maximum 2.35 m; minimum 0.8 m). They correspond with Donkin's *cross-channel* type (1979, p.32) and could also be described as *check-dams* (Stager 1976).
(3) On the slopes above the main valley, terrace walls often have heights between the two extremes of Types 1 and 2 and tend to approximate the contours. They can be classified as *lateral* or *contour terraces* (Donkin 1979, p.32) or *parallel terraces* (Moody and Grove 1990). Whilst they vary considerably in width, the terraces are long and, in proportion to their width, narrow. The most continuous systems are situated on moderate (10-20°) slopes, though discontinuous or fragmented systems are found on steep slopes, especially above side valleys.

To these may be added the isolated small terraces found as ones or twos in relatively remote locations,

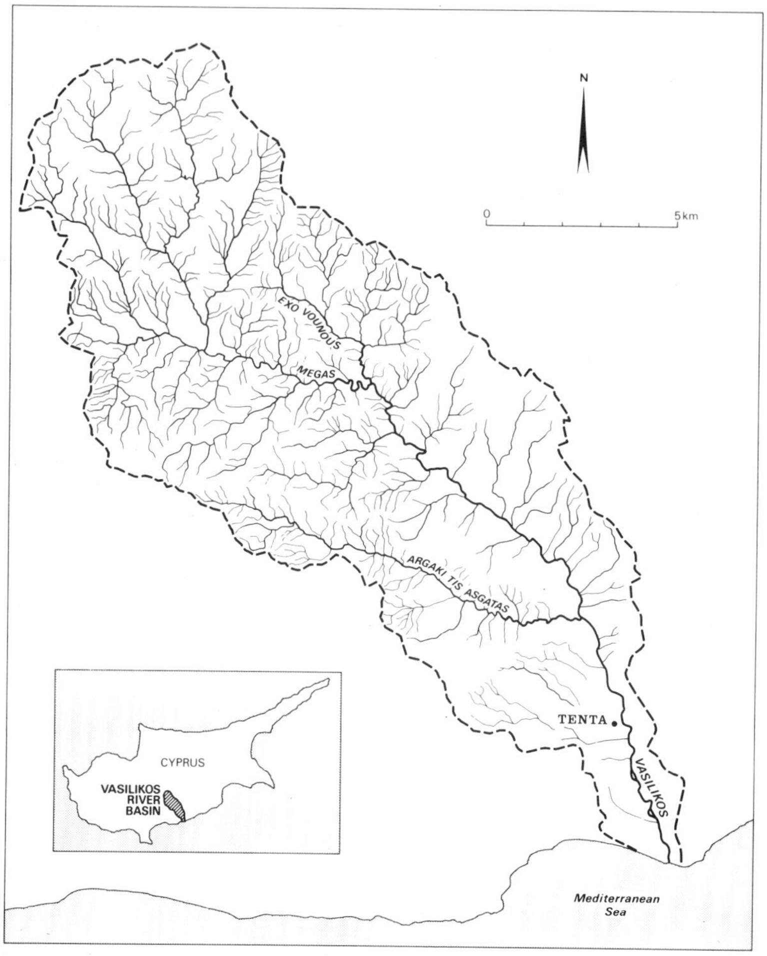


Figure 14.1 Catchment of the Vasilikos River, Cyprus, and the archaeological site of Tenta.

frequently in slight depressions or hollows which occur within areas otherwise characterised by steep gradients (>30°). This appears to be a type also recognised by Spencer and Hale (1961, 6, Type 4).

The existence of man-made terraces raises a number of questions. These include:

(a) Why were they built?
(b) Did their construction meet the initial aims?
(c) When were they built?
(d) What have been the effects of their construction?

**Investigations**

Answers to the questions have been sought through a detailed morphological examination of two sets of terraces falling into the second group recognised above, that is *cross-channel terraces* found in major side valleys. This was done in the belief, shared by archaeologists and geographers, that studies of form provide a reliable basis for making inferences about such matters as process and chronology.

The particular terrace systems were selected for two reasons. First, they form relatively self-contained systems (that is, they were not obviously integrated with terrace systems which covered neighbouring slopes) and, second, they could be examined in a relatively short amount of time.

One set of terraces lies in a side valley which reaches the Vasilikos trough near the Neolithic site of Kalavasos Tenta (hereafter Tenta), the main focus of archaeological investigation in the middle reaches of the main valley (Figure 14.2). Early human occupation here raises an important question for further investigation, namely the possibility of functional and chronological connections between the site and the terrace system. The second terrace system is located on the south side of an area called Mitsinjites, about 1200 m to the north of the first series and provides a useful, neighbouring unit for comparison (Figure 14.3). In both cases, the terrace system consists of a series of flat or gently sloping units of land contained within a valley but defined and retained on the down-slope side by a single stone wall, built without mortar. Heavily silted and presumably disused irrigation channels reached the Tenta system at three points and seem integral to recent exploitation, whilst a more recent and actively used irrigation channel crossed the lower section of the Mitsinjites system taking water elsewhere.

Examination of the two terrace systems followed

standard procedures. Each terrace system was mapped. The heights of the retaining walls were measured in the centre of their length and towards either end. The general angle of the wall in relation to the perpendicular was noted, along with details of the wall's construction. Exposures in the fill were sought and, when found, were usually located where part of the retaining wall had collapsed. Retaining walls were not dismantled. When found, exposures were cleaned and their stratigraphy examined. No samples of sediment were taken. The land use on the terrace itself was noted and the circumference of a sample of olive trees measured with a tape measure. Where vegetation allowed, a search was made for pottery sherds and other cultural materials. Finds were put into plastic bags, labelled and retained for later identification and dating by specialists.

## Results

In answer to the first question posed above, 'Why were they built?' several deductions can be made from the terraces themselves. Examination of a number of 'natural' exposures along the side of some terrace walls suggests that, at least in their upper courses, these particular walls were constructed as the fills were actually accumulating (Figure 14.4). This observation, duplicated for the Negev in Israel by Bruins (1990), suggests that the terraces were created because their builders recognised that a wall across a stream channel would halt the transportation of eroded material and induce alluviation. In this way, not only was a relatively flat area of land created for cultivation but water was also retained within the fill and thus made available for exploitation by plants over a longer period than would have been possible under 'natural' conditions. For it may be argued that the building of walls across a valley indicates a markedly seasonal stream regime which, on the one hand, allowed sufficient time of nil or low flow for construction to take place and, on the other hand, meant that water for plant growth was scarce.

All forms of man-made terrace have the effect of transforming the appearance of the landscape. Not only do they give it a stepped appearance but, strikingly in the case of cross-channel terraces or

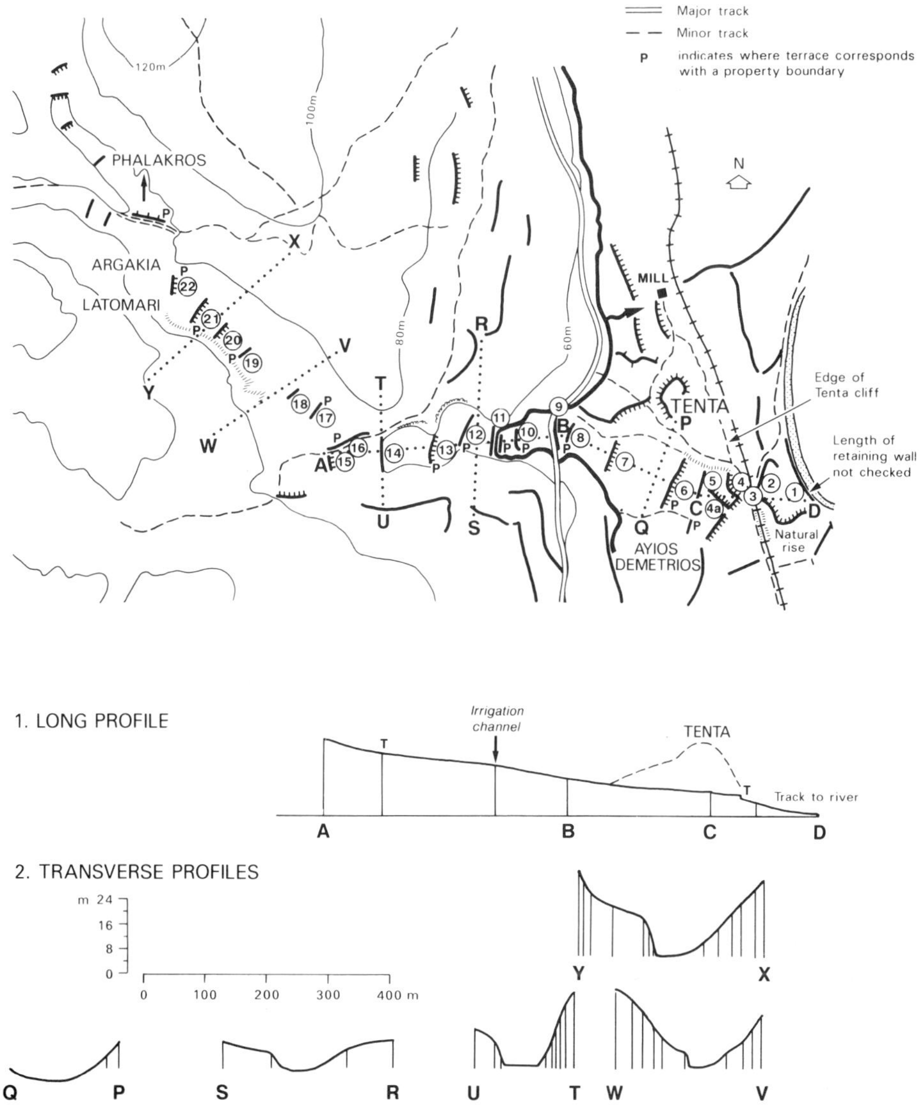


Figure 14.2 Terrace systems in the vicinity of Tenta.

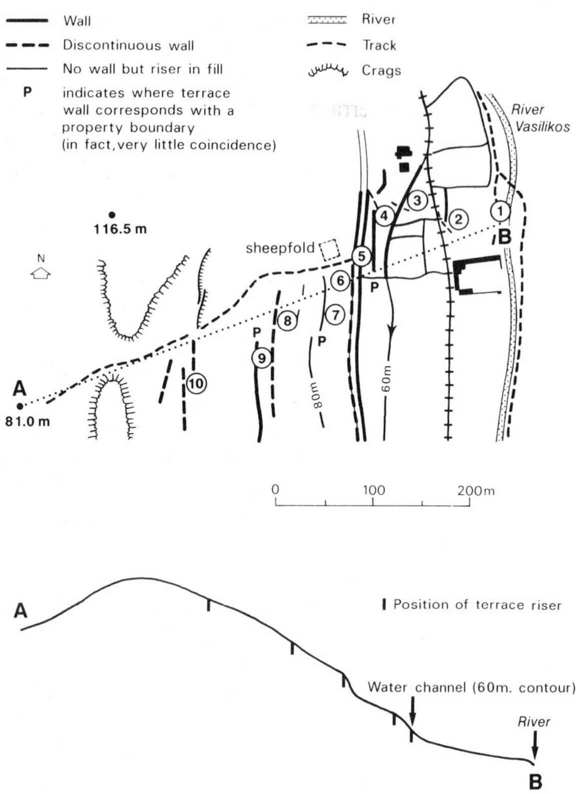


Figure 14.3 Terrace systems in the vicinity of Mitsinjites.

check-dams, they have had a smoothing effect. On the one hand, they have filled up the stream channel and replaced it with a level, or almost level, surface, as has been observed in more dramatic form elsewhere, for example in Italy (Delano Smith 1981). On the other hand, cross-channel terraces have reduced the gradient down the long profile. Recognition of these changes is crucial to the contextualisation of archaeological sites for it is important to know, in the case of land use and resource exploitation, whether the valleys were in a 'natural' state when particular sites, such as Tenta, were occupied or whether they had already been deliberately modified by the construction of cross-channel walls. The importance of correct contextualisation, in turn, puts a premium on dating particular terrace systems, a point to which I will return.

Acceptance of the Zipf principle (1949), that people will attempt to secure their goals through minimum effort, suggests the existence of some pressing need behind terrace creation. We can speculate as to what this might be.

Conservation of soil and water is one possibility. The extension or creation of agricultural lands is another. They are not mutually exclusive. Although wall-building seems relatively laborious, it does not by itself argue for the existence of either population pressure on available agricultural land or a surplus of labour as the driving force behind terrace building. The retaining walls could have been built in the slack seasons of the farming year. In the eastern Mediterranean these used to be in the summer and winter (Pepelasis and Yotopoulos 1962, p.63, Table 2). Two features may militate against the notion that the terrace systems were created, at one moment in time, under the possible aegis of large estates, which might be thought to have possessed both the necessary organisational authority and also the operational necessity to carry out such engineering works. First, the two systems examined are relatively small and conformable with the scale of the two valleys concerned. They are nothing like the 'state terraces' of the Incas where trimmed stone was used and the system is beautifully regulated and carefully adjusted to the natural gradient (Donkin 1979, p.32-33). Nor are they similar to the massive (?) Herodian wall which retains the Ein Khandak system of irrigation terraces near Jerusalem studied by Ron (1966). Second, cadastral maps for the area (1917, revised 1973 and 1980) indicate that terrace walls in the sequence near Tenta coincide with property boundaries in 12 cases (54.4 per cent), whereas the large-estate notion might presuppose either an even greater coincidence (where subdivision of an original holding has taken place) or an even lower degree of correlation where estates covered large tracts of country. The Tenta terrace system thus appears conformable with notions of individual land-ownership and private enterprise. However, the conclusion cannot be insisted upon. The terrace system near Mitsinjites shows little correlation between terrace walls and property divisions. All that can be safely said is that the existence of what appears to be a single system, does not necessarily point to an overall design-plan nor to only one period of construction.

Thus far answers have been attempted for questions (a) and (d). The existence of terrace systems in other side valleys around the Vasilikos Valley argues that, whatever the objectives were, they were successfully met (Question b). Even though the systems are almost certainly of different date and could have been meant to meet different requirements, equifinality indicates widespread agreement on the solution to a particular group of problems, those specified immediately above.

The continued survival of terrace walls and especially the visual signs of maintenance which were clearly evident when the studies were made (including the use of old oil drums to repair one wall), show that the terraces were still thought by their owners to be valuable assets. Signs of recent ploughing on a number of terraces provide supporting evidence. Their agricultural use, however, was diverse. Polyculture was evident on the wider ones, where olives, carobs and the occasional almond or fig tree were undercropped with cereals. Several terraces had been planted with vines, though the stocks were seriously neglected and the vineyards badly overgrown. The smaller and more stony terraces in the higher reaches of the two systems examined in detail were devoted almost exclusively to olives and carobs, though undercropping would obviously be feasible.

**Dating**

The answer to the third question - when were the terrace walls built? - is more complex than those so far attempted. There is no reason to suppose that all the walls are contemporaneous in either system. Indeed, the constructional evidence for several phases of wall heightening are against such a view for the two systems studied. Similar evidence has been reported from the Negev and Sinai (e.g. Bruins 1990) and southern Greece (Wells *et al.* 1990). Piecemeal development seems probable in the Vasilikos Valley, though the initial stages in establishing a system may have taken place at roughly the same date. Interdependence, in the sense that walls in succession up the valley rise out of the fill of the lower terraces, might suggest a relative chronology through the sequence from lowest and oldest to highest and youngest. However, if the terrace walls rest on the bedrock, the natural slope of the valley would seem to rule out the necessity of a uni-directional chronology based on ascending height within the valley. The reverse is equally plausible.

The retaining walls are built from undressed stones, presumably collected from nearby: large cobbles, probably from the bed of the Vasilikos, were also included in some cases. The stones are laid in courses, without mortar, but the construction technique lacks the sophistication of, for example, field walls in northern England which were built to be free-standing and self-supporting. No obvious variations in building style were observed and so the possibility of using that as a means of dating had to be ruled out (see also van Andel and Runnels 1987, p.145).

The walls in each of the two terrace sequences vary in height in a way which is not apparently related to their horizontal position within the sequence but appears to be haphazard. Some walls are higher than others in the same sequence and seem to coincide with the point where the valley steepens down its long profile. Other walls, not so tall, may have been inserted subsequently, either as the accumulations of fill may have necessitated or as the requirements for a larger cropped area could have suggested. The terraces which these walls support might thus be regarded as *secondary*.

Less hypothetical and more absolute means of dating are offered by the trees growing on the terrace flats and by material found in the fills. Olives may be the best arboreal indicators, particularly in the absence of direct dendro-chronological data from cores. The longevity of olives is legendary (Rackham 1981), whilst in the two terrace systems examined there is great diversity in size. On the Tenta sequence olives varied in size from a double-trunked specimen with a circumference of about 9.0 m down to one with a circumference of 2.9 m. The largest single trunk olive trees were about 6 m in circumference. The calculated radii of these trees is thus 1.4, 0.96 and 0.46 m respectively. If we knew the average rate of ring-growth per annum, it would be possible to estimate the age of each tree from its radius (Rackham 1981 hints at this possibility in his work on Crete). For illustrative purposes, let us suppose that an olive tree puts on 1 cm per century. That would give us a range of ages on this particular sequence of terraces of perhaps 960 to 460 years, ignoring the 9 m giant. Any such dates, more correctly derived, would give a *terminus ante quem* for a late phase of fill accumulation.

The fills are not homogeneous, either between terraces or within a section through an individual terrace. Variations in the degree of stoniness, as well as in the size and geological composition of pebbles, are the most noticeable differences between terraces. Sections examined contained not only a preponderance of light-coloured, comparatively fine silt but also lenses of gravel, particularly in the lower levels. Several phases of deposition appear to be suggested. Pottery was also found in various situations, suggesting the possibility of firmer dates.

Only five sherds were collected from the sequence of terraces near Tenta (Figure 14.2). These were found about 0.2 m beneath the surface of a terrace just below the area known as Phalakros (Figure 14.2). However, the fragments were so tiny (mean 2.7 x 2.0 cm) and so worn as to be almost impossible to identify typologically with any certainty. One of them, though,

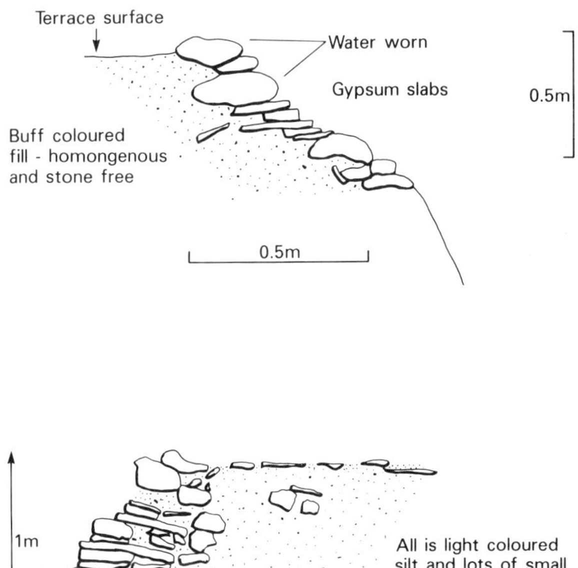


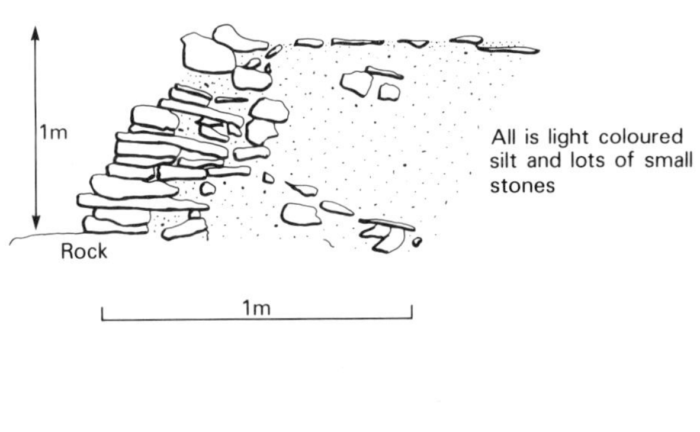


Figure 14.4 Section through the front of two terraces showing the stepped form of the retaining walls.

carried an incised linear pattern and was red in colour. There was sufficient of it to hint that a Middle Bronze Age date might be in order (*c.* 2000–1600 B.C., according to the chronology in Buchholz and Karageorghis 1973, p.137–38). The sherds themselves probably originated in a pottery and lithic scatter a little away from the valley terrace sequence at Phalakros (Cadastral Sheet LV, Plan 20, Plot 316) where Middle and Late (1600–1050 B.C.) Bronze Age pottery types have been identified together with some of Archaic date (*c.* 750–450 B.C., Karageorghis 1981, p.128 and p.158).

More pottery was recovered from the second sequence of terraces near Mitsinjites (Figure 14.3). Collapse of the wall supporting Terrace No. 4 revealed sherds of post-Roman pottery and a tiny amount of Middle–Late Bronze Age pottery in the fill within 0.5 m of the surface. An exposure alongside the road and underneath a retaining wall at Point 6 produced six Middle–Late Bronze Age sherds from a depth of about 0.9 m. A 'natural' section on the edge of Terrace 7 yielded Middle Bronze Age pottery, as well as some sherds of Hellenistic/Roman date and of later glazed ware. Three Middle Bronze Age sherds were recovered from 0.7–0.8 m below the retaining wall of Terrace 8. In all probability, the whole of the material mentioned was washed into the valley from the extensive, multi-period site at Mitsinjites.

Whilst the pottery evidence suggests that the fills retained by the terrace walls may have begun to accumulate in the Middle Bronze Age *c.* 2000–1600 B.C., the typological identifications were not always made with complete confidence. Moreover, our understanding of the processes of sedimentation raises various other possibilities. For example, apparently early cultural material may have been deposited at much later dates, depending on when erosion and transportation began and when terrace walls were built. Reworking of the sediments at one or more dates is also a possibility, whether through ploughing or through renewed erosion and transportation when retaining walls were neglected. Finally, the late date of some of the pottery suggests that fill was accumulating in the early Middle Ages.

## Conclusions

Clearly, without radiocarbon dates, answers to questions about the date and origin of terrace systems can be derived only from a careful examination of the terrace fills, the identification of possible sources for the pottery found in them and the study of the form of the retaining walls. As Dennell (1982) showed, a minute study of terrace fills would produce evidence for the ways in which silt accumulated and provide a reliable basis for inferences not only about the processes involved but also about what he called the life-histories of terraces. There would also be an opportunity to test the conclusion reached by Gibson and Edelstein (1985) from work near Jerusalem that accelerated soil erosion had taken place before the *lateral* terraces in their area had been built and that the fills were created, not by natural processes, but by carrying soil and stones from nearby valleys. A similar possibility, that people dumped material behind terrace walls, was also suggested by Wells *et al.* (1990) for the higher terraces on the Charvati spur in the Berbarti-Limnes survey area near Mycenae in southern Greece. *Transverse* terraces, the type examined in the present paper, are probably different, and may be entirely dependent for their fills on the process of sedimentation. Local, relative chronologies might be derived from the recognition of building phases within the retaining walls. Material, such as sherds, from within the fills could provide terminal dates (though provenance will be a problem), whilst some chronological information about the age of the surfaces (or near surface levels) might be derived from long-lived trees, particularly olives. The excavation of longitudinal sections might tie the phases of sedimentation and wall-building together (Bruins 1990; Dennell 1982), while careful deconstruction of walls might produce more material (sherds, coins) which could allow absolute dates to be advanced. Such work remains to be attempted on terrace systems in the Vasilikos Valley, but the potential is considerable.

## References

Bagnall, P. S. 1960 *The Geology and Mineral Resources of the Pano Lefkara-Larnaca Area.* Geological Survey Department, Cyprus, Memoir No. 5.

Bruins, H. J. 1990 The impact of man and climate on the central Negev and northeastern Sinai deserts during the late Holocene. In: Bottema, S., Entjes-Nieborg, G. and van Zeist, W. (eds.) *Man's Role in the Shaping of the Eastern Mediterranean Landscape.* A. A. Balkema, Rotterdam and Brookfield, 87–99.

Buchholz, H. G. and Karageorghis, V. 1973 *Prehistoric Greece and Cyprus: An Archaeological Handbook.* Phaidon, London.

Delano Smith, C. 1981 Valley changes: some observations from recent field and archive work in Italy. In: Baker, G. and Hodges, R. (ed.) *Archaeology and Italian Society: Prehistoric, Roman and Medieval Studies.* British Archaeological Reports, International Series 102, Oxford, 239–247.

Dennell, R. W. 1982 Dryland agriculture and soil conservation: an archaeological study of check-dam farming and wadi siltation. In: Spooner, B. and Man, H. S. (eds.) *Desertification and Development: Dryland Ecology in Social Perspective.* Academic Press, London and New York, 171–200.

Donkin, R. A. 1979 *Agricultural Terracing in the Aboriginal New World.* Viking Fund Publication 56, University of Arizona Press, Tucson.

Gibson, S. and Edelstein, G. 1985 Investigating Jerusalem's rural landscape. *Levant* 17, 139–155.

Gomez, B. 1982 Observations on the fluvial geomorphology of the Vasilikos Valley. In: Todd, I. A. (ed.) Vasilikos Valley Project: Fourth

Preliminary Report, 1979-80. *Journal of Field Archaeology* 9, 67–74.

Gomez, B. 1987 The alluvial terraces and fills of the lower Vasilikos Valley, in the vicinity of Kalavasos, Cyprus. *Transactions of the Institute of British Geographers*, NS 12, 345–359.

Karageorghis, V. 1982 *Cyprus from the Stone Age to the Romans*. Thames and Hudson, London.

Moody, J. and Grove, A. T. 1990 Terraces and enclosure walls in the Cretan landscape. In: Bottema, S., Entjes-Nieborg, G. and van Zeist, W. (eds.) *Man's Role in the Shaping of the Eastern Mediterranean Landscape*. A. A. Balkema, Rotterdam and Brookfield, 183–191.

Pantazis, T. M. 1977 *The Geology and Mineral Resources of the Pharmakas-Kalavasos Area*. Geological Survey Department, Cyprus, Memoir No. 8.

Pepelasis, A. A. and Yotopoulos, P. A. 1962 *Surplus Labor in Greek Agriculture, 1953-1960*. Center of Economic Research, Research Monograph Series 2, Athens.

Perry, W. J. 1916 The geographical distribution of terraced cultivation and irrigation. *Proceedings of the Manchester Literary and Philosophical Society* 60, 1–25.

Rackham, O. 1981 Land-use and the native vegetation of Greece. In: Bell, M. and Limbrey, S. (eds.) *Archaeological Aspects of Woodland Ecology*. British Archaeological Reports, International Series 146, Oxford, 177–198.

Ron, Z. 1966 Agricultural terraces in the Judean Mountains. *Israel Exploration Journal* 16, 33–49, 111–122.

Spencer, J. E. and Hale, G. A. 1961 The origin, nature and distribution of agricultural terracing. *Pacific Viewpoint* 2, 1–40.

Stager, L. E. 1976 Farming in the Judaean Desert during the Iron Age. *Bulletin of the American Schools of Oriental Research* 221, 145–158.

Todd, I. A. 1977 The Vasilikos Valley Project, 1976. *Journal of Field Archaeology* 4, 375–381.

Todd, I. A. (ed.) 1986 *The Bronze Age Cemetery at Kalavasos Village*. Paul Astroms Forlag, Göteborg.

Todd, I. A. (ed.) 1987 *Excavations at Kalavasos-Tenta*, Vol. 1. Paul Astroms Forlag, Göteborg.

van Andel, T. D. and Runnels, C. 1987 *Beyond the Acropolis: A Rural Greek Past*. Stanford University Press, Stanford.

Wells, B., Runnels, C. and Zangger, E. 1990 The Berbati-Limnes archaeological survey. The 1988 season. *Opuscula Atheniensia* 18, 207–238.

Zipf, G. K. 1949 *Human Behaviour and the Principle of Least Effort*. Cambridge University Press, Cambridge.

# CHAPTER 15

# PALAEOENVIRONMENTAL EVIDENCE FOR AGRICULTURAL SOIL EROSION FROM LATE HOLOCENE DEPOSITS IN THE MONTAGNOLA SENESE, ITALY

C.O. HUNT, D.D. GILBERTSON AND R.E. DONAHUE

*Phases of Holocene valley alluviation have now been described from many countries around the Mediterranean and ascribed either to human activity or to climatic change. It is rare, however, for direct palaeoenvironmental evidence to be reported, thus allowing these conflicting hypotheses to be evaluated. In this paper, we describe palaeoenvironmental evidence from Late Holocene slope and coarse alluvial deposits from the hills of the Montagnola Senese in Tuscany, Italy. In the Farma, Rosia and Feccia valleys, these deposits contain sedimentological, palaeobiological and artifactual evidence for accumulation in sparsely vegetated agricultural landscapes during three phases, provisionally attributed to the eleventh to early twelfth, sixteenth to early seventeenth and late nineteenth centuries AD. These phases appear to compare well with historical and archaeological evidence for periods of intensification of human activity and the expansion of farming in the area.*

## Introduction

Valley alluviation in the Mediterranean coastlands has been a topic of considerable academic interest since Vita-Finzi (1969) proposed that climatic changes caused a major episode of valley sedimentation throughout the region during the Late-Roman to early Post-Roman period. His hypothesis was supported by further research (Vita-Finzi 1975; Bintliff 1977, 1982) though Vita-Finzi himself later (Vita-Finzi 1976) acknowledged that valley sedimentation was to an extent diachronous throughout the Mediterranean region. An increasing body of work (among others Davidson 1971, 1980; Davidson *et al.* 1976; Wagstaff 1981; Bell 1982; Gilbertson *et al.* 1983; Pope and van Andel 1984; van Andel *et al.* 1986, 1990; Chester and James 1991) suggests that alluviation occurred at different times at different localities, and thus in most cases climatic change was unlikely to be as important a causal factor as human disturbance of landscapes. In practically all of this work, however, one vital link in the chain of hypothesis is missing (Bell 1982). Very little direct evidence of environmental conditions during the alluviation episodes has been recorded. Instead, the argument rests on circumstantial evidence and the application of the principle of uniformity to the observation that clearance and cultivation of landscapes today causes soil erosion, which leads to enhanced levels of sediment entering watercourses and thus to the build up of alluvium as rivers are unable to remove all this increased load to the sea.

The link between human activity and alluviation usually rests on arguments deriving from some combination of three types of circumstantial evidence, which can be crudely summarised as follows. First, climatic changes are widely believed to be synchronous over wide areas, and Vita-Finzi (1969) and his followers suggested that if alluviation could be demonstrated to be synchronous, then it must follow that the cause was climatic, since people were unlikely to have had uniform impact over wide areas. Since alluviation can be shown to have occurred at different times in different places, the climatic argument has been deemed to have lost much of its force (for instance, Wagstaff 1981; Bell 1982; van Andel *et al.* 1990). In its simplest form this position neglects, however, the critical concept of geomorphological thresholds (Schumm 1979). The threshold for alluviation might be passed at different times in different places, during a phase of climatic change. The result might well be an uneven pattern of alluviation. Second, pollen or other environmental evidence from lakes or marshes near to the valley system in question, or third, archaeological or historical evidence of enhanced human activity in the surrounding region, will be cited as evidence for clearance episodes at around the time that alluviation occurred and a causal link will be suggested (for instance, Chester and James 1991). Variants on the use of these types of evidence include the work of van Andel *et al.* (1990), who note that 'probably anthropogenic alluviation' in Thessaly predates archaeological though not palynological evidence for clearance and the work of van Andel and Runnels (1987) who presented evidence suggesting that in parts of Greece alluviation occurred during times of landscape abandonment and soil terrace collapse. It is still rare, however, for direct environmental evidence from the valley fill deposits themselves to be cited to support the arguments for human activity being the major cause of Holocene valley alluviation in the Mediterranean coastlands.

In this paper, we describe colluvial and alluvial sequences in the Montagnola Senese, Tuscany, Italy, from the catchment of the river Merse (Figure 15.1). These deposits provide direct environmental evidence for an anthropogenic cause for dramatic phases of aggradation during a number of periods over the last

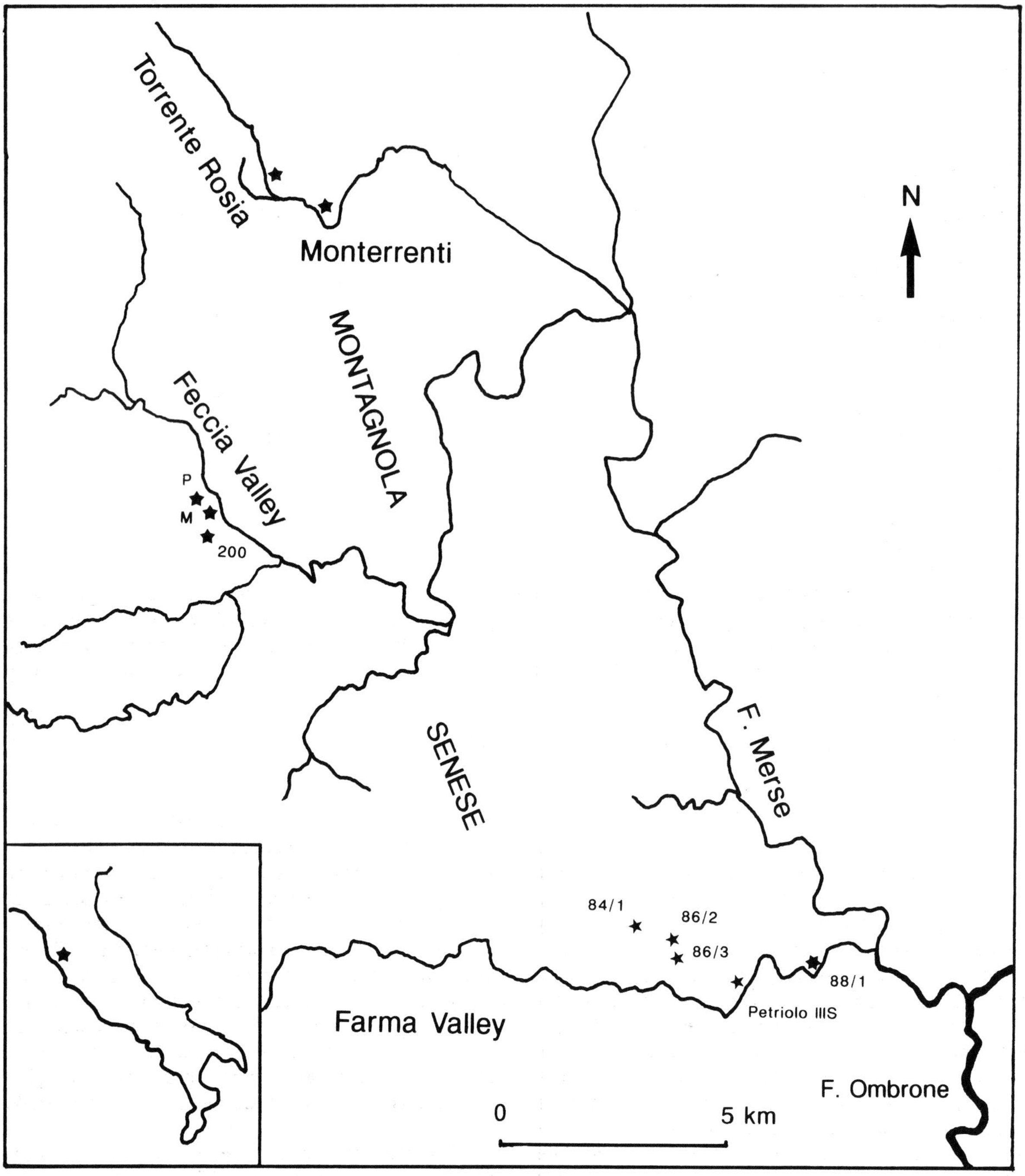


Figure 15.1 Location map, showing the location of field sites mentioned in the text.

| Locality | 1 | 2 | 3 | 4 | 5 | 6 | 7 | 8 | 9 | 10 | 11 | 12 |
|---|---|---|---|---|---|---|---|---|---|---|---|---|
| *Pomatias elegans* | | | | 1 | | | | | 27 | 4 | 4 | 3 |
| *Oxychilus* cf. *draparnaudi* | | | | 1 | | | | | | | | |
| Clausiliidae (undiff.) | | | | | | | | | | 1 | 1 | |
| *Retinella* sp. | | | | | | | | | | 1 | | |
| *Helicodonta* sp. | | | | | | | | | | 1 | | |
| *Ena* sp. | | | | | 1 | | | | | | | |
| *Clausilia bidentata* | | | | | 1 | | | | | | | |
| *Acicula* sp. | | | | | | | | | 7 | | | |
| Helicidae (undiff.) | | | | 1 | | | | | | | | |
| *Cepaea nemoralis* | | | | | | | 1 | | | | | |
| *Helix aspersa* | | | | | 1 | | | | | | | |
| *Cernuella virgata* | | 11 | | 11 | 9 | | | 6 | | | | |
| *Cernuella* sp. | | | | | | | | 10 | | | | |
| *Monacha cartusiana* | | 4 | 1 | | 15 | | | | 2 | | | |
| *Monacha cantiana* | | | | | | | | | 4 | | | |
| *Trochoidea pyramidata* | 33 | | | | | | 9 | | 4 | | | |
| *Trochoidea elegans* | | | | | | | | | 1 | | | |
| *Candidula* sp. | 2 | | 2 | | | 37 | 2 | | | | | |
| *Abida* sp. | | | | | | | 4 | | | | | |
| *Granaria stabilei* | | | | | | | 3 | | 1 | | | |
| *Granaria frumentum* | | | | | | | | | 12 | | | |
| *Rumina decollata* | | | | | 5 | 5 | | 18 | | | | |
| *Truncatellina claustralis* | | | | | | 1 | | | | | | |
| *Chilostoma* sp. | | | | | | | | 4 | | | | |
| *Chondrula tridens* | | | | | 1 | | | | | | | |
| *Eobania vermiculata* | | | | | 2 | | | | | | | |

Habitats: 1 = abandoned eroded field with surface gravel lag and isolated low ruderals; 2 = regenerating grassland after ploughing; 3 = open short grassland grazed by deer; 4, 6 = long grass below wall; 5 = grass and scrub; 7 = wall in light shade; 8 = 'anvil stone' used by thrushes; 9 = *maccia* - broken scrub with some large trees; 10, 11 = walls under dense shade from oak woodland; 12 = dense oak woodland, no understorey vegetation.

Table 15.1 Molluscs found in contemporary habitats in the Montagnola Senese

thousand years.

The Montagnola Senese is basin and range country, with horsts of metamorphosed Mesozoic mudrocks, conglomerates and limestones, interspersed with graben containing fills of Miocene to early Quaternary age (Gilbertson and Hunt 1987). The horsts today are heavily wooded since recent agricultural development has been discouraged by the steep slopes, by poor soils and by the depredations of wild boar. Considerable areas of former agricultural land are now reverting to dense scrub. In the areas discussed in this paper, agricultural terracing is extremely rare, even on the steepest slopes. The once-marshy graben floors were drained in the late nineteenth century and are now cultivated for sunflowers, vines, wheat, sweetcorn, vegetables and rice. Rainfall events, particularly those associated with impressive late summer thunderstorms, can be very severe and we observed ground lowering rates of around 3 mm per year at a number of localities where agricultural soils were exposed.

## Methods

Soil pits were dug by hand and these and natural and artificial sections were measured and drawn and samples taken for palaeobiological analysis. The mollusc diagrams and plant macrofossil counts were constructed using numbers of specimens per 5 kg sample wet-sieved at 0.5 mm, except in the case of Monterrenti, where 25 kg samples were used. Pollen was extracted from sediments by boiling in 5% KOH for 5 minutes, sieving on nominal 7 micron mesh and swirling on a clock glass. 250 pollen grains per sample were counted. A number of samples from modern environments were processed to give comparative material for the mollusc analysis (Table 15.1).

The most problematical aspect of this work is the dating of sediment sequences. The age estimates were made on the basis of the youngest artifact found in any given layer, or by radiocarbon. The pioneering nature of these studies in valley-floor and hillslope sequences, which frequently lack biogenic remains amenable to radiocarbon dating, of necessity forces reliance on age estimates from pottery, or worse, derived pottery. We have had to exercise our judgement upon the reliability of such age estimates, and hence we emphasise the importance of regarding our chronology as provisional.

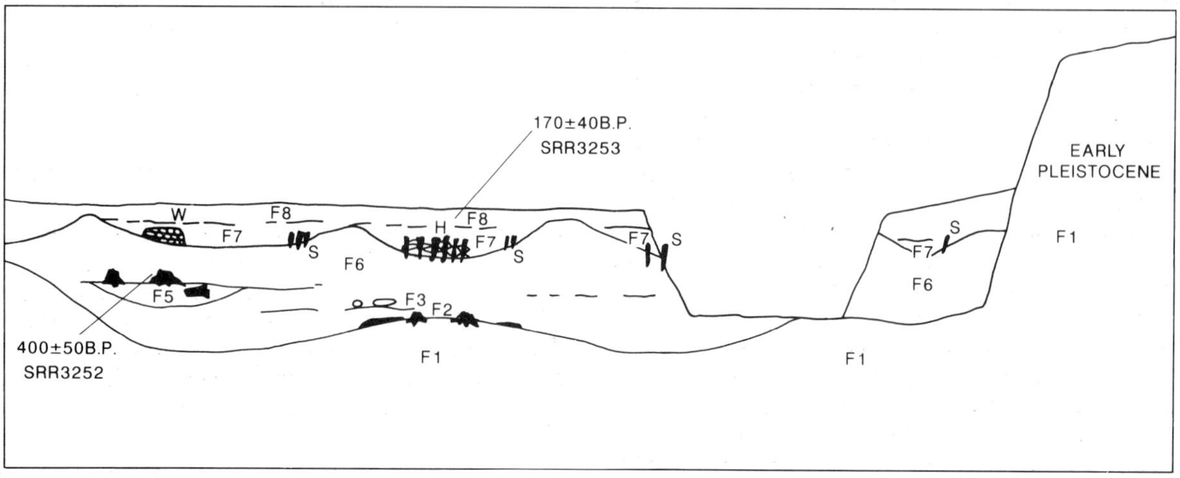


Figure 15.2 Feccia Valley - schematic composite stratigraphical section (not to scale).

## Fluvial deposits

Most of the rivers in the Merse catchment appear to conform to the 'wandering gravel' style of Blacknell (1981), in spite of severe agricultural soil erosion in the area today. Access to the valley floors is limited by dense shrubby and woody streamside vegetation, which seems to act as an efficient 'filter' which adsorbs sediment laden runoff from the valley slopes, thus regulating to some extent the discharge regime.

Post-Medieval fluvial deposits are present on the valley floor as low terraces on the margins of the floodplains of the rivers of the Merse catchment. Most modern floodplains are very narrow and poorly developed, usually in the region of a few tens of metres across and up to 2 m deep, but the terrace deposits are much larger, in the case of the Feccia Valley being up to a kilometre across and 3 m or more deep.

TORRENTE ROSIA. Fluvial deposits containing Medieval to late nineteenth century potsherds and large fragments of masonry were exposed in cutbanks of the Torrente Rosia (Figure 15.1). The sediments are plane-bedded cobbly gravels and massive silty cobbly diamicts. Exposures were very shallow and fragmentary and no lithologies suitable for sampling for palaeobiological investigations were found.

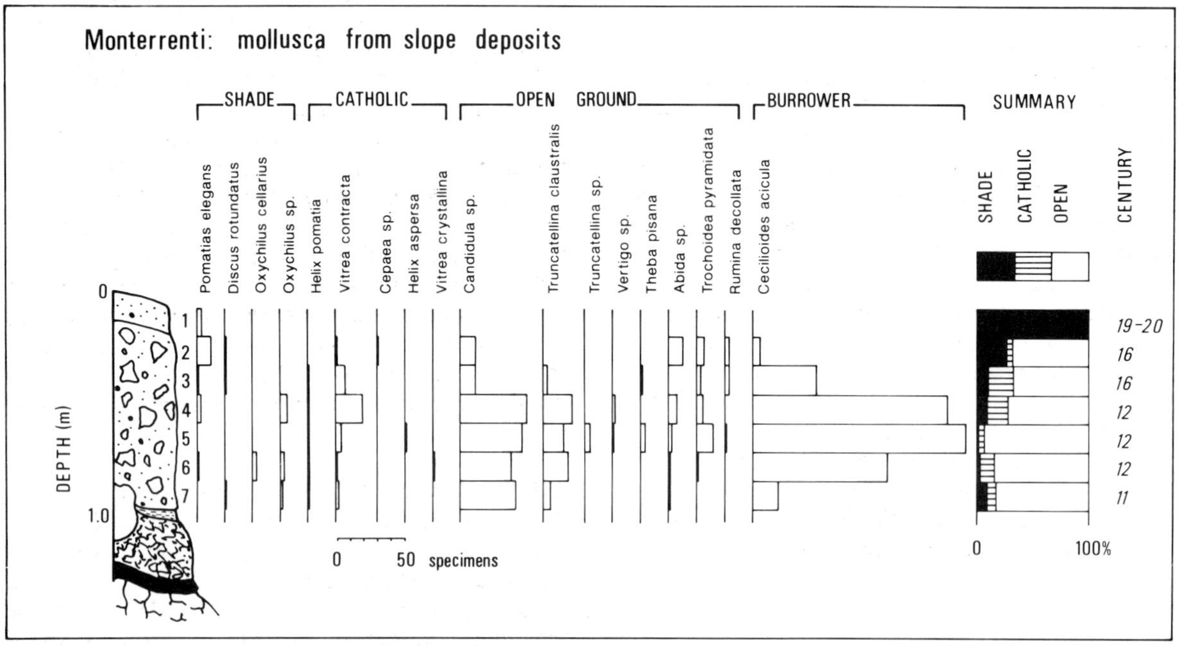


Figure 15.3 Monterrenti – stratigraphy and mollusc diagram

FECCIA VALLEY. A complex sequence of fluvial deposits was exposed in drainage ditches and gravel pits in the Feccia Valley (Figures 15.1 and 15.2). A preliminary description of some of these deposits was given by Gilbertson *et al.* (1983) and a more detailed account will be given elsewhere. The sequence (Figure 15.2) consists of plane and cross-bedded silts, sands and gravels (F8), overlying meander plugs (F7) containing silts, sands and fine gravels with a radiocarbon date of 170±40 BP (SRR-3253) on a fish-weir within the meander plug, then plane and trough cross-bedded gravels (F6) overlying silty meander plugs (F5) with a radiocarbon date of 400±50 BP (SRR-3252) on the outermost layers of a poplar stump that had grown in the top of the meander plug, then further plane-bedded and trough cross-bedded gravels (F3) overlying Lower Pleistocene siltstones (F1). Meandering rivers are generally associated with relatively even discharge regimes, low sediment input, high bank cohesiveness and well-vegetated landscapes, whereas braided rivers, which give rise to trough cross-bedded gravels, are usually associated with 'flashy' discharge regimes, high sediment input, low bank cohesiveness and poorly vegetated landscapes (Briggs and Gilbertson 1980). Mollusc and pollen analysis of unit F5 (Gilbertson *et al.* 1983) suggested that these meander plugs accumulated in landscapes characterised by some woodland and some cultivated land and pasture.

Plant macrofossil analysis and further molluscan analysis provides more information concerning the environment close to the river. The plant macrofossil assemblage from the meander plug unit F5 is characterised by a number of macrofossils from trees and shrubs, a small number from herbaceous taxa and some from marsh and aquatic plants (Table 15.2). The mollusc analyses (Table 15.3) contain very few individuals, but include woodland taxa, a number of catholic taxa which might be expected amongst wet streamside vegetation and aquatic taxa which require permanent flowing water. Samples from the gravel unit F6 did not yield palaeo-environmental evidence. The plant macrofossil assemblages from the meander plug unit F7 include abundant remains from open ground taxa and a few from aquatic, marsh and scrub taxa (Table 15.2). The mollusc assemblages (Table 15.3) are very small but include specimens of one aquatic and several catholic and open ground taxa. The plant macrofossil assemblages from unit F8 contained only open ground and very rare marsh and aquatic taxa (Table 15.2). No molluscs were recovered from unit F8.

FARMA VALLEY. A section (88/1) in the lowest terrace of the Farma (Figure 15.1) showed about 2 m of plane-bedded cobbly gravel with nineteenth century brick, tile and potsherds. No lithologies suitable for palaeobiological sampling were found.

### Slope deposits

The slope deposits appear to form sheets, occurring on all the lower slopes of the Farma Valley and at Monterrenti. In general, the deposits are thicker on areas, such as river terraces, with low gradients. The deposits are generally very poorly sorted and crudely and indistinctly bedded. Rare palaeobiological material suggests accumulation in relatively open, disturbed landscapes.

MONTERRENTI. At Monterrenti (Figure 15.1), slope deposits were exposed by construction work and footpath erosion on the wooded hillside below the small Medieval hilltop settlement. A summary diagram showing the slope deposits is given in Figure 15.3. They consist of dark brown stony sandy diamictons containing abundant mollusc shells and potsherds, overlying Pleistocene deposits and Triassic limestone. Seven samples were taken from the stony diamicts.

Dating evidence at Monterrenti rests on potsherds, with two sherds in the lowest sample, and between ten and twenty sherds in the others. The lowest sample contained potsherds only of eleventh century date, while the three samples overlying this had sherds of both eleventh and twelfth century date. The next two samples contained material of twelfth to early seventeenth century age and the highest sample contained sherds of twelfth to late nineteenth century age.

The mollusc assemblages from Monterrenti are divisible into three groups (Figure 15.3) which, from the evidence of the ceramics, reflect three phases of deposition, during the eleventh to twelfth centuries, the early seventeenth century and the late nineteenth century. The intervening periods are not represented by sediments. The assemblages from the lowest four samples are characterised by abundant 'open ground' taxa (*Candidula, Truncatellina, Abida* and *Trochoidea pyramidata*), some 'catholic' taxa, mostly *Vitrea*, together with rare specimens of shelter-demanding taxa (*Pomatias elegans, Discus rotundatus* and *Oxychilus* spp.). These assemblages probably reflect areas of grassy vegetation, patches of exposed ground and some scrub or rock-rubble habitats. The middle assemblages, of probable sixteenth or early seventeenth century age, are dominated by 'open ground' taxa, particularly *Candidula* sp., *Trochoidea pyramidata, Rumina decollata* and Chondrinidae, with some shelter-demanding taxa. These assemblages probably reflect fairly bare ground with patches of rock rubble or scrub. The highest assemblage, of probable late nineteenth century age, contains only the species *Pomatias elegans*, usually associated with light or disturbed woodland but also found (M. Bell, written communication 1991) among olive groves in the Tuscan Hills.

FARMA VALLEY. Slope deposits are widely distributed in the Farma Valley and were sampled in a number of temporary and permanent exposures including quarries, roadcuts, and the prehistoric archaeological excavations at Petriolo IIIS. The slope deposits are between 0.2 m and over 4.0 m thick, the deepest deposits infilling what appear to have been gullies on relatively gentle slopes, the thinnest being

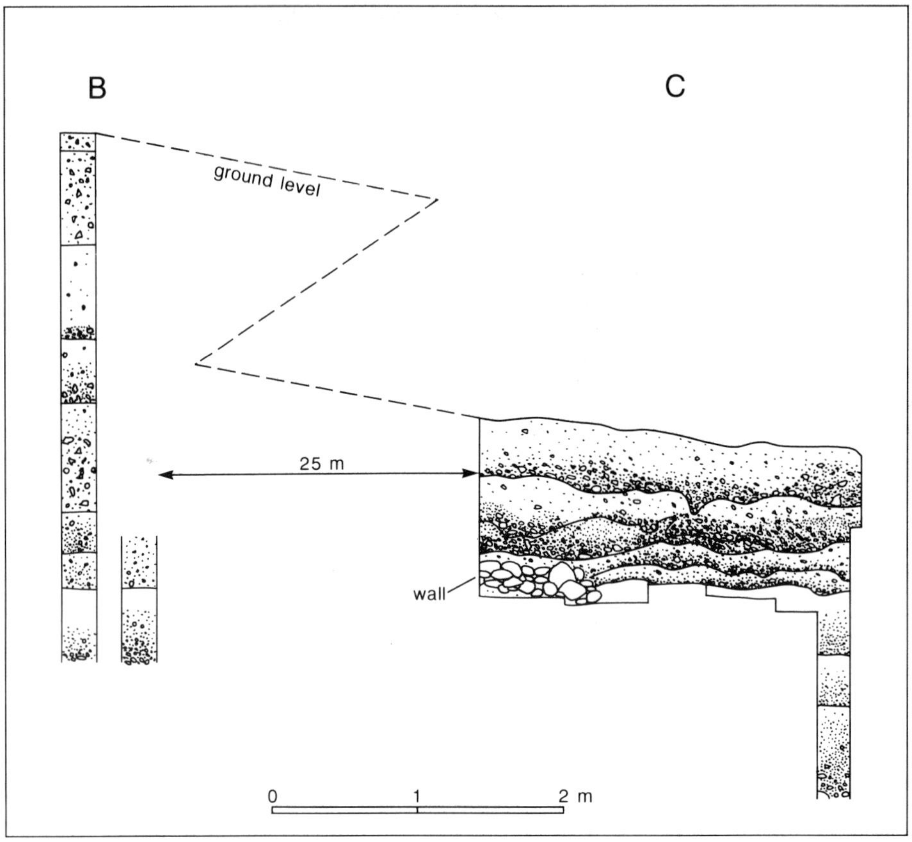


Figure 15.4 Farma Valley – cross section through trenches B and C at Petriolo IIIS, showing colluvial deposits and the buried wall.

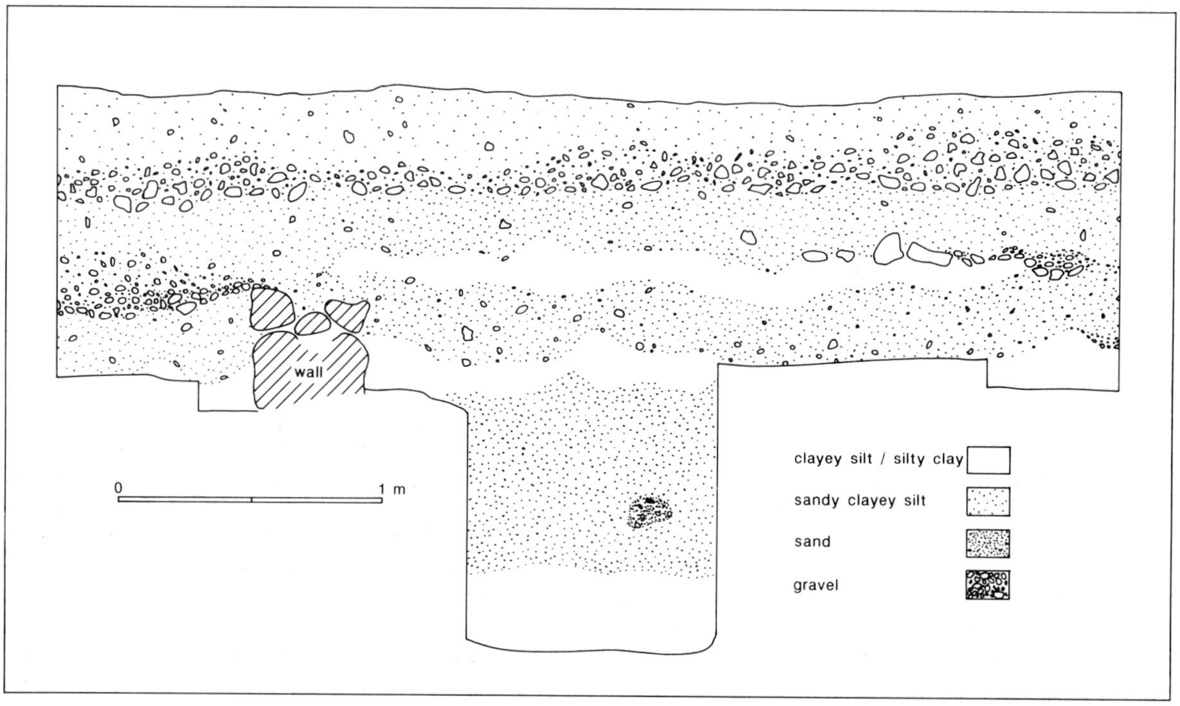


Figure 15.5 Farma Valley – cross section through trench C at Petriolo IIIS, showing colluvial deposits and the buried wall.

| Unit | F5 | F7 | | | F8 | |
|---|---|---|---|---|---|---|
| Sample | M2.5 | P2.4 | 200.1 | 200.3 | 200.4 | 200.5 |
| Taxon | | | | | | |
| TREES/SCRUB | | | | | | |
| Rosaceae (a) | 1 | 1 | | | | |
| Rosaceae (w) | + | | | | | |
| *Cornus* sp. (n) | 1 | | | | | |
| *Cretaegus* sp. (sp) | 1 | | 2 | | | |
| *Carpinus betulus* (n) | 1 | | | | | |
| *Quercus* sp. (w) | + | | | | | |
| *Alnus* sp. (w) | + | | | | | |
| *Populus* sp. (w) | + | | | | | |
| *Rosa* sp. (a) | | 2 | | | | |
| *Ribes* sp. (sp) | | 1 | 2 | | | |
| *Ribes* sp. (a) | | | 1 | | | |
| *Corylus avellana* (n) | | | 1 | | | |
| HERBS | | | | | | |
| Gramineae (s) | 1 | 1 | | | | |
| Gramineae/Cyperaceae (st) | + | | + | | + | + |
| Papilionaceae (s) | 1 | | | | | |
| *Sinapis* sp. (a) | | 10 | 4 | 4 | 5 | 3 |
| *Lepidum ruderale* (a) | | 3 | | 1 | | |
| *Polygonum persicaria* (a) | | 2 | 7 | 2 | | |
| *Rumex* spp. (a) | | 6 | 10 | | 2 | |
| *Thalictrum* aff. *lucidum* (a) | | | 1 | | | |
| Cruciferae (a) | | | 1 | | | |
| Compositae (fl) | | | | | 1 | |
| Chenopodiaceae (a) | | | | | 2 | 1 |
| WETLAND | | | | | | |
| *Carex* spp. (n) | 2 | 1 | 1 | | 1 | |
| *Scirpus* sp. (n) | | 2 | | | | |
| *Ranunculus* aff. *planatifolius* (a) | | | 1 | | | |
| *Ranunculus* aff. *hederacus* (a) | 3 | 7 | 10 | 4 | 1 | 1 |
| *Chara* sp. (oo) | 2 | | 3 | | | |

a = achene; w = wood; sp = spine; st = stem; oo = oogonium; n = nut (let); fl = floret; s = seed.

Table 15.2 Plant macrofossils from the Feccia Valley.

on steep, convex slopes. A range of lithologies is present, including silty diamicts, clast-supported gravels, laminated and massive silty sands, and massive and indistinctly laminated clayey silts. Clasts are angular to subrounded. In the diamicts and gravels, most large clasts are orientated with their long axes pointing downslope. Most sections show crude, elongately lenticular bedding running approximately parallel to the ground surface (Figures 15.4 and 15.5). Beds often fine upwards but occasional large clasts are found in even the most fine-grained horizons. Individual beds vary between 0.25 m and 0.75 m thick. Beds often show signs of weak pedogenesis - darker colouration, some development of weak columnar or blocky textures - in their upper parts.

The diversity of lithologies present in the slope deposits suggests that a number of different mechanisms emplaced them. The poor sorting and crude downslope bedding and fabric of the diamicts suggests that they are most probably the result of deposition from fairly viscous debris flows. Some gravel and boulder sized clasts seem to have concentrated at the base of some of the debris flows, as a 'lag' deposit. Other gravels filling narrow channels incised into the underlying deposits are most probably the result of gullying and rill flow. The massive silty sands and the massive and laminated clayey silts were probably laid down by sheet floods.

| Unit | F5 | | | F7 | |
|---|---|---|---|---|---|
| Sample | M2.5 | P1.1 | P2.4 | 200.1 | 200.3 |
| Taxon | | | | | |
| SHELTERED HABITAT | | | | | |
| *Pagodulina* sp. | 1 | | | | |
| *Aegopinella* sp. | | 1 | | | |
| CATHOLIC | | | | | |
| *Cochlicopa* sp. | 1 | | | | |
| *Vertigo* sp. | 1 | | | 1 | |
| Helicidae (indet.) | 1 | | | | |
| *Trichia* sp. | 1 | 3 | | | |
| *Vitrea* sp. | | | | 1 | 1 |
| OPEN HABITAT | | | | | |
| *Candidula* sp. | | | | 3 | |
| WETLAND | | | | | |
| *Vallonia pulchella* | 2 | | | | |
| *Lymnaea truncatula* | 3 | 1 | | | |
| AQUATIC | | | | | |
| *Valvata piscinalis* | 1 | 1 | | | |
| *Bythinia tentaculata* | 1 | 1 | 1 | | |

Table 15.3 Molluscs from the Feccia Valley

The slope deposits regularly contain fragments of pottery and tile, though no great concentrations of material have been found. The fragments are up to several centimetres across and are always heavily abraded, as would be expected from manuring scatters. Over 100 sherds were examined and a number of pottery fabrics ranging up to, but probably not later than, the early seventeenth century have been recognised. A small, carefully faced dry stone wall was found at a depth of about 0.80 m in the slope deposits in the excavations at Petriolo III South (Figures 15.4 and 15.5). The wall trended west-north-west - east-south-east (downslope). It measured 0.40 m wide and 0.40 m high, but displaced stones found near it suggested that it had once been higher. A fragment of late Medieval tile was found immediately below the wall. Modern plough soils overlying the slope deposits contain potsherds of seventeenth to twentieth century age, in addition to older material.

Molluscs were found in two sections in the Farma Valley (Figures 15.6 and 15.7), where slope deposits were developed on calcareous substrates. Two ecological groups of molluscs are present. The dominant group includes species typical of exposed, dry habitats, such as *Candidula* sp., *Jaminia quadridens, Trochoidea pyramidata, Granaria* sp. and other Chondrinideae. The second group consists of taxa typical of sheltered habitats, including *Pomatias elegans, Helix pomatia, Oxychilus* sp. and an unidentified Clausiliid.

Pollen analysis of the slope deposits was generally unsuccessful, most samples yielding only recycled palynomorphs, fungal hyphae, thermally altered fragments of wood, plant cuticle fragments (mostly of Graminaceae type) and very occasional highly corroded pollen grains, usually Gramineae, Compositae (Liguliflorae), *Plantago, Pinus* and fern spores. A richer and better preserved pollen assemblage (Table 15.4) was recovered from a weak palaeosol, developed on slope deposits and sealed by the small wall and further slope deposits at Petriolo III South. This assemblage was dominated by Gramineae (29%) and weed pollen, but also contained 26% tree and shrub pollen and 21% fern spores. Also of note are the algal microfossils.

Pollen analysis was also successful on the upper part of boulder-rich slope deposits and on the peaty sediments overlying them at section 84/1. The slope deposits at this locality are derived from a slightly cleaved arkosic sandstone and are very acid. Pollen preservation was excellent and the pollen diagram (Figure 15.8) appears well stratified and not subject to problems of downwash or differential preservation. Three pollen assemblages may be distinguished, all dominated by Ericaceae pollen. The lowest assemblage (A) is marked by low incidences of tree pollen and the presence of a number of taxa typical of disturbed and exposed habitats, including *Plantago lanceolata, Artemisia, Hippophaë* and Hypericaceae. The second assemblage (B) also contains low incidences of tree pollen, but is distinguished by a different assemblage of weeds, with the disappearance of the disturbed ground taxa and the incoming of *Galium*. Cereal pollen is present in one sample. The

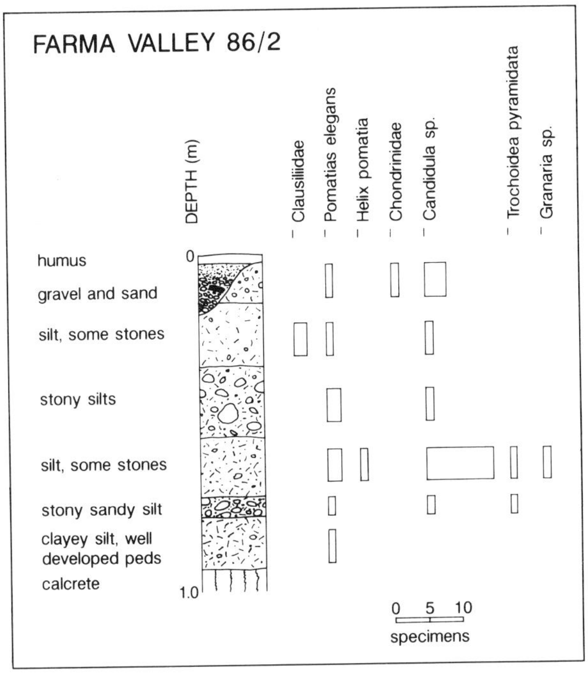


Figure 15.6 Farma Valley - molluscan diagram from colluvial deposits at site 86/2.

| Species | % |
|---|---|
| HERBS | |
| Gramineae | 28.7 |
| Compositae (Liguliflorae) | 14.9 |
| Cyperaceae | 6.0 |
| *Plantago lanceolata* | 5.0 |
| *Spergula* type | 3.0 |
| *Lamium* type | 3.0 |
| Cereal | 2.0 |
| Compositae (Tubuliflorae) | 2.0 |
| *Anagallis* | 1.0 |
| *Solanum nigrum* | 1.0 |
| *Galium* | 1.0 |
| Chenopidiaceae | 1.0 |
| *Rumex* | 1.0 |
| SHRUBS | |
| *Corylus* | 3.0 |
| *Juniperus* | 2.0 |
| Ericaceae | 1.0 |
| *Rhamnus* | 1.0 |
| TREES | |
| *Quercus* | 8.0 |
| *Pinus* | 8.0 |
| *Tilia* | 1.0 |
| FERNS | |
| *Pteridium* | 10.9 |
| Filicales (undiff.) | 8.0 |
| ALGAE | |
| *Spirogyra* sp. | 1.0 |
| *Cosmarium* sp. | 1.0 |

Table 15.4 Pollen from the palaeosol below the wall, Pit C, Petriolo III South.

highest assemblage (C) is marked by higher incidence of *Quercus* and by the presence of *Fagus*, *Ilex* and *Juniperus*. Nevertheless, weed pollen is still present in this assemblage and *Hippophaë*, an indicator of disturbed ground, returns.

**Discussion**

Evidence from the fluvial deposits points to three phases of Holocene coarse sediment aggradation, separated by two phases when meandering channels incised slightly into the previously aggraded sediment. The latest phase of gravel aggradation appears to match the latest phase of slope deposit accumulation identified, of late nineteenth century date. Coarse alluvial deposition of unit F6 in the Feccia Valley appears to have started soon after 1550 and ended before the late eighteenth century, while slope sedimentation at Monterrenti and in the Farma Valley probably took place in the sixteenth and early seventeenth century. There is no dating evidence for the earliest fluvial aggradation phase, but the earliest phase of slope deposit accumulation at Monterrenti appears to have been in the eleventh and twelfth centuries AD.

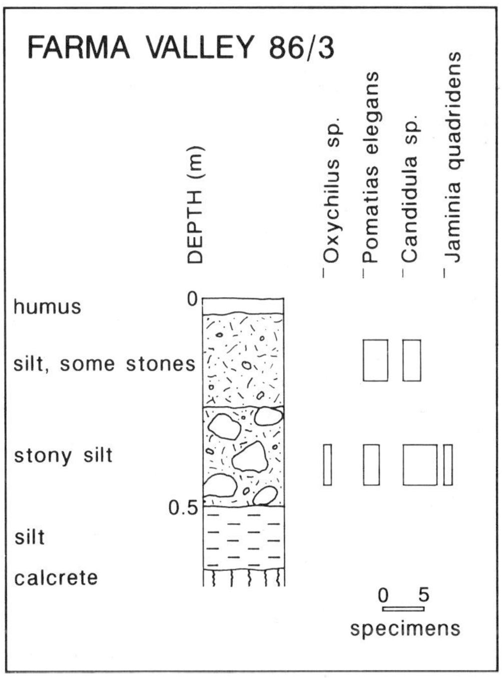


Figure 15.7 Farma Valley - molluscan diagram from colluvial deposits at site 86/3.

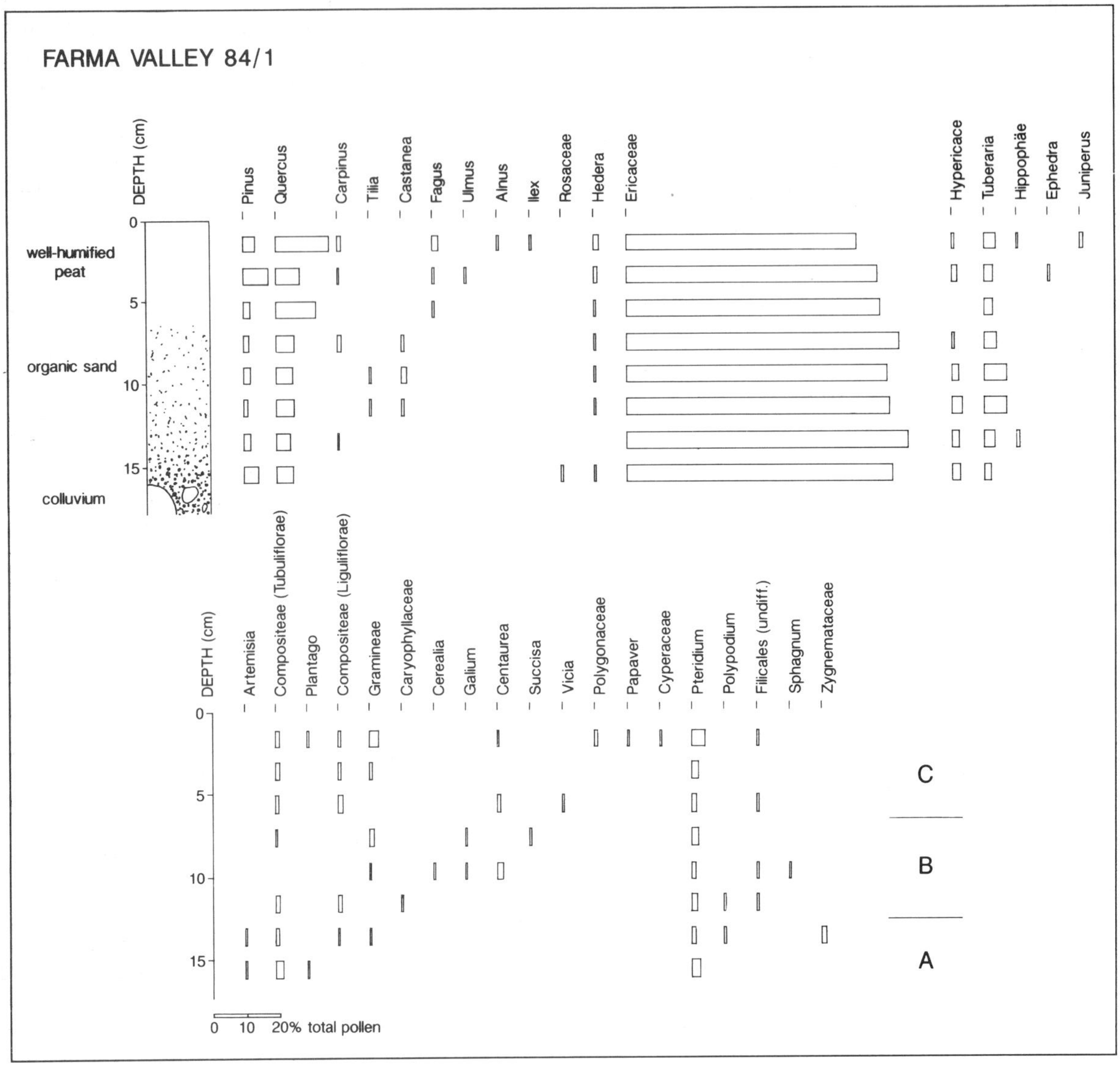


Figure 15.8 Farma Valley - pollen diagram from soil profile overlying colluvium at site 84/1.

The slope deposits seem to have been emplaced by a number of processes, including debris flows and shallow overland flow. These processes, and debris flows in particular, are promoted by steep slopes, lack of vegetation, short periods of abundant water supply and a source providing debris with a muddy matrix. Similarly, exposed soils would be necessary for rain and overland flow to have sufficient erosive effect to lead to large-scale gully and hillwash deposits. A largely devegetated landscape would thus be necessary to promote large-scale mudflows and slopewash. The aggradation of coarse fluvial deposits, particularly on braided floodplains, is similarly linked with 'flashy' discharges and high sediment supply, both characteristic of relatively devegetated landscapes (Briggs and Gilbertson 1980).

The palaeobiological information is variable in quality, but in the fluvial deposits there is a lack of evidence for arboreal or shrubby vegetation contemporary with the aggradation of the large coarse sediment units, though there is restricted evidence for herbaceous vegetation. Briggs *et al.* (1990) have commented on the low probability of deposition of mollusc shells in similar coarse gravel depositional environments.

There is, however, relatively abundant palaeobiological evidence from the approximately coeval slope deposits, both from molluscs and from palynology. In both cases, there are clear indications that the slope deposits accumulated in an environment characterised by patches of bare exposed ground, areas of short, sparse herbaceous vegetation and some light scrub or rock rubble. In particular, the palynological data, with indicators of exposed, disturbed ground such as *Plantago lanceolata, Hippophaë, Artemisia, Spergula,* Chenopodiaceae and Hypericaceae suggest areas of unstable bare soil. Cereal pollen at Petriolo III and 84/1 might suggest some arable farming. The high count for *Pteridium* at Petriolo may be consistent with overgrazing, though it may reflect only preferential preservation of these resistant spores in the soil profile. Other parts of the valley, such as 84/1, were probably covered with scrubby degraded vegetation, in this case heathers. Some trees were also still present, most probably in the upper part of the valley where the studies of Ferrarini and Marracini (1979) do not show disturbance of the forest cover. The later pollen record at 84/1 shows the partial regeneration of the vegetation after disturbance, with first the stabilisation of soils under a cover of tall herbs and scrub and later the spread of trees, including oak, beech, holly and juniper.

Between the episodes of coarse fluvial sedimentation, the meander plugs give an indication of reduced 'flashiness' of discharge and reduced sediment input into the fluvial system. The palaeobiological evidence from unit F5 in the Feccia Valley suggests significant valley floor afforestation, though there is also evidence for open habitats and agriculture, particularly from the palynology (Gilbertson *et al.* 1983). In unit F7, there is no compelling evidence for valley floor woodland, but there is certainly evidence for shrubby vegetation and small trees. There is, however, also clear evidence for disturbed, open habitats, which suggests that the Feccia floodplain was not completely abandoned for agricultural use at this time.

There are differences between the palaeobiological 'signal' from the different phases of activity, and from the different periods of 'recovery'. At Monterrenti, the eleventh-twelfth century slope deposits appear to have accumulated in a relatively grassy, well-vegetated landscape, whereas the sixteenth/seventeenth century slope deposits at Monterrenti and in the Farma Valley appear to have accumulated in more exposed, and thus perhaps more eroded, landscape. Fieldwalking data suggests a major change in manuring patterns with much larger areas manured in the sixteenth century, probably more intensively than in the Medieval period (Barker and Symonds 1984; G.W.W. Barker *pers. comm.* 1987). This may reflect a change in techniques within an agricultural landscape otherwise unchanged, but might also be the result of an expansion in the area farmed. In North Lazio, pollen analysis from the Lago di Monterosi (Bonatti 1970) and Lagaccione (Hunt *in* Barker and Rasmussen 1987) suggests major clearance and increase in the area cultivated around the sixteenth century. The nineteenth century episode seems to have been less widespread and dramatic in its impact, with relatively little accumulation of slope deposits and comparatively fine-grained alluviation in the Feccia Valley. Likewise, the palaeobiological data from meander plug unit F5 in the Feccia Valley suggest a greater degree of reafforestation during the earlier 'recovery phase' than does the data from the later unit F7.

Archaeological evidence found by the Sheffield-Siena Monterrenti Project and the Michigan State University Palaeolithic Archaeology Program suggests that the rather marginal uplands of the Montagnola Senese were subject to only intermittent intensive use during the last thousand years. Field walking and archaeological excavation suggest that major phases of activity occurred in the late eleventh and twelfth centuries, the sixteenth century and the later nineteenth century (Barker and Symonds 1984; G.W.W. Barker *pers. comm.* 1987; R. Colten *pers. comm.* 1988). Historical accounts, for instance Cardini (1986) identify these periods as the times of economic boom in Tuscany. The evidence described above suggests that these phases of economic activity were accompanied by agricultural penetration of marginal upland habitats, which led to episodes of enhanced erosion and consequent colluviation and coarse gravel alluviation.

Comparison with the available palaeoclimatic information for Europe since the eleventh century, as summarised by Lamb (1982) shows that the known climatic events are out of phase with the episodes of sedimentation identified here. The eleventh and twelfth centuries were a time of general warmth in Europe and relatively high rainfall in the Mediterranean. Climatic deterioration occurred in Europe from the thirteenth century, although the fourteenth and fifteenth centuries were mild and equable in southern Europe. The sixteenth and seventeenth centuries were a time of climatic severity in Europe and of very variable climate, with severe droughts and floods in southern Europe. The eighteenth century and early nineteenth century were generally warm, though the later eighteenth century was very variable, with some cold years. The later nineteenth century and earlier twentieth century were cool, before the warming of the 1940s. The eleventh/twelfth century sedimentation therefore accompanied benign climatic conditions, while the sixteenth/early seventeenth century episode took place during a complex time of climatic severity and the nineteenth century episode took place under cool, but not extreme conditions. It is therefore unlikely that these long term climatic events were the immediate cause of the pattern of sedimentation observed in the Montagnola Senese. The link with short duration extreme climatic events is untestable in the Montagnola Senese, given the low resolution of the dating so far achieved, but it seems to be clear that if such events occurred, they have had significant erosive effect only during periods when human activity had largely denuded the landscape of its protective cover of vegetation.

## Conclusion

This study, though limited, provides relatively unambiguous direct sedimentary and palaeo-biological evidence linking phases of increased human activity, most probably largely agricultural, with phases of enhanced sedimentation in the Montagnola Senese. The episodes of sedimentation are clearly out of phase with long term climatic events in southern Europe. If short term extreme climatic events caused some individual sedimentation events (as they no doubt did on occasion) it seems clear that they were only effective during periods when human activity had reduced vegetation cover and made the landscape vulnerable to their effects. In this case, therefore, the link between anthropogenic activity and alluviation is more compelling than the link between climate and alluviation. The position may be different elsewhere, but further research is now needed to clarify the complex relationship between human activity, climate and sedimentation.

**Acknowledgements**

This study owes much to the kindness of Dott. G. Roncaglia and Dr. H. Patterson, who identified the potsherds, Dr. D. Harkness for the radiocarbon dates, Prof. G. Barker and Mr. R. Colten, for discussion of the fieldwalking evidence, Dr. S. Mitchell for historical discussion, the Provincia di Siena for hospitality and numerous students from the universities of Sheffield, Siena and Michigan State who took part in the surveys on which this account is based. We thank the Editors and an anonymous referee for their constructive comments on the manuscript.

**References**

Barker, G. W. W. and Rasmussen, T. 1987 The archaeology of an Etruscan *Polis*: a preliminary report on the Tuscania Project (1986 and 1987 seasons). *Papers of the British School at Rome* 46, 25–42.

Barker, G. W. W. and Symonds, J. 1984 The Monterrenti Survey, 1982-1983. *Archeologia Medievale* 11, 278–289.

Bell, M. 1982 The effects of land-use and climate on valley sedimentation. In: Harding, A.F. (ed.) *Climatic Change in Later Prehistory*. Edinburgh: Edinburgh University Press, 127–142.

Bintliff, J. L. 1977 *Natural Environment and Human Settlement in Greece.* British Archaeological Reports, Supplementary Series 28, Oxford.

Bintliff, J. L. 1982 Climatic change, archaeology and Quaternary science in the eastern Mediterranean region. In: Harding, A. F. (ed.) *Climatic Change in Later Prehistory*. Edinburgh: Edinburgh University Press, 143–161.

Blacknell, C. 1981 Morphology and sedimentary features of point bars in Welsh gravel bed rivers. *Geological Magazine* 118, 181–192.

Bonatti, E. 1970 Pollen sequence in the lake sediments. *Transactions of the American Philosophical Society* 60, 26–31.

Briggs, D. J. and Gilbertson, D. D. 1980 Quaternary processes and environments in the Upper Thames Basin. *Transactions of the Institute of British Geographers* NS 5, 53–65.

Briggs, D. J., Gilbertson, D. D. and Harris, A. L. 1990 Taphonomy in a braided river environment and its implications for studies of Quaternary cold-stage river deposits. *Journal of Biogeography* 17, 623–637.

Cardini, F. 1986 *Tuscany.* SCALA, Firenze.

Chester, D. K. and James, P. A. 1991 Holocene alluviation in the Algarve, Southern Portugal: the case for an anthropogenic cause. *Journal of Archaeological Science* 18, 73–87.

Davidson, D. A. 1971 Geomorphology and prehistoric settlement on the Plain of Drama. *Revue de Géomorphologie Dynamique* 20, 22–26.

Davidson, D. A. 1980 Erosion in Greece during the first and second millennia B.C. In: Cullingford, R. A., Davidson, D. A. and Lewin, J. (eds.) *Timescales in Geomorphology.* New York: Wiley, 143–158.

Davidson, D., Renfrew, C. and Tasker, C. 1976 Erosion and prehistory on Melos: a preliminary note. *Journal of Archaeological Science* 3, 219–227.

Ferrarini, E. and Marraccini; L. 1979 Pollini fossili in depositi lacustri della Valle della Farma (Toscana Meridionale). *Atti Della Societa Toscana di Scienze e Nature, Memorie* ser. B 85, 29–34.

Gilbertson, D. D., Holyoak, D. A., Hunt, C. O. and Paget, F.N. 1983. Palaeoecology of late Quaternary floodplain deposits in Tuscany: the Feccia Valley near Frosini. *Archeologia Medievale* 10, 340–350.

Gilbertson, D. D. and Hunt, C. O. 1987 An outline and synthesis of the geoarchaeological development of the southern Montagnola Senese, Tuscany. *Archeologia Medievale* 14, 349–408.

Lamb, H. H. 1984 *Climate, History and the Modern World.* London: Methuen.

Pope, K. O. and van Andel, T. H. 1984 Late Quaternary alluviation and soil formation in the Southern Argolid: its history, causes and archaeological significance. *Journal of Archaeological Science* 11, 281–306.

Schumm, S. A. 1979 Geomorphic thresholds: the concept and its applications. *Transactions of the Institute of British Geographers* NS 4, 485–516.

Van Andel, T. J., Runnels, C. N. and Pope, K. O. 1986 Five thousand years of land use and abuse in Southern Argolid, Greece. *Hesperia* 55, 1, 103–128.

Van Andel, T. J. and Runnels, C. 1987 *Beyond the Acropolis: A Rural Greek Past.* Stanford: Stanford University Press.

Van Andel, T. J., Zangger, E. and Demitrack, A. 1990 Land use and soil erosion in prehistoric and historical Greece. *Journal of Field Archaeology* 17, 379–396.

Vita-Finzi, C. 1969 *The Mediterranean Valleys.* Cambridge: Cambridge University Press.

Vita-Finzi, C. 1975 Late Quaternary alluvial deposits in Italy. In: Squyres, D. J. (ed.) *Geology of Italy.* Tripoli: Earth Sciences Society of the Libyan Arab Republic.

Vita-Finzi, C. 1976 Diachronism in Old World alluvial sequences. *Nature* 263, 218–219.

Wagstaff, J. M. 1981 Buried assumptions: some problems in the interpretation of the 'Younger Fill' raised by recent data from Greece. *Journal of Archaeological Science* 8, 247–264.

# CHAPTER 16

# EPISODES OF EROSION AND STABILITY SINCE THE LATE-GLACIAL: THE EVIDENCE FROM DRY VALLEYS IN KENT

R.C. PREECE

*An outline of the environmental history of Holywell Coombe, Folkestone, is given, based on an intensive borehole survey and detailed study of representative sections. Waterlogging of basal levels has led to the preservation of datable materials, allowing the establishment of a firm radiocarbon chronology, largely by dating charcoal and seeds by accelerator mass spectrometry. Volumetrically, Late-glacial deposits account for over 60% of the valley infill and in places it is possible to recognize two distinct interstadial horizons. The lower deposits accumulated in marshes between 12-13,000 BP, whereas the upper interstadial horizon is represented by a soil that has yielded dates clustering around 11,500 BP. These interstadial horizons are separated by units of chalk gravel that were emplaced between 11,800-11,500 BP. Above the soil are further units of chalk gravel that are thought to have formed during the Younger Dryas (Loch Lomond Stadial).*

*The Holocene sequence is represented primarily by deposits of calcareous tufa in the early part and chalky silt ('hillwash'), following forest clearance in the Neolithic and early Bronze Age. Two palaeosols within the hillwash reflect periods of slope stability, during the Bronze Age and Iron Age.*

*This sequence is compared with others in Kent and elsewhere.*

## Introduction

The stratigraphy of the infills of chalkland dry valleys (coombes) in south-east England has provided much of the direct evidence relating to the landscape history of this region. The development of coombes has been the subject of protracted debate that is still not fully resolved. Two schools of thought have emerged. One school explains their origin by 'normal' fluvial erosion and spring-sapping (e.g. Fagg 1923; Sparks and Lewis 1957; Small 1965), whereas the second favours development under periglacial conditions characterized by intense physical weathering, freeze-thaw processes, active solifluction and seasonal melt-waters at a time when the Chalk was rendered impermeable by permafrost (e.g. Reid 1887; Kerney *et al.* 1964). There is growing support for a compound explanation which maintains that many coombes originated and evolved through both fluvial erosion and spring-sapping, although they have been greatly influenced by periglacial activity (e.g. Jones 1981).

A major impediment to a full understanding of the origin and antiquity of many coombes is the fact that most of the valley sediments are oxidized and lack datable materials. This largely results from the fact that many of the valley bottoms now lie above the level of the water-table. Although it is clear that many coombe valleys, particularly those studied in Kent, are floored by deposits of Devensian Late-glacial age, none has so far yielded a detailed chronology.

The discovery that the thick basal Late-glacial sediments at Holywell Coombe, Folkestone (Figure 16.1) are waterlogged and, in places, have preserved organic materials, offered an opportunity to establish a detailed record of the landscape history that could be tied to a firm radiocarbon chronology. The importance of this site was first appreciated in the late 1960s when some trial pits were excavated for the Institute of Geological Sciences (now British Geological Survey) in connection with proposals to build a Channel Tunnel. Work on samples taken at that time has already been published elsewhere (Kerney *et al.* 1980). This work demonstrated that a fine Late-glacial and Holocene molluscan succession was present within the colluvial infill and indeed this locality was selected as the regional type site for a series of mollusc zones covering this period. As a

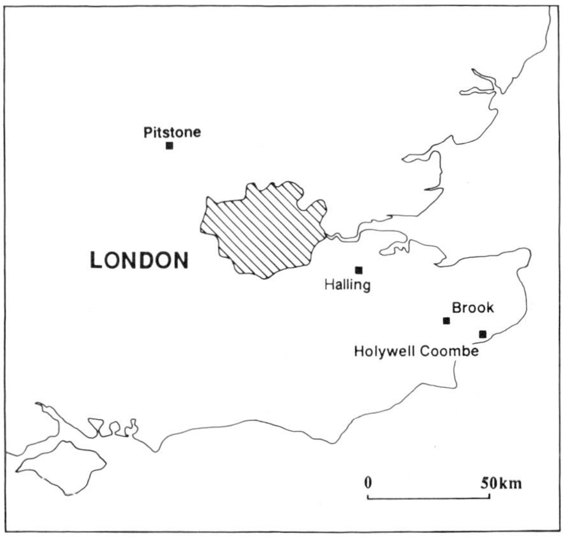


Figure 16.1 Location map of the main sites mentioned in the text.

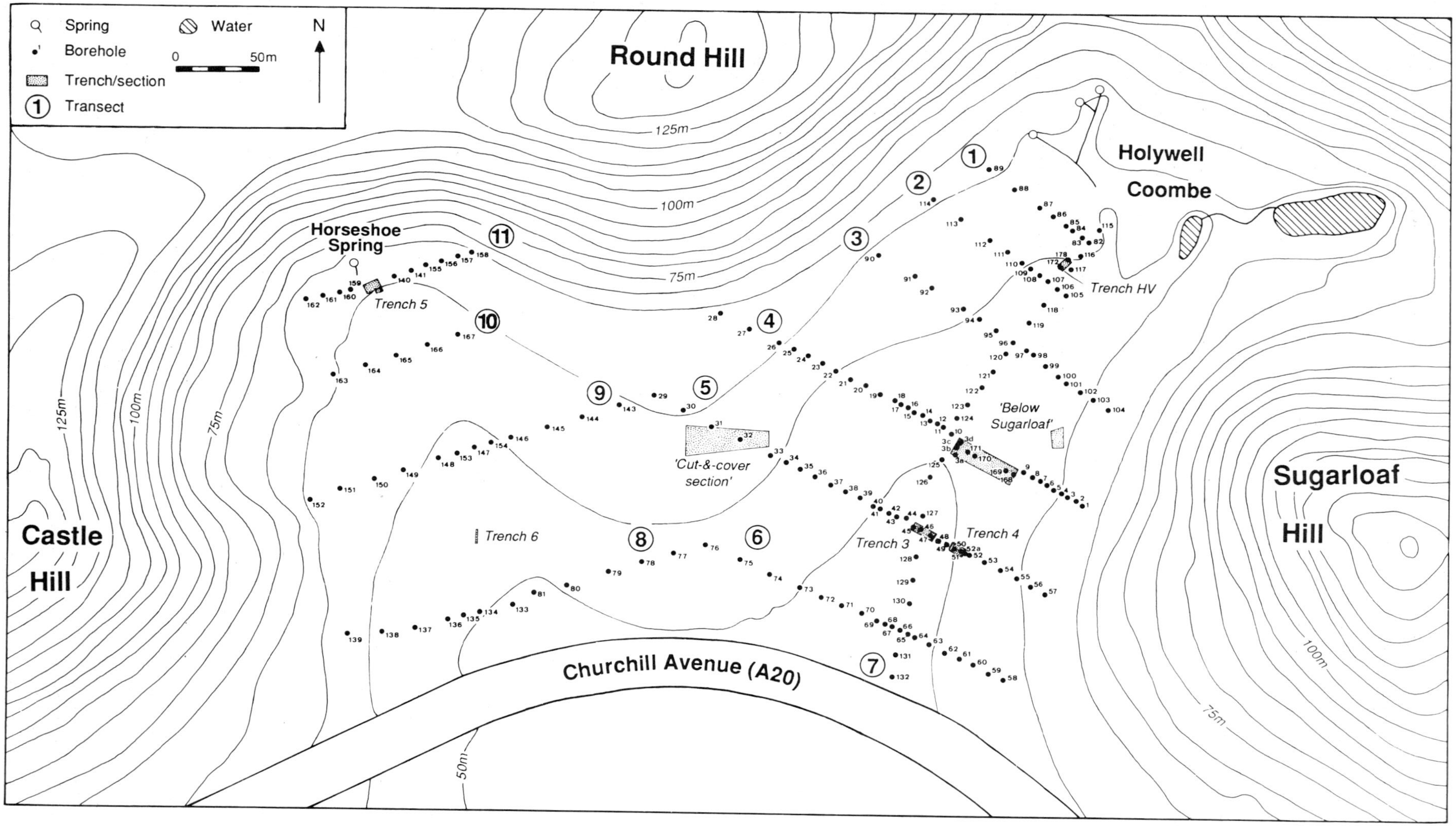


Figure 16.2 Location map showing the position of the boreholes (numbered dots) and trenches.

result of this research, recommendations made to the Nature Conservancy Council (now English Nature) led to the scheduling of the area as a Site of Special Scientific Interest (SSSI). The resurrected plans to proceed with the construction of the Channel Tunnel threatened to destroy this important SSSI, and in July 1987 a major rescue excavation, funded by Eurotunnel, was initiated. This paper summarizes some preliminary conclusions resulting from this new work. Full details will be published elsewhere.

**Location and setting of the site**

Holywell Coombe (NGR TR 220379) is the name given to the valley between Castle Hill and Sugarloaf Hill, immediately north of Folkestone, Kent (Figure 16.2). In fact, this area is occupied by two small valleys, Holywell Coombe proper, that skirts Sugarloaf, and an unnamed western limb flanking Castle Hill, extending down-valley from Horseshoe spring. Both valleys have been cut into the Cretaceous Gault Clay, above which there is an important sequence of colluvial deposits covering a large part of the Late-glacial and most of the Holocene. The stratigraphy essentially consists of a tripartite sequence of (1) basal solifluction debris and associated deposits (Late-glacial) consisting largely of re-deposited Gault Clay, Greensand and Chalk, (2) calcareous tufa (early to mid Holocene), and (3) hillwash (post-Neolithic), but there is much lithological variation, particularly within the Late-glacial (Kerney *et al.* 1980).

**The present project**

The present project began with a detailed borehole survey, aimed primarily at identifying stratigraphically important areas and pinpointing the location of organic deposits. The boreholes were made commercially using a mechanised, petrol-driven, percussion corer with each drive recovering some 1-1.5 m of sediment. The maximum diameter of the corer was 10 cm, but this was reduced to about 5 cm during the deeper drives. The location of each borehole was carefully planned according to a pre-determined grid so that the interval between boreholes was closest in the valley axes (every 5 m or occasionally every 2.5 m), with greater intervals of 10 m or 20 m on the valley sides (Figure 16.2). At the end of the survey, a total of over 180 boreholes had been carefully logged. This detailed survey has provided a reasonably clear picture of the general morphology and spatial disposition of the sedimentary infill of the valleys.

The solifluction deposits are derived from the slopes of Castle Hill, Round Hill and Sugarloaf Hill and from each source they form wedges, thinning downslope. The Round Hill wedge has filled a small valley on the scarp face whereas that emanating from Castle Hill has filled one half of what was once a double valley near Horseshoe spring from the west. The wedge derived from Sugarloaf Hill has partially infilled Holywell Coombe from the east-south-east causing it to be substantially narrowed and realigned from north-east to north-north-east. There is only a thin veneer of soliflluct underlying the valley axes and in places it may be lacking entirely. Although the thickness of soliflluct thins dramatically downslope, there appears to be no appreciable change in thickness down-valley. The margins of solifluction units appear to be eroded and the present valley is in places incised into soliflluct.

Scars are present at various points along the Chalk escarpment, such as on the flanks of Castle Hill and Sugarloaf Hill and it is clear that landslipping has affected much of this area. Further evidence of this has emerged from sub-surface data. In three separate borehole transects at the foot of Sugarloaf Hill (eastern ends of transects 4, 5 and 6; Figure 16.2), the surface of the Gault Clay rises anomalously in a form suggestive of the toes of one or more landslips. Marsh deposits in the hollow immediately in front of these toes (Trench 4) date from the early part of the Late-glacial (12-13,000 BP), providing a minimum age for the landslip. This landslipping clearly contributed to an earlier re-alignment of the valley axis.

The main tufa deposits are confined to the valley axes and generally thin down-valley. A few other localized sources have been identified in areas upslope from the axes and these probably relate to independent spring-heads.

In contrast to the Late-glacial soliflucts, the hillwash forms thin wedges that thicken both downslope and down-valley. The hillwash is very thin on the interfluves and upper slopes, and is the product of slope erosion and valley floor deposition resulting from anthropogenic forest clearance. Deposition is concentrated in various topographic hollows along the valleys which appear to act as effective sediment traps. It would appear that the hollows closest to the contributing slopes are infilled first and only subsequently are the more distal areas supplied. This could explain the differences in basal radiocarbon dates (Preece 1991) and in the number of palaeosols recognized within the hillwash.

Using the borehole data, estimates of the volumes of each major sediment type have been calculated. These estimates relate only to the area covered by the borehole grid (Figure 16.2). Sediments of Late-glacial age are volumetrically by far the most important group, accounting for over 60% of the total. Next are the two deposits of calcareous tufa which, although spatially limited to the valley axes, together account for about 15%. The hillwash never exceeds 2 m in thickness and forms a mantle making up nearly 12% of the sedimentary infill of this area. The modern ploughsoil which would have truncated the upper levels of the hillwash accounts for a slightly lower figure of about 10%.

Having pinpointed several critical areas in the boreholes, it was desirable to re-examine these areas in open section. A series of deep trenches was consequently dug with a hymac excavator. As these trenches extended below the water-table they required shoring, pumping and, in a few cases, well-pointing. Small gaps were deliberately left between each retaining trench sheet to enable critical examination of the sediments and to allow systematic serial sampling.

| | |
|---|---|
| 0–6000 BP | Renewed activation of slope processes following forest clearance. Periods of stability during the early Bronze Age and Iron Age reflected by soil horizons. Zones e and f. |
| 10–6,000 BP | Cessation of slope processes. Development of hazel scrub. Formation of calcareous tufa in the two valleys. Temperate climate. Zones a to d. |
| 10–11,200 BP | Active erosion of valley slopes. Deposition of chalk gravels. Arctic-alpine communities. Cold climate. Zone z. |
| 11,200–11,500 BP | Stability. Soil formation. Minor tufa deposition towards head of Holywell Coombe. Zone z. |
| 11,500–11,800 BP | Active erosion of valley sides. Deposition of thick units of chalk gravel by solifluction. Climatic deterioration. |
| 11,800–13,100 BP | Stability. Deposition of marsh sediments in hollows. Birch-juniper scrub. Temperate climate. Zone y. |

Table 16.1 Summary of the environmental history of Holywell Coombe. The radiocarbon chronology is based on more than two dozen determinations; more complete details are given in Preece (1991). Respective mollusc zones (Kerney *et al.* 1980) are also indicated.

**The Late-glacial succession**

In several places in Holywell Coombe it was possible to subdivide the Late-glacial sediments into successive units. The general environmental sequence is summarized in Table 16.1.

The earliest sediments recognized were interstadial marsh deposits that usually rested directly on the surface of the Gault or occasionally on a veneer of Chalk gravel. These were discovered at several points in Holywell Coombe. No evidence of any pre-interstadial cold stage sediments could be located, except possibly for this gravel veneer. The marsh deposits yielded an impoverished molluscan fauna, dominated by *Pupilla* and *Vallonia*, ascribable to zone y (*sensu* Kerney *et al.* 1980). These accumulated in the period between 12-13,000 BP (Preece 1991). They have also yielded remains of tree birches and a rich thermophilous insect fauna indicating that summer temperatures were broadly similar to those of today.

Towards the top of the marsh deposits, dated to around 12,000 BP, there are changes in the composition of the insect faunas that indicate cooling. The marsh deposits (humic silts) are usually overlain by chalk gravel, that in places can be several metres thick (Figure 16.3). This gravel would seem to represent the culmination of this cooling trend and to reflect a phase of active solifluction from the slopes of Sugarloaf Hill. These chalk gravels are unfossiliferous or virtually so, and it is therefore impossible to gain a clearer insight into the precise thermal environment at the time of their formation. Radiocarbon dates from the adjacent units indicate that the gravel was emplaced between about 11,500 and 11,800 BP.

In several places, particularly on the valley sides, a well developed palaeosol has formed an important marker horizon. This was developed in fine chalk silts at the top of the gravel units and could be recognised as a distinct grey horizon in several borehole transects. Molluscan assemblages from this palaeosol are more diverse (with *Abida, Trichia* and *Arianta*) than those from the underlying marsh deposits and clearly belong to zone z (*sensu* Kerney *et al.* 1980). Several radiocarbon dates clustering around 11,500 BP have been obtained from this unit (Preece 1991). The palaeosol is in turn overlain by a further unit of chalk muds (and gravels). These have also yielded Late-glacial molluscan faunas ascribable to zone z. At one location, near the portal entering Castle Hill, a basal coarse chalk gravel passed into a thin unit of chalk mud yielding remains of arctic-alpine plant and animal communities, including macroscopic remains of *Betula nana* and *Dryas octopetala* itself. These units are thought to have accumulated by solifluction and as snow-melt deposits during the Younger Dryas (Loch Lomond Stadial).

**The Holocene succession**

The early to mid Holocene succession is composed chiefly of calcareous tufa, which in places exceeds 3 m in thickness. This commonly occurs as a soft, pure white sediment ('sinter') with some iron-staining particularly towards the base of the unit. As this was deposited by springs issuing from the two valleys, its occurrence is confined primarily to the valley axes. In places the tufa is mixed with much plant debris, including wood fragments and hazel-nuts. In these

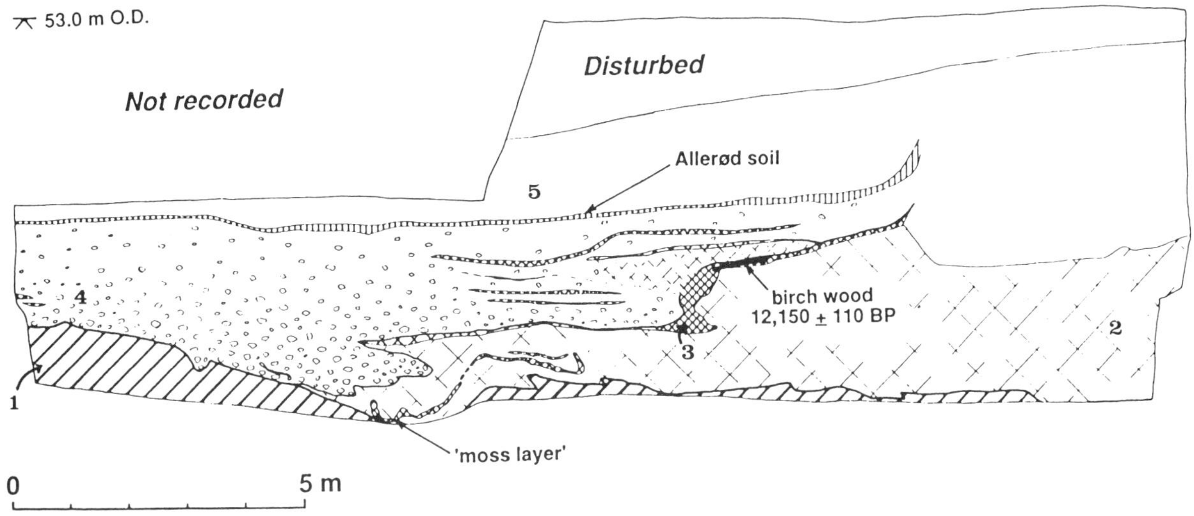


Figure 16.3 Late-glacial section 'Below Sugarloaf Hill' (Figure 16.2).
1. Gault Clay with sharp periglacially scoured surface.
2. Humic chalk muds ('Bølling Interstadial') yielding molluscan assemblages assignable to zone y (*sensu* Kerney *et al.* 1980) and a thermophilous insect fauna.
3. Organic detritus mud forming the surface of unit 2. Beetles suggest cooler climate than unit 2.
4. Chalk gravel (soliflucted).
5. Chalk silts (gravelly in places).
Note the superposition of the 'Allerød soil', unit 4 (? Older Dryas) and unit 2 ('Bølling Interstadial').

situations the tufa becomes distinctly coarser in texture and nodular (onchoidal) facies may be found. Towards the head of Holywell Coombe, tufa formation on a minor scale (lenses up to 10 cm thick), occurred in the latter part of the Late-glacial (about 11,500 BP). However, the main phase of tufa formation began about 10,000 BP.

Biostratigraphical analyses (palaeobotanical, malacological and entomological) show a progressive replacement of open country communities at the beginning of the Holocene by those typical of woodland. It is difficult to date the cessation of tufa formation with any precision, since the upper levels invariably occur above the water-table and so lack datable materials. A date of about 6000 BP seems probable on the basis of the biostratigraphy (cf. Kerney *et al.* 1980).

A unit of chalky silt ('hillwash') forms the upper part of the succession. This rests directly on Late-glacial deposits on the valley sides and over the interfluves, but in the axes it mantles the tufa. Two palaeosol horizons have been recognized in this unit. The earliest occurs at the base and is the most prominent. This has yielded decorated pot sherds belonging to the early Bronze Age, as well as some Neolithic pottery and flint-work. A single microlith was excavated from the base of the palaeosol. The upper palaeosol is much less distinct and only became obvious after the main trench section had been left exposed for over a week, allowing differential drying of this slightly more clay-rich unit. Artifacts from this level were primarily of Iron Age date. Magnetic susceptibility measurements show significant enhancement at these two horizons and there are marked increases in the abundance of shells and artifacts. There are also significant increases in the frequency of shade-demanding taxa, implying the retardation of slope erosion and the regeneration of scrub.

The dating of these palaeosols has proved problematic. Charcoal from the lowest palaeosol yielded a date of 5620±90 BP (OxA-2091), whereas a shell date (using *Arianta arbustorum*) from the same horizon proved to be somewhat younger, 4470±90 BP (OxA-3224). Both dates are older than the Early Bronze Age date suggested by the archaeology. Greater consistency was obtained from dating the upper palaeosol where the determination based on charcoal, 3515±80 BP (OxA-2090), was statistically indistinguishable from one based on *Arianta* shells, 3430±90 BP (OxA-3223). As before, both these dates were older than the archaeological evidence which suggested an Iron Age date.

**Discussion**

The foregoing outline of the environmental history of Holywell Coombe has identified a number of phases of soil erosion and slope stability and it is worth considering whether any of these may have regional significance.

The basal marsh deposits accumulated under temperate conditions during the early part of the Late-glacial, between 12-13,000 BP. Activation of slope processes sometime after about 11,800 BP led, in places, to the accumulation of several metres of chalk gravel. A major period of pedogenesis around 11,500 BP implies a temporary cessation of these slope processes. Renewed erosion of the valley slopes ensued, leading to the deposition of a further unit of chalk gravel.

This sequence of events conforms to the classical model of Late-glacial stratigraphy which recognises two interstadial horizons (Bølling and Allerød) separated by two cold phases, the intervening Older Dryas period and the Younger Dryas period following the Allerød. Studies in Britain, mostly in upland regions of the north and west, have generally failed to distinguish two distinct interstadials and have led to the abandonment of this terminology. Instead, a single interstadial, variously termed the 'Windermere' or 'Late-glacial Interstadial', was recognised, followed by a cold episode called the Loch Lomond Stadial (Coope and Pennington 1977; Lowe and Gray 1980). Much debate surrounded the definition of the boundaries between these stages, as the palaeobotanical response apparently lagged behind that of the insects and lithological boundaries were found to be unreliable guides to climatic history. The general conclusion was that there was 'little direct evidence for continued periglacial activity in the Interstadial' save for some minerogenic inwash into some Scottish lake basins (Lowe and Walker 1984: 338).

The stratigraphical evidence from valley-fills, such as Holywell Coombe, does not support the concept of a unified interstadial. Kerney (1963; 1965) reached the same conclusion nearly thirty years ago. In Kent there is not only a clear lithostratigraphic separation, but there are also important biostratigraphic differences between the two interstadial horizons. The diversity of the molluscan fauna from the upper ('Allerød') horizon is greater than that of the lower. Several taxa with southern modern ranges account for some of this increased diversity. Kerney (1963) took this to be an indication of climatic amelioration. The new work at Holywell Coombe, particularly the beetle evidence (G.R. Coope, pers. comm.), has demonstrated that warmer temperatures were actually attained in the earlier ('Bølling') levels rather than in the later interstadial horizons. This matches the story now obtained from many sites in Britain and elsewhere (Coope 1975; 1977). The real difference from most other British sites concerns the existence of an intervening cold episode. I believe that the differences result from the fact that these Kentish sites are all situated on Chalk bedrock. As Chalk is extremely susceptible to frost-shattering (e.g. Williams 1980), even the most short-lived cold oscillation is likely to produce substantial bodies of solifluction, particularly at the foot of slopes as steep as those of Sugarloaf Hill. In other words the climatic signal will be amplified in areas of frost-sensitive bedrock and will give only a muted response (i.e. insignificant lenses of minerogenic inwash) in areas of hard Palaeozoic bedrock. It is also possible that we are witnessing the effects of a climatic gradient across the British Isles with the more maritime western locations experiencing significantly different regimes to those of south-east England. However, it is interesting to note that some sites in western Britain have also recently been shown to have evidence of 'climatic deterioration prior to 11400 yr BP and a subsequent short-lived recovery some 100 years later based on fluctuations in the *Betula* curve' (Walker and Harkness 1990: 141).

It is worth noting that in some sections (e.g. the main section, Figure 16.4), the thickness of chalk gravel emplaced *between* the two interstadial horizons far exceeds that produced during the Younger Dryas (Loch Lomond Stadial), which is generally considered to have been very much colder. As mentioned earlier, gravel units dating from this period scarcely reach the valley axis of Holywell Coombe. The reasons for this are not entirely clear. The other point to remember is the fact that it is only possible to distinguish between the two gravel units in profiles where the 'Allerød soil' is preserved.

Is the sequence of events seen at Holywell Coombe typical of other dry valleys in Kent and elsewhere? The obvious site that has been studied in similar detail is the Devil's Kneadingtrough near Brook, 16 km north-west of Folkestone (Kerney *et al.* 1964). Much of this valley was floored with shattered chalk debris that extended as fans onto the Gault Clay plain beyond the escarpment. Here these debris fans overlie marsh deposits ascribed to the 'Allerød Interstadial', and the intense period of erosion responsible for their formation has been taken to be the Younger Dryas (Loch Lomond Stadial). Unlike the situation at Holywell Coombe, no evidence for any Late-glacial interstadial deposits was recognized within the Devil's Kneadingtrough itself.

At Brook (Figure 16.1), the key section in the Late-glacial deposits was provided by Borehole III, located in the middle of a more southern lobe of debris (Kerney *et al.* 1964). In this borehole, basal chalk gravel passed into a light grey calcareous mud (330-338 cm), which was overlain by a grey, highly calcareous quartzitic silt with fine plant debris (295-300 cm) and by an organic detritus mud (290-295 cm). This last unit was covered by chalk muds (277-290 cm). The organic sequence culminated with a thin marsh soil with charcoal (274-278 cm), which in turn was covered by a thick series of chalk muds and silts assigned to the Younger Dryas. A radiocarbon date of 11,900±160 BP (Q-618) was obtained from the organic detritus mud. This date has been regarded as being 'Allerød' in age. I think that this is not so for two reasons. First, the date itself is too old and significantly earlier than other dates obtained from the 'Allerød soil' in Holywell Coombe, which cluster

Figure 16.4 Main section at Holywell Coombe showing the dated 'Allerød soil' (1), the mid Holocene tufa (2) thinning upslope and the hillwash with two well-developed palaeosols (3 and 4), that have yielded early Bronze Age and Iron Age artifacts respectively. Some Neolithic material was recovered from the base of unit 3. Chalk gravels extend for a further 2 m below the 'Allerød soil'. An organic detritus mud, dated to 11,820±140 BP (Q–2720), was present towards the base of the Chalk gravels just above the Gault Clay below the far left of the picture.

around 11,500 BP (Preece 1991). Second, the molluscan faunas below this level belong to zone y (*sensu* Kerney *et al.* 1980) whereas those above it are ascribable to zone z. Although no molluscs were reported from the detritus mud itself, it would seem likely from the date that this is a pre-Allerød marsh deposit equivalent to those seen in Holywell Coombe which have *all* produced zone y molluscan assemblages. The true Allerød horizon would seem to be represented by the thin 'marsh soil' 15 cm higher in the sequence. The condensed nature of this profile, compared with others in Holywell Coombe, may result from its more distal position, over 1 km from the Chalk escarpment. The important conclusions regarding the dating of the emplacement of the overlying gravels remain largely unaffected by this reinterpretation. It does suggest that a lithostratigraphic separation of two distinct interstadial horizons is a widespread phenomenon in Kent.

Kerney *et al.* (1964) calculated that about a third of the chalk eroded from the Devil's Kneadingtrough was still present within a radius of about 1 km from its source. As the authors point out, even this astounding figure is likely to be a serious under-estimate, since a considerable amount of calcium carbonate is likely to have been carried out of the area in suspension or in solution or lost by subsequent weathering. This erosion was thought to have been accomplished entirely during the Younger Dryas (Loch Lomond Stadial) but it would seem from the foregoing reinterpretation that this does include a minor pre-Allerød component.

At Holywell Coombe it was not possible to map the fans of solifluction debris to their terminal limits, since these would have extended across what is now the A20 and into the area now occupied by an industrial estate. It is therefore not possible to provide comparable volumetric estimates that encompass the sediments in full measure. One fact is clear. The volume of sedimentary infill of the present valley falls a long way short of filling the cavities of the coombes. This raises the interesting question of the ultimate origin of the coombe and the absence of any pre-Late Devensian sediments. The volumetric discrepancy between the sedimentary infill and the size of the valley does suggest that the coombe has a pre-Late Devensian origin, but it is impossible to be more precise. Although the earliest Pleistocene sediments filling both the Devil's Kneadingtrough and Holywell Coombe are Late-glacial in age, other valleys such as that at Halling, in the Medway Valley (Kerney 1971), do have Full-glacial sequences below typical Late-glacial profiles.

Outside Kent, Pitstone in Buckinghamshire (Figure 16.1) is probably the best studied site covering this period. Evans (1966) described both transverse and longitudinal sections through the infill of two scarp-face coombes exposed by quarrying. Above a basal unit of unbedded chalk rubble (coombe rock) were deposits of weakly stratified chalk muds and fine rubbles within which a prominent palaeosol ('Allerød' soil) had formed. This unit was overlain by coarse chalk and flint gravels of Younger Dryas age (Loch Lomond Stadial). In places this buried soil had a double profile with a lower, more weakly developed horizon occurring intermittently a few centimetres below the main upper unit. Evans (1966) thought that this lower horizon may have formed before the cessation of active downwashing and that there may have been a temporary cold phase between the two units. The molluscan faunas recovered from both horizons are composed of species of open country, and both belong to zone z. In other words, this soil relates only to events during the 'Allerød' and not to any earlier interstadial episode.

An independent stratigraphical terminology has been proposed for 'soil stratigraphic units'. Following Evans' (1966) detailed work at Pitstone, Rose *et al.* (1985) formally named the palaeosol ('Allerød soil'), the 'Pitstone Soil', which they regarded as having a characteristic profile. Some accelerator dates have recently been obtained from charcoal extracted from the Pitstone Soil (Evans 1986). These gave quite a scatter of ages, but the most reliable date was said to be 10,900±130 BP (OxA-415). Although it is likely that the Pitstone Soil is the equivalent of other 'Allerød soils' studied in Kent, there are some notable differences. First, at many sites in Kent, the equivalent soil does not show a double profile. Second, the dates from the Kentish soils are significantly older than the published dates from Pitstone.

Clearly there is scope for much further work on the stratigraphy and dating of valley infills and it remains to be seen to what extent the story emerging from Kentish valleys holds good for other parts of Britain.

**Acknowledgements**

I would like to thank all my collaborators on the Holywell Coombe project, particularly Drs. D.R. Bridgland, G.R. Coope and M.J. Sharp and members of the Canterbury Archaeological Trust. Special thanks are due to Eurotunnel who have funded much of this new work. The two editors, Martin Bell and John Boardman, and an anonymous referee, kindly commented on an earlier draft.

**References**

Coope, G. R. 1975 Climatic fluctuations in north-west Europe since the last interglacial, indicated by fossil assemblages of Coleoptera. In: Wright, A. E. and Moseley, F. (eds.) *Ice Ages ancient and modern. Geological Journal Special Issue* No. 6, 153–168.

Coope, G. R. 1977 Fossil coleopteran assemblages as sensitive indicators of climatic changes during the Devensian (last) cold stage. *Philosophical Transactions of the Royal Society of London* B, 280, 313–340.

Coope, G. R. and Pennington W. 1977 The Windermere Interstadial of the Late Devensian. *Philosophical Transactions of the Royal Society of London* B, 280, 337–339.

Evans, J. G. 1966 Late-glacial and Post-glacial subaerial deposits at Pitstone, Buckinghamshire. *Proceedings of the Geologists' Association* 77, 347–364.

Evans, J. G. 1986 Radiocarbon dates from the Pitstone Soil at Pitstone, Buckinghamshire. In: Gowlett, J. A. J. and Hedges, R. E. M. (eds.) *Archaeological Results from Accelerator Dating*. Oxford University Committee for Archaeology, Monograph 11, 91–93.

Fagg, C. C. 1923 The recession of the Chalk escarpment and the development of Chalk valleys in the regional survey area. *Proceedings of the Croydon Natural History and Scientific Society* 9, 93–112.

Jones, D.K.C. 1981 *The geomorphology of the British Isles: Southeast and southern England*. Methuen, London and New York.

Kerney, M. P. 1963 Late-glacial deposits on the Chalk of south-east England. *Philosophical Transactions of the Royal Society of London* B, 246, 203–254.

Kerney, M. P. 1965 Weichselian deposits on the Isle of Thanet, East Kent. *Proceedings of the Geologists' Association* 76, 269–274.

Kerney, M.P. 1971 A Middle Weichselian deposit at Halling, Kent. *Proceedings of the Geologists' Association* 82, 1–11.

Kerney, M. P., Brown, E. H. and Chandler, T. J. 1964 The Late-glacial and Post-glacial history of the Chalk escarpment near Brook, Kent. *Philosophical Transactions of the Royal Society of London* B, 248, 135–204.

Kerney, M. P., Preece, R. C. and Turner, C. 1980 Molluscan and plant biostratigraphy of some Late Devensian and Flandrian deposits in Kent. *Philosophical Transactions of the Royal Society of London* B 291, 1–43.

Lowe, J. J. and Gray, J. M. 1980 The stratigraphic subdivision of the Lateglacial of North-west Europe. In: Lowe, J. J., Gray, J. M. and Robinson, J. E. (eds.) *Studies in the Lateglacial of North-west Europe*. Pergamon, Oxford and New York, 157–175.

Lowe, J. J. and Walker, M. J. C. 1984 *Reconstructing Quaternary Environments*. Longman, London and New York.

Preece, R. C. 1991 Accelerator and radiometric radiocarbon dates on a range of materials from colluvial deposits at Holywell Coombe, Folkestone. In: Lowe, J. J. (ed.) *Radiocarbon dating: recent applications and future potential. Quaternary Proceedings* 1, 45–53.

Reid, C. 1887 On the origin of dry Chalk valleys and of Coombe Rock. *Quarterly Journal of the Geological Society of London* 43, 364–373.

Rose, J., Boardman, J., Kemp, R. A. and Whiteman, C. A. 1985 Palaeosols and the interpretation of the British Quaternary stratigraphy. In: Richards, K. S., Arnett, R. R. and Ellis, S. (eds.) *Geomorphology and soils*. London: George Allen and Unwin, 348–375.

Small, R. J. 1965 The role of spring sapping in the formation of chalk escarpment valleys. *Southampton Research Series in Geography* 1, 3–29.

Sparks, B. W. and Lewis, W. V. 1957 Escarpment dry valleys near Pegsdon, Hertfordshire. *Proceedings of the Geologists' Association* 68, 26–38.

Walker, M. J. C. and Harkness, D. D. 1990 Radiocarbon dating the Devensian Lateglacial in Britain: new evidence from Llanilid, South Wales. *Journal of Quaternary Science* 5, 135–144.

Williams, R. B. G. 1980 The weathering and erosion of chalk under periglacial conditions. In: Jones, D. K. C. (ed.) *The shaping of southern England*. Institute of British Geographers Special Publication, No. 11, 225–248.

# CHAPTER 17

# PATTERNS AND RATES OF CONTEMPORARY SOIL EROSION DERIVED USING CAESIUM–137: MEASUREMENT, ANALYSIS AND ARCHAEOLOGICAL SIGNIFICANCE

T. A. QUINE AND D. E. WALLING

*Contemporary soil erosion on agricultural fields has been investigated using the Caesium–137 ($^{137}Cs$) technique. Using this technique a single site visit can yield point estimates of rates of erosion and deposition, the spatial resolution of which is limited only by sampling density. These rates are retrospective, representing the net effect of soil redistribution over the past 30 to 35 years. It is therefore possible to (1) map patterns of erosion and deposition across the landscape, (2) establish gross, net and mean erosion rates over defined areas, and (3) estimate sediment delivery ratios for such areas. The high spatial resolution of the erosion rate data has also permitted analysis of topographic controls upon rates of erosion and deposition.*

*It is possible to identify both direct and indirect applications of such studies of contemporary soil erosion to archaeological investigations. Direct applications which are considered include the implications of the measured erosion rates for archaeological site survival on agricultural land, the influence of spatially-varied erosion rates upon the distribution of artifacts identified in field survey, and the potential for immediate retrospective measurement of rates of erosion on archaeological sites. Indirect applications considered relate to the inference of past erosion rates, with particular reference to sediment delivery ratios and sediment budgets.*

## Introduction

Patterns and rates of contemporary soil erosion have been recognised both as a topic of interest for the geomorphologist and as an area of concern to the environmentalist (e.g. Boardman *et al.* 1990). They are also of vital importance to the archaeologist investigating the surviving archaeological record and interpreting past people–land interactions. This paper will describe the recently developed caesium–137 ($^{137}Cs$) technique for soil erosion measurement, the data which it can provide, and potential archaeological applications of both the method and the data.

## The caesium–137 technique

Since the early work of Rogowski and Tamura (1965; 1970), $^{137}Cs$ has been used to trace sediment movement in a wide range of environments (Figure 17.1) and the $^{137}Cs$ technique has become established as a valuable tool in soil erosion research (e.g. Campbell *et al.* 1986; De Jong *et al.* 1983; Ritchie *et al.* 1974; Walling *et al.* 1986; Zhang *et al.* 1990). As the basis of the technique is well documented (Campbell 1983; Walling and Quine 1990a, 1991a; Wise 1980) the description below is limited to a short summary of the salient features.

$^{137}Cs$ is an artificial radionuclide with a half-life of 30.17 years which may be used as a sediment tracer. Measurable quantities of $^{137}Cs$ were released into the stratosphere as a result of atmospheric testing of thermonuclear weapons from the mid-1950s to the 1980s. The $^{137}Cs$ was distributed globally through the stratosphere and deposited as fallout, mostly in association with precipitation. Although significant regional variation exists in the total fallout input (Cawse and Horrill 1986), deposition over an extended period of time led to essentially uniform local distributions of the radionuclide. However, the particular value of $^{137}Cs$ lies in the rapid and strong adsorption of the radionuclide by soil particles, particularly the clay fraction, in most environments (Bachhuber *et al.* 1982; Klobe and Gast 1970; Livens and Baxter 1988; Livens and Loveland 1988; Squire and Middleton 1966). Despite the close affinity of $^{137}Cs$ for the clay fraction, effective 'capture' of $^{137}Cs$ may be expected in most soils regardless of texture, because the rates of $^{137}Cs$ fallout deposition were very low and even small proportions of fine material within the soil profile provide sufficient adsorption capacity (Quine and Walling 1992). Subsequent lateral redistribution of $^{137}Cs$ takes place in association with sediment particles as a result of erosion, transport and deposition processes. Therefore, measurements of radiocaesium redistribution may be used to investigate the erosion and deposition of soil particles.

Estimation of soil erosion rates using $^{137}Cs$ is based upon measurement and analysis of the spatial distribution of caesium inventories, which are expressions of the $^{137}Cs$ activity within the soil profile per unit surface area. The reference inventory for the location of interest is established by sampling an undisturbed, uneroded location, where it may be assumed that the amount of $^{137}Cs$ in the soil will be equal to the total amount of fallout less natural loss by radioactive decay. This reference inventory forms the baseline for identification of erosion and deposition and, therefore, must be established with great care. Where possible it should be checked against independent measures of atmospheric fallout such as

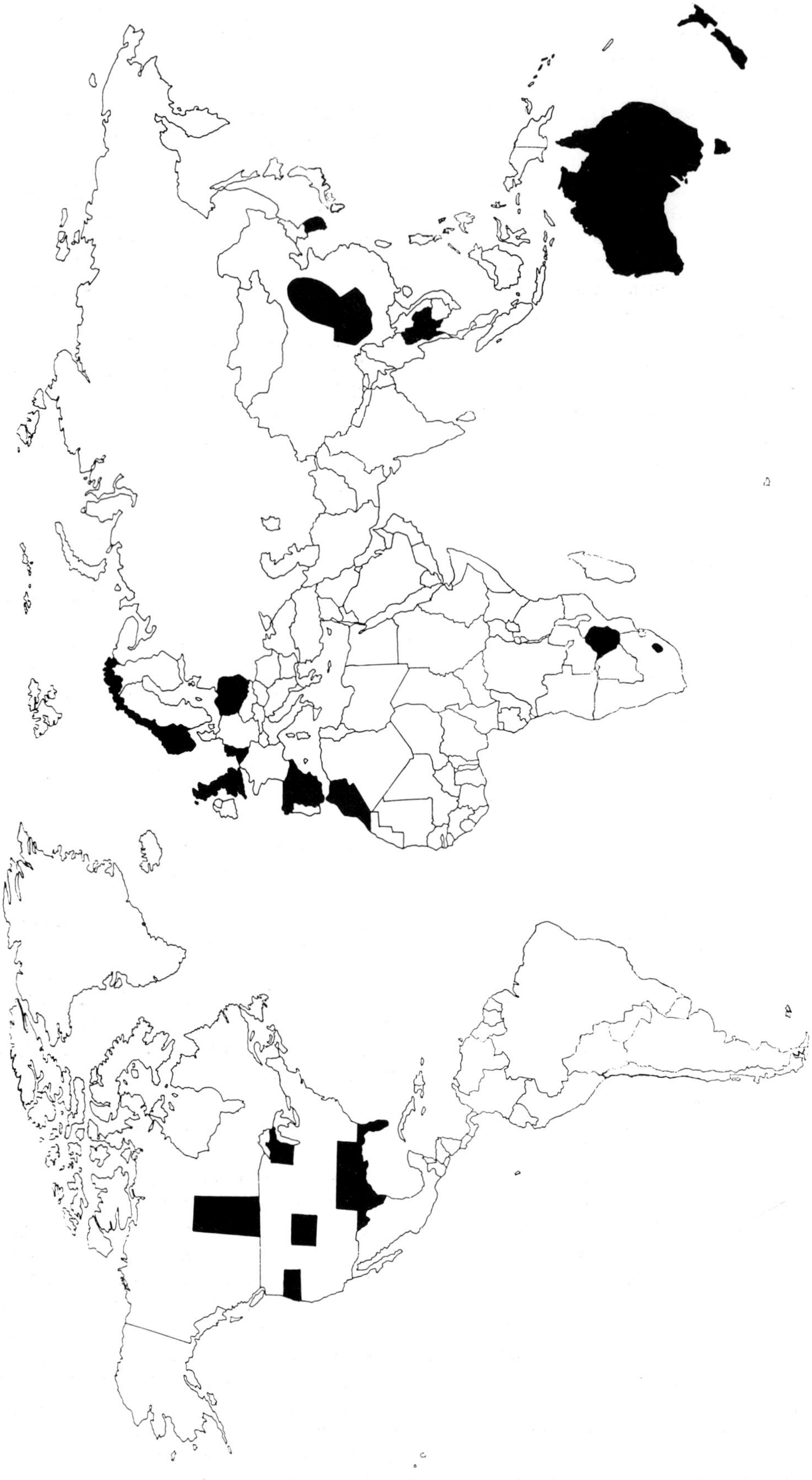

Figure 17.1 World map illustrating the range of environments in which the $^{137}Cs$ technique has been used successfully.

those carried out by the United Kingdom Atomic Energy Authority (e.g. Cawse and Horrill 1986). Inventories from the study field or catchment are compared with this reference inventory; where the point inventory is less than the reference inventory erosion is indicated, and where the point inventory exceeds the reference inventory deposition is evidenced. Therefore, the spatial distribution of $^{137}Cs$ inventories may be used to infer patterns of soil redistribution. Rates of soil redistribution are estimated by establishing calibration relationships between $^{137}Cs$ loss and gain, and soil loss and gain (Walling and Quine 1990b). At each sampling point, measurement of the radiocaesium inventory may be used to calculate a point rate of soil redistribution and it is possible to plot spatial distributions of soil erosion and aggradation rates based upon the $^{137}Cs$ data (Figure 17.2). The spatial resolution of such distributions is defined by the sampling density which may be adapted according to the resolution required. It is also possible to calculate a variety of whole-field parameters by integrating the point values of erosion and deposition. These parameters include:

*Gross Erosion Rate*: the total mass of eroded soil divided by the total area

*Net Erosion Rate:* total erosion minus total aggradation divided by the total area

*Mean Erosion Rate*: the mean rate of erosion from those areas of the field subject to net soil loss

*Sediment Delivery*: the ratio of net erosion to gross erosion

In addition to its use for soil erosion studies, $^{137}Cs$ may also be employed in other sediment tracing applications. These include the examination of rates and patterns of floodplain sedimentation (Walling and Bradley 1989; Walling *et al.* 1992). The approach is based upon comparison of measured $^{137}Cs$ inventories on the floodplain with the local reference inventory. Where sedimentation has taken place the sampling site will have received $^{137}Cs$ both from atmospheric fallout and from deposition of $^{137}Cs$-labelled sediment eroded from the upstream catchment and transported by the river during overbank flood events. The rate of sedimentation will therefore be reflected in the degree to which the inventory for the sampling point exceeds the reference inventory (i.e. the excess $^{137}Cs$). Conversion of the $^{137}Cs$ data to actual rates of sedimentation may be based upon analysis of the depth distribution of $^{137}Cs$ or on the magnitude of the excess $^{137}Cs$ (Walling *et al.* 1992; Walling and He 1992). Although $^{137}Cs$-labelled sediment will be dominated by the clay and silt fractions, it is possible to adapt the calibration procedure to account for variation in the particle size distribution of deposited sediment by establishing the relationship between excess inventory and depth of sediment deposition at a range of localities, over the study area, characterised by different particle size distributions. $^{137}Cs$ measurements may therefore be used directly to examine patterns and relative rates of fine sediment deposition on floodplains and, after conversion, to map actual rates of floodplain sedimentation (e.g. Figure 17.3). The values obtained are again average values for the period extending from the onset of significant fallout (mid-1950s) to the time of sampling.

The major constraints upon the application of the $^{137}Cs$ technique are the expense of the gamma spectrometry equipment and the long count times required for measurement of $^{137}Cs$ (usually 8 hours per sample). However, in the context of soil erosion studies, the $^{137}Cs$ technique offers the following suite of advantages (2 to 7 also apply to floodplain studies):

1. Relates to all erosive processes.
2. Both erosion and deposition may be estimated.
3. Identification of both patterns and rates.
4. Provision of long-term rates (*c.* 35 years).
5. High spatial resolution (determined by sampling density).
6. Whole study area may be examined with minimal disturbance.
7. Only a single site visit is required.

The particular relevance of these advantages, in the context of the archaeological significance of contemporary erosion, will be highlighted in the following discussion of both direct and indirect applications of the $^{137}Cs$ technique and the data derived from its use.

### Archaeological applications of contemporary erosion rate data

Two broad groups of potential archaeological applications for data obtained from studies of contemporary soil erosion may be identified. First, contemporary rates may be utilised directly in the assessment of the survival and integrity of the archaeological record. Secondly, contemporary data may be employed indirectly in the estimation and interpretation of past erosion rates. These two broad groups encompass a range of applications which will be considered in the following sections.

### Survival of the archaeological record

Whilst elevated rates of agricultural erosion have been recognised with concern by archaeologists, the true impact upon the survival of the archaeological record has been difficult to establish due to the paucity of quantitative data. The capacity of the $^{137}Cs$ technique to provide both quantitative estimates of rates of soil redistribution and evidence of the spatial distribution of erosion and deposition greatly facilitates examination of this problem. Within the broad theme of survival of the archaeological record three areas can be identified in which $^{137}Cs$-derived data are of particular significance. First, the existence of spatial variability in erosion rates has important implications for the interpretation of archaeological survey data based upon surface artifact distributions. Secondly,

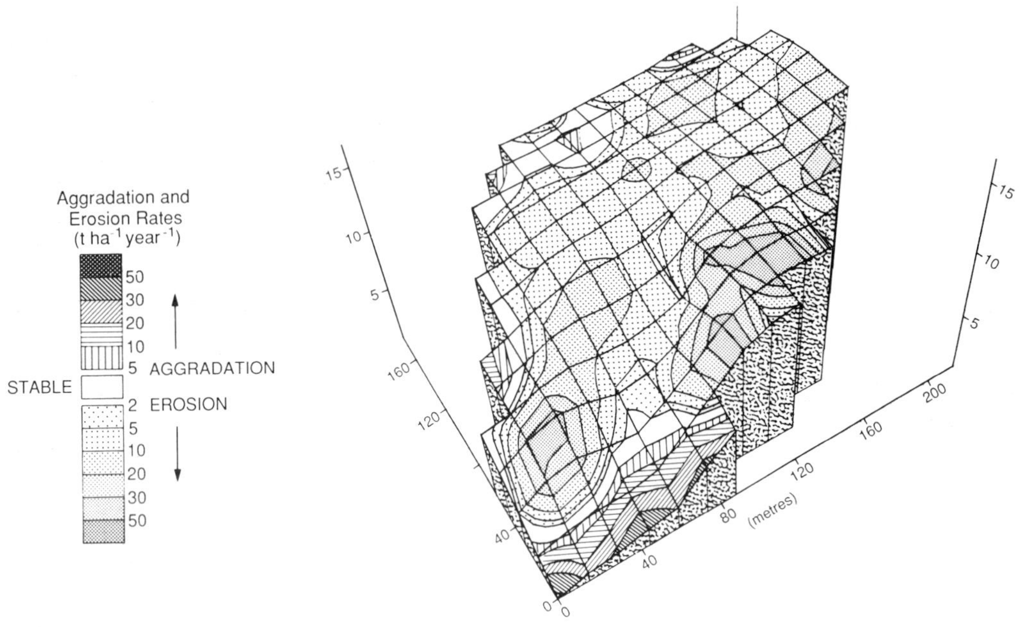


Figure 17.2 Erosion and deposition rates, derived from $^{137}Cs$ measurements, within the field at Dalicott Farm, Shropshire (based on Walling and Quine 1991a).

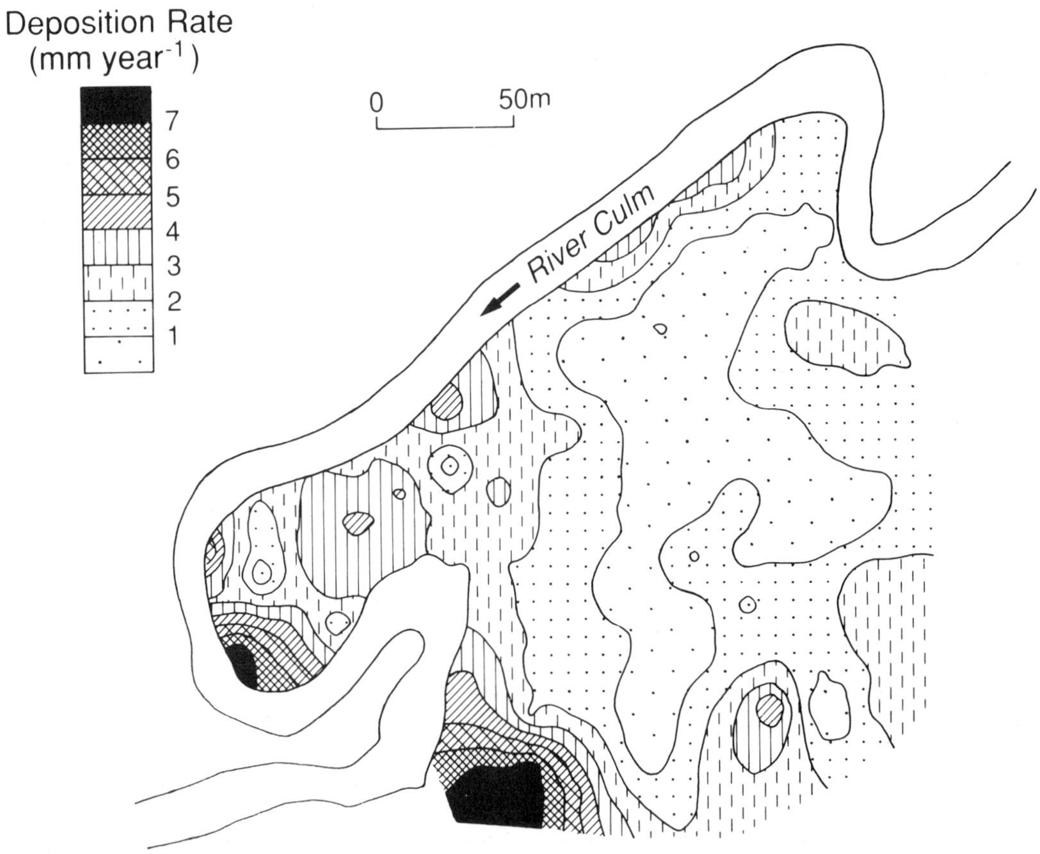


Figure 17.3 The spatial pattern of floodplain sedimentation, derived from $^{137}Cs$ measurements, over a study reach of the River Culm, Devon (based on Walling *et al.* 1992).

the existence of within-field deposition can lead to underestimation of the local severity of erosion when assessments are based upon net rates of soil loss from a field or catchment. Recognition of this is essential to realistic evaluation of risks to sites of interest. Thirdly, the capacity for one-off erosion surveys using $^{137}Cs$ provides the potential for site-specific investigations to be incorporated into cultural resource management strategies. Such site-specific data could also be used to evaluate site formation models which would otherwise remain untestable. The following discussion will examine these three areas of direct application of $^{137}Cs$-derived erosion rate data to archaeological investigations.

*Spatial variability of erosion - implications for survey*

The results of a $^{137}Cs$-based survey of soil erosion, described elsewhere (Walling and Quine 1990a, 1991a), have highlighted the local spatial variability of both erosion and deposition on agricultural land in the UK. As shown in Figure 17.2, soil redistribution rates may vary from –20 (erosion) to +20 (deposition) t $ha^{-1}$ $year^{-1}$ over distances of less than 100 metres. Such variations have important implications for surface distributions of artifacts recorded in field survey. Allen (1991) has provided a valuable summary of some of the potential impacts of erosion and deposition upon surface distributions of artifacts, with special reference to the chalk downlands of south-east England. These impacts include the influence of soil erosion and deposition upon concentrations of artifacts in the plough zone, and therefore at the soil surface, and differential artifact displacement by high energy erosion events.

Contemporary soil erosion data may contribute to an understanding of the extent of these impacts upon the archaeological record. Examination of $^{137}Cs$-derived patterns of erosion will allow identification of zones suffering high average rates of soil loss (e.g. in excess of 20 t $ha^{-1}$ $year^{-1}$) which are likely to have been subject to episodic extreme events capable of artifact displacement. However, the most important contribution of $^{137}Cs$-derived erosion rate data to the understanding of spatial surveys is the provision of information on patterns of erosion and deposition and their implications for artifact incorporation into the plough layer. As soil erosion leads to surface lowering, artifacts buried at a uniform depth will be first incorporated into the plough zone and exposed at the soil surface in areas exhibiting the highest rates of erosion. Such areas of high rates of erosion are often partially topographically controlled (Quine and Walling 1992), with the consequence that patterning of surface artifact distributions may be created as a result of the impact of erosion upon subsurface distributions. Such 'spurious' patterns could bear no relationship to the true artifact distributions and could lead to erroneous interpretations of the extent and location of sites of archaeological significance. This is illustrated for a hypothetical landscape segment in Figure 17.4. Figure 17.4a shows a hypothetical sub-surface artifact distribution which, for the sake of simplicity, is assumed to be located at an initial uniform depth below the base of the plough zone. Figure 17.4b shows a simple modelled distribution of soil erosion and deposition, based upon topographic variables. Figure 17.4c illustrates the initial pattern of the resultant surface distribution of artifacts incorporated into the plough zone through surface lowering by erosion. In the absence of artifact displacement by extreme events, this 'spurious' pattern would be reinforced as continued removal of fine sediment from areas suffering from erosion would lead to concentration of artifacts in the plough layer, which in turn would result in higher frequency artifact exposure at the surface of the soil in eroding areas (Allen 1991). Where high energy erosion events lead to displacement of coarse material further 'spurious' patterns may develop in the artifact distribution through concentration of artifacts in depositional zones which are often located at the base of slopes (Figure 17.4b).

It is clearly imperative that the impact of spatially discontinuous soil erosion should be taken into consideration when examining patterns of artifacts identified by surface survey. A first approximation of this impact may be made by consideration of potential topographic controls upon erosion. However, it should be recognised that both the rates of erosion and the potential controls will vary according to the soil type. Furthermore, in a recent investigation of topographic controls on patterns of soil erosion and deposition based on $^{137}Cs$-derived estimates of soil redistribution, Quine and Walling (1992) found that less than 50% of the variation in rates of erosion and deposition could be explained by reference to selected topographic variables alone, and in many cases the degree of explanation was much lower. Despite this, it was possible to produce a tentative identification of sensitive areas on a range of soil types (Table 17.1). Such sensitive areas may provide initial guidelines for the evaluation of potentially 'spurious' patterns in the surface distribution of artifacts.

| Soils | Zones |
|---|---|
| Sandy }<br>Chalky}<br>Silty} | Crests and spurs, upper slopes, high angle |
| Clay | Valley bottoms |

Table 17.1 Tentative identification of areas susceptible to erosion (based on Quine and Walling 1992)

*Rates of Erosion - Implications for site survival*

The identification of areas sensitive to erosion is also of fundamental importance in the assessment of site

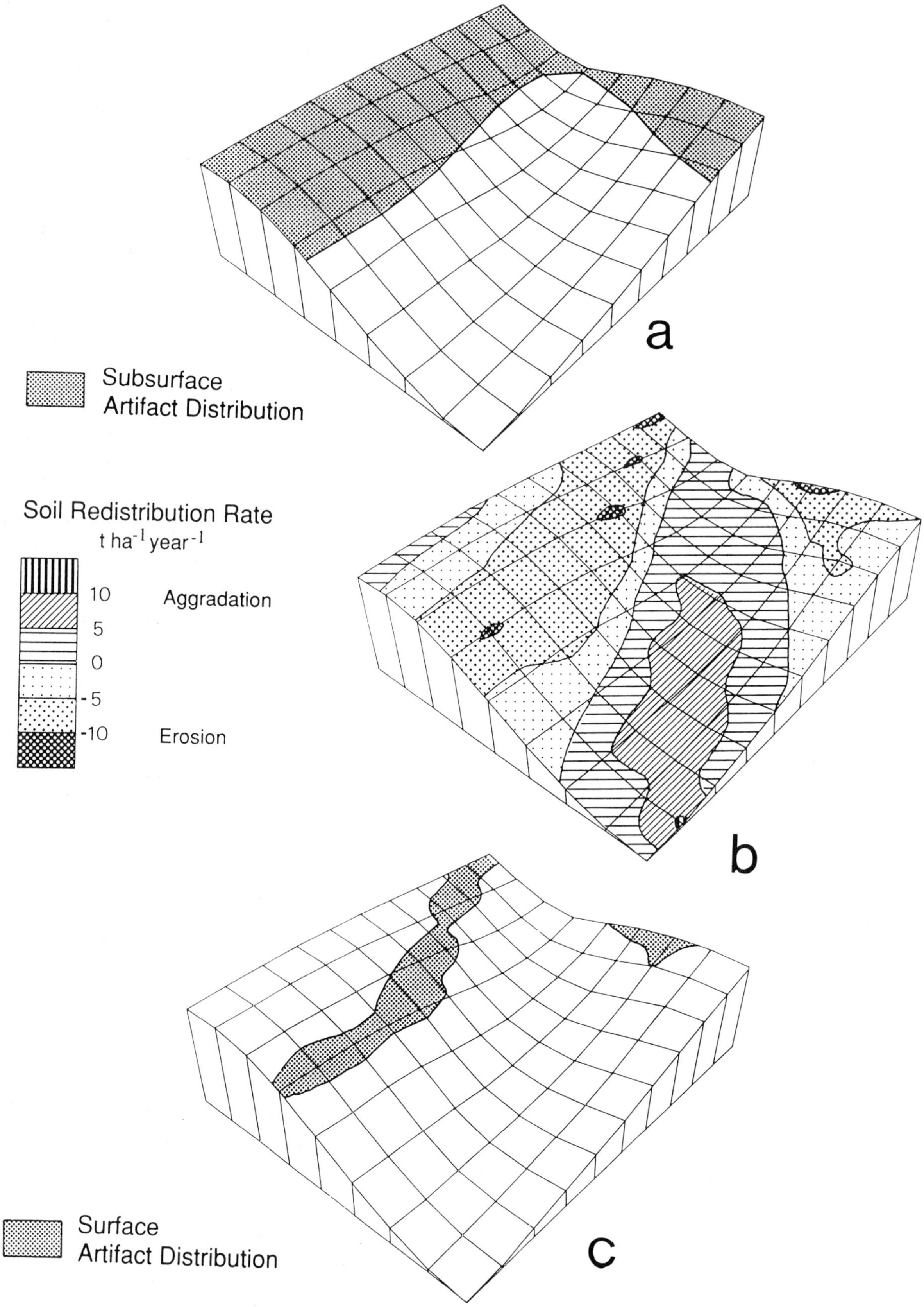


Figure 17.4 The impact of differential erosion upon surface artifact distributions for a hypothetical landscape.
a. Hypothetical sub-surface artifact distribution.
b. Modelled spatial pattern of erosion and deposition.
c. Resultant surface distribution of artifacts.

survival, especially on agricultural land. The enhanced susceptibility of crests and spurs to erosion on sandy, silty and chalky soils is of particular significance to archaeologists, because such locations may be preferred sites for earthworks such as hill forts and enclosures. This reinforces the need to consider the most appropriate measure of soil erosion in the evaluation of site risk.

The most frequently quoted and, therefore, easily accessible measures of erosion are net or gross rates (as defined above). These are appropriate for the applications for which they are measured, such as assessment of wide-scale impacts upon productivity or potential environmental damage. However, in the assessment of risk to a specific locality, such as an archaeological site on agricultural land, it is more meaningful to examine long-term mean rates of soil loss from eroding areas within the landscape, and in some cases to examine the worst affected areas. This is possible using $^{137}$Cs-derived data (Quine and Walling 1991; Walling and Quine 1991a) and the various erosion rate measures obtained by the authors for 13 sites examined in the UK are listed in Table 17.2. The rates for the sites investigated will tend to represent the upper end of the spectrum for the soil types studied, because the locations were chosen on the basis of a known propensity to erode. However, the data highlight the importance of using the most appropriate measure of erosion rate, since in all but one case (the most severe) the mean rate of erosion from eroding areas within the study fields is more than double the net rate, and in many cases the increase is in excess of three times. Examination of the worst affected areas within the study fields reinforces the importance of spatial variation, because in most cases the minimum rate occurring over the worst 10% of the field area was almost twice the mean rate for the total eroding area.

Consideration of topographic context, soil type and land use will allow the assessment of the true risk of destruction through soil erosion which is faced by archaeological sites. The data presented in Table 17.2 highlight the danger of relying on general landscape based rates in the evaluation of site-specific risk.

### *Rates of site erosion*

Published data for rates of erosion may provide general guidelines for the assessment of risk to sites on agricultural land. However, it may be necessary to obtain more detailed information concerning rates of erosion on a specific site, for example in the development of cultural resource management strategies. The $^{137}$Cs technique is uniquely suited to this task, as it is possible to obtain retrospective estimates of soil erosion over an area of interest on the basis of a single site visit. The spatial resolution of such a survey is defined by the sampling strategy and may therefore be site-specific and problem-orientated. Furthermore, measurements are generally made on cores about 7 cm in diameter and 50 cm in depth, consequently causing minimal disturbance to the site. Although the studies discussed above were carried out on agricultural land, it is possible to use the method on uncultivated areas subject to widespread erosion, as may be the case where natural vegetation provides a low percentage cover for all or part of the year.

Detailed erosion surveys may be valuable for site interpretation as well as site preservation. An understanding of contemporary erosion and deposition is vital to the development of effective models of site formation, and may contribute to the prediction of properties such as the past extent of a site. This may be illustrated by consideration of tell formation. A number of models have been devised to predict the initial form and extent of tells on the basis of their present shape and predicted geomorphological processes (e.g. Davidson 1976; Rosen 1986). Rosen (1986), for example, suggests that the present asymmetric form of many tells is a product of differential erosion, and contrasting processes of slope retreat. Measurement of patterns of contemporary soil erosion would facilitate partial testing of such models. Although contemporary rates could not be applied directly across the spectrum of changing environmental conditions in the past, it would be possible to relate contemporary rates to known current land-use, climatic and edaphic parameters and thereby test the feasibility of model predictions derived for estimated past combinations. The long-term (*c.* 35 years) retrospective rates established by the $^{137}$Cs technique would be particularly appropriate for such model testing.

## **Estimating and interpreting past erosion rates**

The estimation of past erosion rates is often based, necessarily indirectly, upon the evidence of deposition whether colluvial (Bell 1986; Allen 1988) or alluvial (Evans 1990). However, a key factor in any such calculation is the estimation of the relationship between the quantities eroded and the quantities deposited. Until recently, quantification of this relationship has been hampered by the absence of empirical data. However, recent work with the $^{137}$Cs technique has begun to provide the necessary information.

### *From colluvial sediment to erosion*

When estimating past erosion from volumes of colluvial sediment the simplest method is to divide the volume of deposited sediment by the upslope length (e.g. Bell 1986). However, this does not take into account two important factors, namely, the sediment delivery ratio and the relative extent of contributing areas and depositional areas.

The sediment delivery ratio provides a means of quantifying the relationship between the amount of sediment eroded within an area and the amount of sediment exported. This relationship has proved extremely difficult to measure, but it might be expected that sediment delivery ratios would reflect soil texture. Once entrained, finer particles should be carried further than coarser particles and, if all other factors were constant, finer soils should therefore be

| Soil / Site | Erosion rate (t ha$^{-1}$ year$^{-1}$) Net | Gross | Mean | Worst 10% (Minimum) |
|---|---|---|---|---|
| *Brown Sand* | | | | |
| Hole Norfolk | 3.0 | 6.3 | 8.1 | 16.5 |
| Dalicott Shropshire | 6.5 | 10.2 | 12.6 | 20.0 |
| Rufford Nottinghamshire | 10.5 | 12.2 | 13.8 | 28.0 |
| *Brown Rendzina* | | | | |
| Manor House Norfolk | 2.4 | 6.3 | 9.4 | 18.2 |
| Lewes Sussex | 1.4 | 4.3 | 6.2 | 10.8 |
| *Brown Calcareous Earth* | | | | |
| West Street Kent | 4.3 | 7.7 | 11.1 | 23.3 |
| *Brown Earth* | | | | |
| Mountfield Somerset | 2.2 | 4.6 | 6.1 | 10.3 |
| Yendacott Devon | 1.9 | 5.3 | 7.3 | 12.9 |
| *Argillic Brown Earth* | | | | |
| Wootton Herefordshire | 2.8 | 6.4 | 8.1 | 14.8 |
| Fishpool Gwent | 1.9 | 5.1 | 8.1 | 15.3 |
| *Calcareous Pelosol* | | | | |
| Higher Dorset | 3.1 | 5.2 | 6.7 | 13.6 |
| Brook End Bedfordshire | 1.2 | 3.6 | 5.7 | 8.7 |
| Keysoe Bedfordshire | 0.6 | 2.2 | 4.0 | 7.0 |

Table 17.2 Erosion rate measures for 13 fields, derived from $^{137}Cs$ measurements

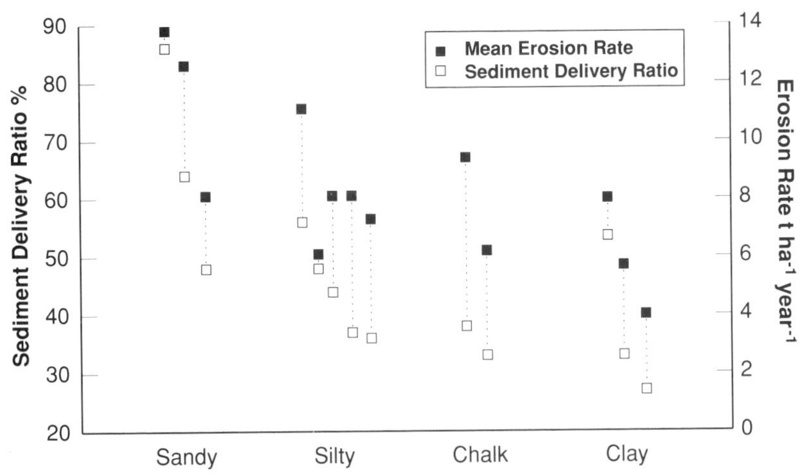


Figure 17.5 Sediment delivery ratios and mean erosion rates for 13 fields in southern UK.

characterised by higher sediment delivery ratios than coarser soils. However, the energy of the erosional system must also be taken into consideration, because this will also influence transport capacity. A fully developed rill system could be expected to generate higher rates of erosion and greater transport distances than discontinuous overland flow. The importance of these considerations may be seen in Figure 17.5 which shows the sediment delivery ratios for the 13 sites studied in the $^{137}Cs$-based soil erosion survey referred to above. The highest sediment delivery ratios from the sites studied were found on the coarsest soils, and the ratios for the sites on silty soil were higher than two of the three sites on clay soil. Examination of the pattern of mean erosion rates, also shown in Figure 17.5, shows a close correlation with the sediment delivery ratios. This suggests that, at the sites studied, the energy of the erosional system has exerted a greater control over the export of sediment from the fields than the texture of the soils.

The preceding discussion emphasises the problems in estimating representative sediment delivery ratios. Further examination of a wider range of fields on various soil textures may allow the identification of trends in the relationships between delivery ratios, topography and erosion rates. This would be very valuable in the estimation of past sediment delivery ratios. However, even the limited data set presented in Figure 17.5 highlights the importance of considering the sediment delivery ratio in the estimation of erosion from deposited sediments. The data indicate that the percentage of eroded sediment represented by infield deposition varies from 14 to 73 (the percentage infield deposition = 100 minus the percentage sediment delivery ratio). Estimates of erosion from the fields based directly on the deposited sediment would have been under-estimates, but more significantly the under-estimate would have ranged from a factor of 1.4 to 7.

The second important factor which should be considered in the calculation of erosion from deposited sediment is the relative extent of contributing areas and receptor areas. While providing a useful approximation on simple uniform slopes such as erosion plots the assumption that the contributing area is equal to the upslope length multiplied by the width of the receptor area is unsatisfactory on the complex slopes which characterize the landscape. In such situations the contributing area is better defined by employing morphological boundaries to establish the local catchment (e.g. Boardman and Hazelden 1986). This may be achieved relatively easily by reference to large-scale topographic maps.

*Estimating erosion from alluvium*

The previous section has demonstrated that a significant percentage of eroded sediment is exported from the field or geomorphic unit in which the erosion took place. Once exported from the immediate locality, much of this sediment may be transported through the fluvial system and in turn a proportion of this could be deposited on floodplains. This has led to the study of alluvial sequences as an aid to elucidating past erosional histories. To be successful, such studies demand an understanding of the relationship between sediment entering a river, stored in the channel, deposited on floodplains and exported from the system. Until recently quantification of this relationship has been almost impossible. Consequently, Evans (1990) recognized the potentially conservative nature of his nationwide estimates of past erosion based upon the current distribution of alluvium in England and Wales, but he was able to suggest that the equivalent of 40–80 mm of eroded topsoil is stored as alluvium within the river systems of that area.

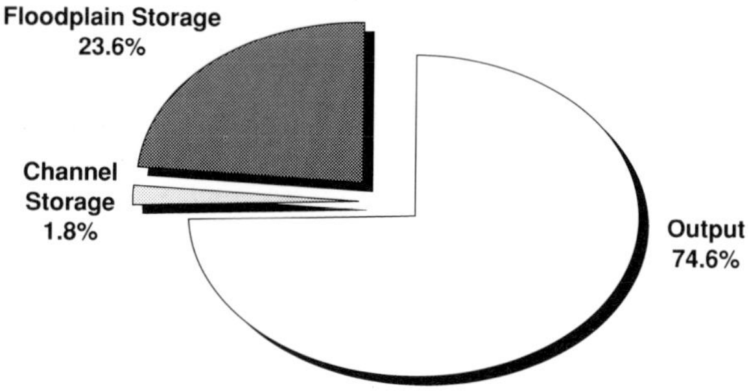


Figure 17.6 A tentative suspended sediment budget for the River Severn, 1986–1989, based upon redistribution of Chernobyl-derived radionuclides.

However, recent studies using sediment yield data and the $^{137}$Cs technique have begun to provide quantitative evidence of the relationship between the sediment entering the river, storage and export. In a study of the lower reaches of the River Culm in Devon, Lambert and Walling (1987) measured conveyance losses of 28% of the total load through channel and floodplain deposition. Because the majority of the conveyance loss resulted from sediment deposition on the floodplain, it may be suggested that the mass of fine sediment transported from this part of the system was approximately equal to 2.5 times the mass of floodplain deposition, i.e. alluvium. In a broader scale study of the River Severn, Walling and Quine (1991b) have produced a tentative sediment budget for the period 1986–1989, based upon redistribution of Chernobyl-derived radiocaesium in association with sediment. The budget is illustrated in Figure 17.6, and suggests that the suspended sediment load transported by the River Severn through the gauging station at Upton on Severn was approximately 3 times the amount deposited on the upstream floodplains. The similarity of the figures obtained using different methods at different scales suggests that the results may be of the correct order of magnitude. If these figures are used in the estimation of past erosion based upon the volume of stored alluvium in England and Wales, calculated by Evans (1990), then the maximum surface lowering represented by the alluvium would be increased from 80 mm to 280–320 mm.

*Contemporary analogues*

Although it is clearly impossible to use the $^{137}$Cs technique to measure rates of erosion prior to 1950, it may be possible to obtain an insight into the erosional impact of 'primitive' agricultural systems through contemporary analogues. Such analogues may be identified in contemporary agricultural systems in developing countries, and in experimental projects such as that at Butser Iron Age Experimental Farm, Hampshire, UK (Reynolds 1988). Current studies by the authors in Lesotho and China are investigating erosional patterns in terraced and strip field agricultural systems, where basic technology is still employed. Although the climate and environment of these study areas are very different from prehistoric Europe, the response of these landscapes to erosion processes may provide insights into the influence of traditional conservation strategies. Similar studies being undertaken in areas of Eastern Europe, sheltered from the recent agricultural intensification seen in the west, may provide closer parallels with archaeological contexts under investigation. As mentioned above, it may also be possible to apply the $^{137}$Cs technique to the investigation of soil erosion patterns and rates at experimental sites such as Butser. Where experiments have been continued over a number of years and prior land-use impacts may be considered as minimal, application of the $^{137}$Cs technique could provide data which would otherwise be unattainable.

**Erosion assessment using caesium-137 and its archaeological relevance**

Studies of contemporary soil erosion using the $^{137}$Cs technique have both highlighted the particular value of the method and yielded data of relevance to the archaeologist. The $^{137}$Cs technique has a unique combination of advantages over other methods of erosion assessment. These advantages include some of special relevance to archaeology, for instance, the potential for studying whole areas with minimum disturbance to the site, and the capacity to provide long-term (*c.* 35 years) estimates of soil redistribution rates and spatial patterns. It has therefore been possible to identify a number of archaeological applications of $^{137}$Cs-derived erosion rate data, particularly in investigating the survival of the archaeological record, and in improving estimates of past erosion rates based upon colluvial and alluvial deposits. In addition to these existing applications, it has been suggested that the $^{137}$Cs technique could be used in detailed site-specific erosion surveys to aid either the development of cultural resource management strategies or the testing of models of site formation. A further potential archaeological application has been identified in the investigation of erosion in contemporary agricultural contexts which are analogous to archaeological examples. This is already underway in China and Lesotho, however parallel studies of erosion at experimental sites such as Butser could prove worthwile.

**Acknowledgements**

The $^{137}$Cs-based erosion measurements at sites in the UK were undertaken as part of the NERC/NCC 'Agriculture and the Environment' initiative, the studies of radionuclide redistribution in the River Severn catchment and floodplain sedimentation in Devon were also undertaken as part of studies funded by NERC. This financial support, and that provided to one of the authors (TAQ) by the Lloyd's of London Tercentenary Foundation Fellowship scheme, are gratefully acknowledged.

**References**


Allen, M. J. 1988 Archaeological and environmental aspects of colluviation in south-east England. In: Groenman-van Waateringe, W. and Robinson, M. (eds.) *Man-Made Soils*. BAR, International Series 410, Oxford, 67–92.

Allen, M. J. 1991 Analysing the landscape: a geographical approach to archaeological problems. In: Schofield, A.J. (ed.) *Interpreting Artefact Scatters: Contributions to Ploughzone Archaeology*. Oxbow Monograph 4, Oxford, 39–57.

Bachhuber, H., Bunzl, K., Schimmack, W. and Gans, I. 1982 The Migration of $^{137}Cs$ and $^{90}Sr$ in multilayered soils: results from batch, column, and fallout investigations. *Nuclear Technology* 59, 291–301.

Bell, M. G. 1986 Archaeological evidence for the date, cause and extent of soil erosion on the chalk. In: Burnham, C.P. and Pitman, J.I. (eds.) *Soil Erosion. SEESOIL* 3, 72–83.

Boardman, J. and Hazelden, J. 1986 Examples of erosion on brickearth soils in east Kent. *Soil Use and Management* 2, 105–108.

Boardman, J., Foster, I. D. L. and Dearing, J. A. (eds.) 1990 *Soil Erosion on Agricultural Land*. Wiley, Chichester.

Campbell, B. L. 1983 Applications of environmental caesium-137 for the determination of sedimentation rates in reservoirs and lakes and related catchment studies in developing countries. In: *Radioisotopes in Sediment Studies. International Atomic Energy Technical Document* 298, 7–30.

Campbell, B. L., Loughran, R. J., Elliott, G. L. and Shelly, D. 1986 Mapping drainage basin sediment sources using caesium-137. In: Hadley R. F. (ed.) *Drainage Basin Sediment Delivery I.A.H.S. Publication* 159, 437–446.

Cawse, P. A. and Horrill, A. D. 1986 *A survey of caesium-137 and plutonium in British soils in 1977*. AERE Harwell report R–10155, H.M.S.O.

Davidson, D. A. 1976 Processes of tell formation and erosion. In: Davidson, D. A. and Shackley, M. L. (eds.) *Geoarchaeology*. Duckworth, 255–266.

De Jong, E., Begg, C. B. M. and Kachanoski, R. G. 1983 Estimates of soil erosion and deposition for some Saskatchewan soils. *Canadian Journal of Soil Science* 63, 607–617.

Evans, R. 1990 Soil erosion: its impact on the English and Welsh landscape since woodland clearance. In: Boardman, J., Foster, I. D. L. and Dearing, J. A. (eds.) *Soil Erosion on Agricultural Land*. Wiley, Chichester, 231–254.

Klobe, W. D. and Gast, R. G. 1970 Conditions affecting cesium fixation and sodium entrapment in hydrobiotite and vermiculite. *Soil Science Society of America, Proceedings* 34, 746–750.

Lambert, C. P. and Walling, D. E. 1987 Floodplain sedimentation: a preliminary investigation of contemporary deposition within the lower reaches of the River Culm, Devon, UK. *Geografiska Annaler* 69A, 393–404.

Livens, F. R. and Baxter, M. S. 1988 Chemical associations of artificial radionuclides in Cumbrian soils. *Environmental Radioactivity* 7, 75–86.

Livens, F. R. and Loveland, P. J. 1988 The influence of soil properties on the environmental mobility of caesium in Cumbria. *Soil Use and Management* 4, 69–75.

Quine, T. A. and Walling, D. E. 1991 Rates of soil erosion on arable fields in Britain: quantitative data from caesium-137 measurements. *Soil Use and Management* 7, 169–176.

Quine, T. A. and Walling, D. E. 1992 Use of caesium-137 measurements to investigate relationships between erosion rates and topography. In: Allison, R. J. and Thomas, D. S. G. (eds.) *Environmental Sensitivity*. Wiley, Chichester (in press).

Reynolds, P. J. 1988 *Butser Ancient Farm 1987*. Butser Ancient Farm Project Trust.

Ritchie, J. C., Spraberry, J. A. and McHenry, J. R. 1974 Estimating soil erosion from the redistribution of $^{137}Cs$. *Soil Science Society of America, Proceedings*, 38, 137–139.

Rogowski, A. S. and Tamura, T. 1965 Movement of cesium-137 by run-off, erosion and infiltration on the alluvial Captina silt loam. *Health Physics*, 11, 1333–1340.

Rogowski, A. S. and Tamura, T. 1970 Erosional behaviour of cesium-137. *Health Physics*, 18, 467–477.

Rosen, A. M. 1986 *Cities of Clay - the Geo-archaeology of Tells*. University of Chicago Press, Chicago.

Squire, H. M. and Middleton, L. J. 1966 Behaviour of $^{137}Cs$ in soils and pastures - a long term experiment. *Radiation Botany*, 6, 413–423.

Walling, D. E. and Bradley, S. B. 1989 Rates and patterns of contemporary floodplain sedimentation: a case study of the River Culm, Devon, UK. *Geojournal*, 19, 53–62.

Walling, D. E. and He, Q. 1992 Interpreting caesium-137 profiles in lake sediments. In: McManus, J. and Duck, R. (eds.) *Geomorphology and Sedimentology of Lakes and Reservoirs*. Wiley, Chichester.

Walling, D. E. and Quine, T. A. 1990a Use of caesium-137 to investigate patterns and rates of soil erosion on arable fields. In: Boardman, J., Foster, I. D. L. and Dearing, J.A. (eds.) *Soil Erosion on Agricultural Land*. Wiley, Chichester, 33–53.

Walling, D. E. and Quine, T. A. 1990b Calibration of caesium-137 measurements to provide quantitative erosion rate date. *Land Degradation and Rehabilitation* 2, 161–175.

Walling, D. E. and Quine, T. A. 1991a The use of caesium-137 measurements to investigate soil erosion on arable fields in the UK: potential applications and limitations. *Journal of Soil Science*, 42, 147–165.

Walling, D. E. and Quine, T. A. 1991b Fluvial

redistribution of Chernobyl radionuclides. *NERC News*, July 1991, 22–25.

Walling, D. E., Bradley, S. B. and Wilkinson, C. J. 1986 A caesium-137 budget approach to the investigation of sediment delivery from a small agricultural drainage basin in Devon, U.K. In: Hadley, R. F. (ed.) *Drainage Basin Sediment Delivery, I.A.H.S. Publication* 159, 423–435.

Walling, D. E., Quine, T. A. and He, Q. 1992 Investigating contemporary rates of floodplain sedimentation. In: Carling, P. A. and Petts, G. E. (ed.) *Lowland Floodplain Rivers: Geomorphological Perspectives*. Wiley, Chichester, 165–184.

Wise, S. M. 1980 Caesium-137 and lead-210: a review of techniques and some applications in geomorphology. In: Cullingford, R. A., Davidson, D. A. and Lewin, J. (eds.) *Timescales in Geomorphology*. Wiley, Chichester, 109–127.

Zhang, X., Higgitt, D. L. and Walling, D. E. 1990 A preliminary assessment of the potential for using caesium-137 to estimate rates of soil erosion in the Loess Plateau of China. *Hydrological Sciences Journal* 35, 267–276.

# CHAPTER 18

# SOIL MICROMORPHOLOGICAL EVIDENCE OF ANCIENT SOIL EROSION

RICHARD I. MACPHAIL

*The methodology of investigating ancient soil erosion through soil micromorphology is demonstrated by a detailed study of Ashcombe Bottom, Sussex. The evidence for (mainly Neolithic, Chalcolithic and Beaker) soil erosion induced by ancient vegetation clearance and cultivation is discussed from 16 United Kingdom and Italian (Ligurian) sites, representative of flatland, low slope, sloping and receiving site elements of the landscape. All soils became potentially erodible when soil disruption by clearance exposed subsoils to rainsplash and gravity movement. Neolithic and Beaker cultivation seems to have produced highly erodible soils, possibly because organic matter levels became reduced, and soil stability diminished. The resulting pore infillings (e.g. coatings and infills) and compaction (e.g. aggregate sorting and settling) may have led to accelerated erosion. Further experimental and site data is required to investigate the effects of clearance and the influence of manuring (possibly introduced in the Early Bronze Age) on soil erodibility.*

## Introduction and methods

Present rates of soil erosion are being measured on many substrates (Burnham and Pitman 1986) as discussed by Boardman and Evans in this volume. Some estimated rates of past soil erosion have also been investigated by Bell and Allen in this volume, but *details* of these rates and information on the mechanisms of past soil erosion (Allen 1991) are far less accessible. Yet exact data from the past is crucial, if global Holocene soil changes, which for instance relate to climate, man and erosion, are to be fully understood and quantified. Reliable information from the past would also permit studies of many modern soils (i.e. eroded and colluvial profiles), which are themselves the result of past erosion, to be placed within the context of time. A comprehension of past mechanisms and factors affecting soil erosion also needs to be added to the equation, if predictions for the future are to be based upon past rates.

The combination of archaeology, which provides datable soils from known cultural contexts, and soil micromorphology has produced useful data (Kwaad and Mücher 1977; 1979; Macphail *et al.* 1987). Investigations have employed field study (Hodgson 1974), the impregnation of soil samples and their manufacture into thin sections (Murphy 1986; Guilloré 1985), the description of these thin sections (Bullock *et al.* 1985) and the interpretation of their microfabrics (Courty *et al.* 1989). Often bulk samples have also been analysed through the techniques of grain size, organic carbon, pH, calcium carbonate and magnetic susceptibility (Avery and Bascomb 1974; Tite and Mullins 1971; Longworth and Tite 1977), although data are not reported here.

These studies have provided direct information on ancient soil erosion, or allowed inferences to be made about the potential of ancient soils to erode under given past landuses. The case study presented here of Ashcombe Bottom, near Lewes, Sussex, demonstrates the methodology. Other micromorphological evidence of soil erosion mainly caused by Neolithic, Chalcolithic and Beaker vegetation clearance and cultivation from additional sites in the United Kingdom and from Eastern Liguria, Italy, is discussed.

## Ashcombe Bottom

Studies of the *in situ* Beaker period cultivated (features interpreted as ard marks) colluvial soil ('Beaker Palaeosol'), and earlier buried soils in subsoil hollows at Ashcombe Bottom (Allen 1984; 1988; 1991; Macphail *et al.* 1990) may permit some elucidation of the processes that induced phases of soil erosion at this site.

### *Parent material, woodland clearance and erosion of the early to mid-Holocene soils*

In general, the site comprises a metre or so of prehistoric colluvium overlying chalky Combe Rock. The base of Trench B (Figure 18.1) also exposed a probable Late Devensian deposit (sample C; position not illustrated) of finely bedded (water lain), fine sand-size rounded chalk and silt-size quartz. Soil micromorphology (Figure 18.2) shows the chalk clasts to have been weathered and some micritic cementation has also taken place. Both Trench B (Figure 18.1) on the east side of the valley floor, and Trench A on the west side of the valley floor, revealed non-calcareous silty clay-infilled solution hollows (Samples A and B) which had been preserved as a post-Combe Rock palaeosol (Table 18.1). A hiatus in the pedological sediment record is caused by the truncation of this palaeosol, and the probable re-establishment of woodland in overlying weakly calcareous colluvium and in areas of the Combe Rock. This weakly calcareous colluvium and the Combe Rock contain

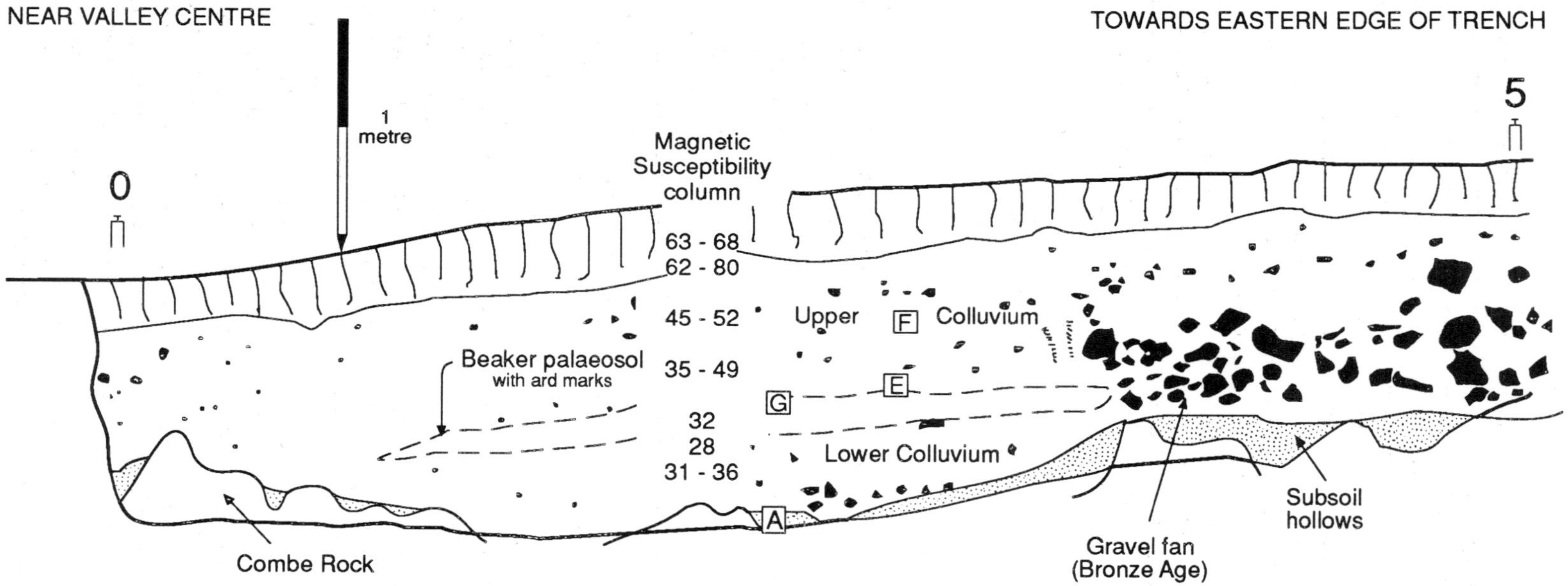


Figure 18.1 Valley bottom part of Trench B, Ashcombe Bottom, Sussex, showing location of soil micromorphological samples 'A', 'E', 'F' and 'G', and approximate position of magnetic susceptibility column (after Allen 1984; pers. comm.) (magnetic susceptibility $10^{-8}$ SI Kg).

probable tree-hollows, with flint artifacts, and a woodland molluscan fauna of post 5,500 BP (Kerney 1977; Allen 1984). Only the earlier, post-Combe Rock subsoil hollows were sampled (samples A and B).

These soils have a spongy fabric with many partially closed and closed vughs (Figure 18.3). The soils are made up of discrete patches of silt/silt loam and clay, with void infills and coatings of clay and/or fine silt. Flint and other residual siliceous material are also present. Sample B is dominated by a clay fabric and contains obvious fragments of highly birefringent fine clay coatings (papules; Bullock and Murphy 1979), indicating an original argillic brown earth (Avery 1980: Typical Luvic Brown Soil; Avery 1990) character (Bt horizon) for this soil. The silty soil fragments only contain about 30–40% reddish brown fine material. In contrast, sample A contains less clay-rich fragmented Bt horizon material. Rather, the fabric is dominated by a brown to yellowish brown silt loam, the pores within which are coated and infilled by dusty to impure clay. Very coarse textural features with compound layering are present, and may include fine organic matter. Much of the fabric has been affected by both black ferro-manganese impregnation and reddish ferruginous impregnation, whereas some of the loamy material is dark because it may be residual of Ah horizon soil. *In situ* ferruginised roots are present in both samples.

Sample A is much more porous than sample B, and most pores have textural coatings (Figure 18.4). The microfabrics of the truncated palaeosols are interpreted as being those of the early to mid-Holocene decalcified argillic brown earth which had formed in a variety of local parent materials, including probable loess (cf. Weir *et al.* 1971), calcareous fluvial deposits and Combe Rock. For example, sample A contains much silt that may in part originate from the decalcification of the calcareous fluvial deposits, found in the base of the same trench (Trench B), which have a quartz silt component (sample C, Figure 18.2). As there is no evidence in either of the trenches examined that the soils in these subsoil hollows are anything but post-glacial phenomena (Allen 1984; pers. comm.), the disrupted fabrics are unlikely to be of a cryogenic origin (Bullock and Murphy 1979), and thus must have been caused by another agency of disturbance. In addition, the soil fragments present in the subsoil hollows appear to be from Ah, Eb(A2) and Bt horizons, suggesting that they are from the on-site soil profile, whereas palaeosols from the last interglacial would more probably have a heterogeneous mixture of *rounded* soil papules, embedded grains etc., of a very dominantly B horizon character and be of a mixed provenance (Bullock and Murphy 1969; Kemp 1986). They are also unlikely to contain much Ah horizon material. The subsoil hollow microfeatures also do not relate to the action of large burrowing fauna (Courty *et al.* 1989, p. 140), but a number of studies (Macphail *et al.* 1987; Courty *et al.* 1989, p. 286; Macphail and Goldberg 1990; Macphail *et al.* 1990) have shown that tree-throw and clearance activity can produce these features (i.e. mixed fragments representative of all horizons, and infilling of fissures around these fragments by clay and silt) of soil profile disruption (cf. Table 18.2, examples 1–3).

Without recourse to clay mineralogical data, decalcification of chalky material seems to have been the main origin of the clay in the subsoil Bt horizon (samples A and B), with probable additions of clay through weathering and particle translocation from the fluvial deposits and loessic upper stratum. Certainly the soil on the west side of the valley (sample B) is more clay rich, even though the overlying colluvium is remarkably silty and acid, perhaps representing a residual deposit of locally reworked loess (Allen, pers. comm.).

There is an inference that if soil profile disruption was caused by clearance, then this was of early Neolithic date (cf. Offham Hill 2.5 km to the east, Drewett 1977; Allen, pers. comm.). Fragmented soil from the clay-rich Bt, the loessic Bt and B(t), and loamy and more organic Eb (A2) and Ah horizons were mixed, the subsoil hollows becoming focal points for the drainage of precipitation (Veneman *et al.* 1984; Schaetzl 1990), which slaked broken soil and washed it into fissures to produce textural coatings and infillings. Moist conditions and the probable preferential accumulation of leaf litter also encouraged both faunal and floral activity (cf. Balksbury Camp, Macphail and Goldberg 1990) such as rooting. The last is recorded by ferruginised root pseudomorphs and could be evidence of revegetation.

Within the subsoil hollows examined, however, the last most important process was the complex deposition of textural features (see Figures 18.3 and 18.4). These have been perforated by fine rooting, and in places disturbed/eroded, and new coatings have accumulated (Figure 18.4). In the first instance, these textural features could be ascribed to loose soil washing-in through time into the subsoil hollows, whereas fine rooting and minor disturbance of the textural features may relate to the re-establishment of (post 5,500 BP) woodland on the site, as recorded by the microfabrics of these infilled subsoil hollows. If primary clearance of the valley floor was of an early Neolithic date, and this was followed by woodland regeneration, then this follows the pattern of Neolithic landuse as recorded, from molluscan evidence, across Sussex (Thomas 1982; Allen, pers. comm.).

The 'lower' colluvium (*sensu* Avery 1990), however, which overlies the tree hollows and truncates the early to mid-Holocene palaeosol, may relate to renewed clearance (and possible cultivation) in the late Neolithic which induced soil erosion (Allen 1984). This later period of soil instability (Table 18.1) could be the source of the last phase of compound, dusty to silty, soil pore coatings and infillings in the subsoil hollows (Figure 18.4).

*Beaker cultivation*

A Beaker palaeosol (Table 18.1), which featured ardmarks, is itself a colluvial soil (see below) forming in and over the lower colluvium. Some 10 cm of this

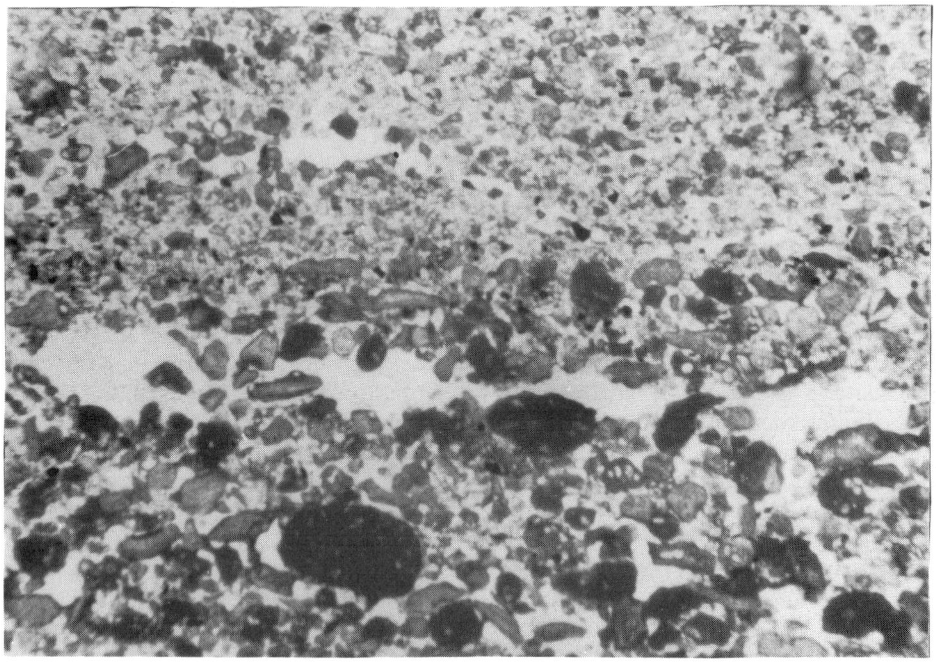

Figure 18.2 Ashcombe Bottom, base of Trench B; lamina bedded coarse silt and fine sand size quartz, and fine sand size chalk fragments; weak micritic cementation; Late Devensian fluvial deposit. Plane polarised light (PPL), frame length is 5.3 mm.

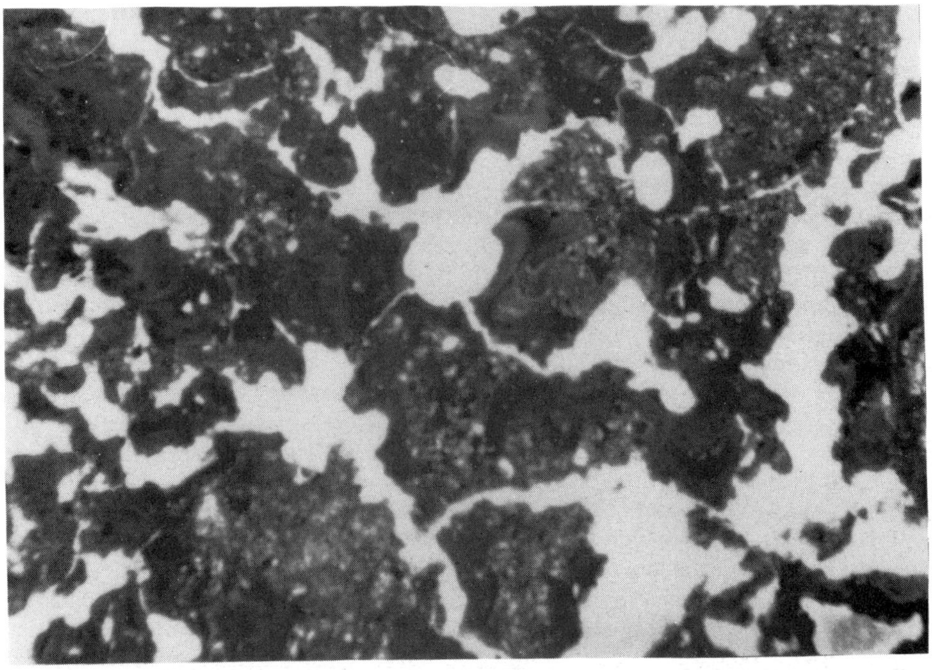

Figure 18.3 Ashcombe Bottom, Trench B, sample A; post-Combe Rock subsoil hollow (Figure 18.1); highly porous Bt horizon (cracks are artifacts of thin section manufacture), comprising biologically worked silt and clay soil, with coarse porosity infilled and coated by multiphase compound textural features forming closed and partially closed vughs (centre top and centre right); early to mid-Holocene argillic brown earth disrupted by clearance. PPL, frame length is 5.3 mm. (Bt horizon in sample B is more dense and clay rich.)

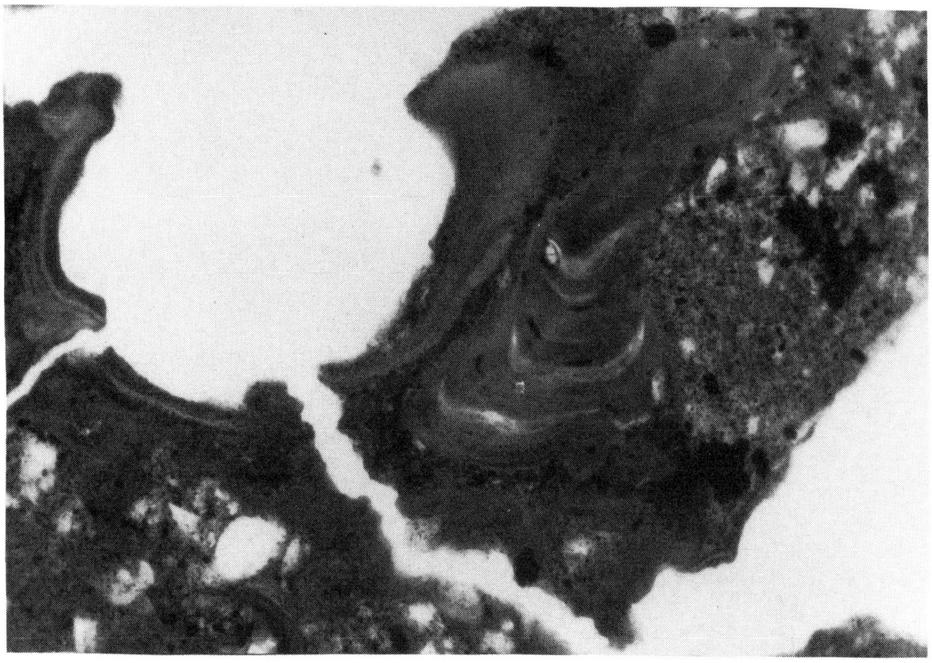

Figure 18.4 Detail of Figure 18.3, showing partially manganese stained (previously humic?) silty clay soil, coated by at least two phases of microlaminated dusty to silty clay coatings. Two periods of soil disturbance (early? and late? Neolithic) may be recorded. PPL, frame length is 0.33 mm.

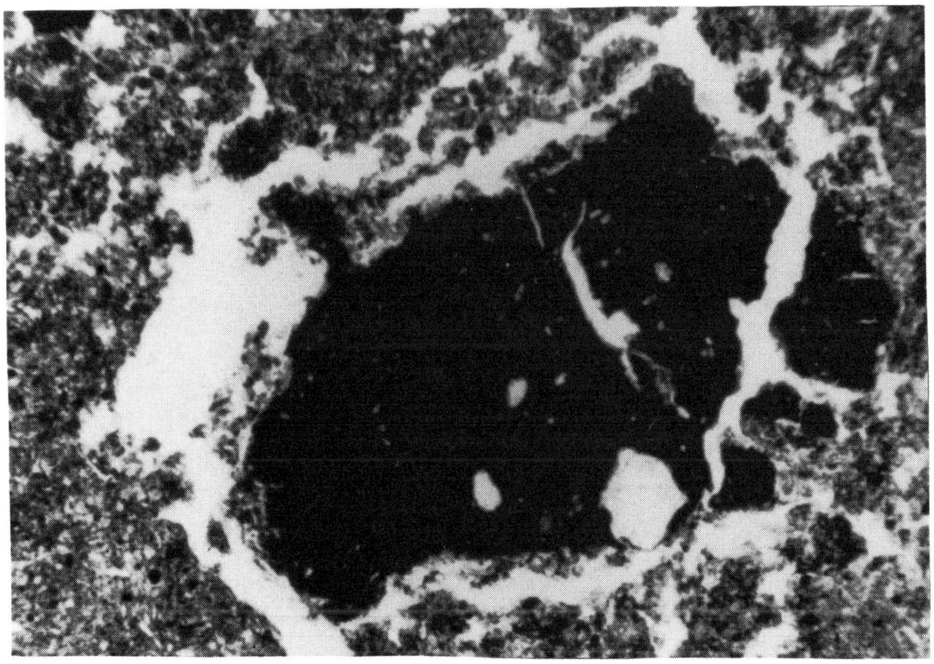

Figure 18.5 Ashcombe Bottom, Trench B, sample G (Figure 18.1) from Beaker palaeosol; eroded coarse fragment of Bt horizon, relic of the early to mid-Holocene argillic brown earth, within silty colluvium. PPL, frame length is 5.3 mm.

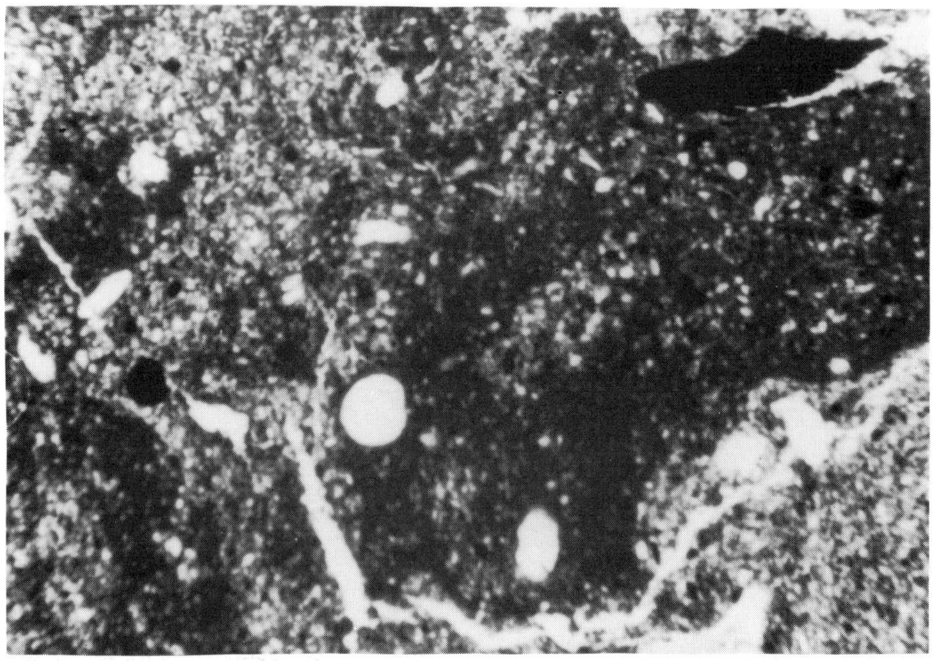

Figure 18.6 Ashcombe Bottom, Trench B, sample G (Figure 18.1) from Beaker palaeosol; within dense ard-ploughed colluvium, a probable crust fragment of surface slaked soil, comprises dusty clay and silty soil elements, and contains vesicles (circular voids) and coarse and fine charcoal. PPL, frame length is 5.3 mm.

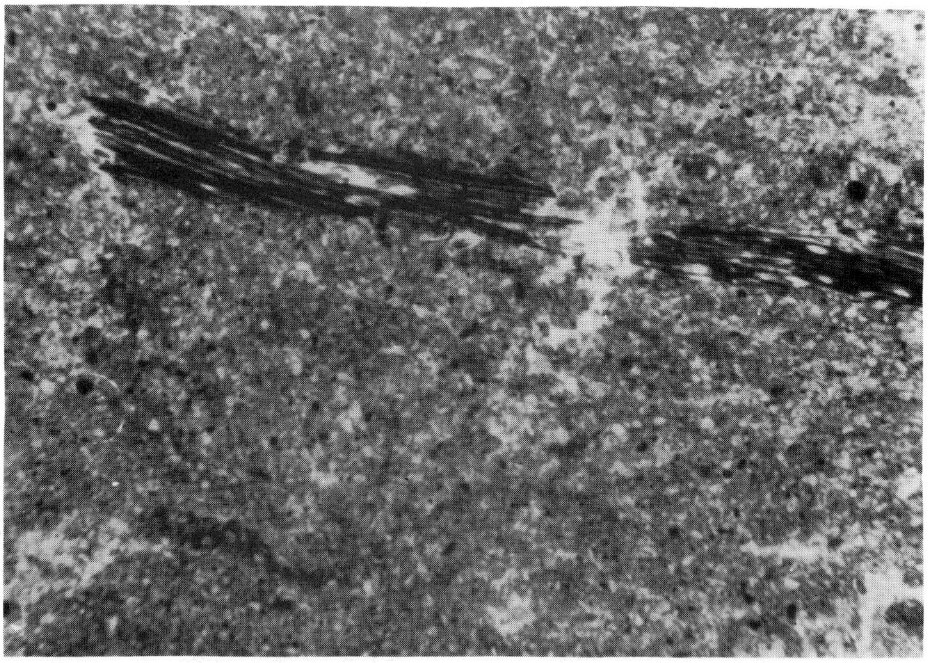

Figure 18.7 Ashcombe Bottom, Trench A, sample D (Iron Age to Roman colluvium); compacted silty colluvium contains large fragment of possible straw (ferruginised), as possible evidence of ploughing-in of chaff/manuring. PPL, frame length is 5.3 mm.

9. Iron Age to Roman cultivation and continued erosion and colluviation ('upper' colluvium).

8. Bronze Age cultivation, erosion and high energy deposition of erosive chalk gravel fan ('upper' colluvium).

7. Beaker occupation, cultivation, erosion and colluviation (Beaker 'palaeosol').

6. Secondary clearance (and cultivation?) of presumed late Neolithic date, leading to the removal of most of the calcareous colluvium, truncation of the underlying palaeosol, and deposition of the 'lower' colluvium.

5. Woodland regeneration, forming tree-hollows in the Combe Rock and in the weakly calcareous colluvium.

4. Primary clearance of inferred early Neolithic date, disruption of woodland argillic brown earth (palaeosol), and deposition of weakly calcareous colluvium (now only patchily preserved).

3. Early to mid-Holocene woodland and development of argillic brown earth (palaeosol) (still extant in solution/subsoil hollows).

2. Late Devensian periglacial activity (involutions, deposition of loess, fluvial activity).

1. Probable Devensian deposition of Combe Rock.

Table 18.1 Suggested sequence of prehistoric events at Ashcombe Bottom (Allen 1984, pers. comm.)

horizon was examined in sample G, whereas its upper boundary was investigated from sample E (Figure 18.1). A number of microfeatures relating to the effects of cultivation were noted (Macphail *et al.* 1990). The upper colluvium (pottery dates of Iron Age to Roman; Allen, pers. comm.) was examined from samples F (Figure 18.1) and D, the latter from Trench A on the west side of the valley. Samples G and E, and D and F are silty soils and contain a variety of included soil fragments (Mücher 1974). Many appear to be sharp-edged and eroded from the clay-rich Bt horizon of the post-Combe Rock early-to-mid-Holocene palaeosol (Figures 18.3 and 18.5). In sample D especially, other fragments could originate from silty Bt horizon material formed more directly from loess, and these occur as weakly integrated inclusions, in contrast to the more coherent, sharp-edge clay-rich fragments. The occurrence of seemingly more fragile silty fragments in Trench A, thus perhaps indicates a very local source for some of the silty colluvium on the west side of the valley.

A further type of common soil fragment is more dusty brown in colour with many very fine birefringent clay fragments. It sometimes contains fine charred organic matter, coarse wood charcoal and has a vesicular porosity (Figure 18.6). The last is indicative of trapped air in a water saturated soil. These dusty brown soil inclusions may originate from surface crusts (layered silts and clays) (Boiffin and Bresson 1987; Courty, pers. comm.), perhaps formed in ard furrows, which were reworked/eroded into the plough soil (Figure 18.8). The birefringent clay fragments are probably relic of previous crusts and occur as fine papules. Dark rounded soil fragments, which are poorly birefringent and red under oblique incident light, are pieces of burned soil. Their presence, alongside charcoal in the crust fragments, could testify to arable agriculture (also on-site occupation; Allen 1984) employing burning. In the later (upper) colluvium (sample D) a fragment of 0.5 cm long ferruginised straw-like material (Figure 18.7) occurs, which in addition to ferruginised root fragments, may indicate the ploughing-in of chaff(?) and possible manuring. Thus, the Beaker palaeosol contains relic fragments of the clay-rich post-Combe Rock palaeosol, loessic Bt, and cultivation crust and burned soil, with charred organic matter and charcoal inclusions, whereas later colluvium may even contain residual straw(?).

The colluviums have a compact fabric (see Figure 18.7), brought about by aggregate sorting (Dexter 1976; 1979), in which weak intercalations had formed as the result of 'internal slaking' (Jongerius 1970; 1983). Perhaps because the soil surface was often sealed by crusts, there were only infrequent patches of soil that had dusty clay void coatings (most noted in D), that were more likely to be the result of more open porosity and surface slaking. Many coarse peds had been broken up presumably by ard impact, and biological activity had been encouraged (Macphail *et al.* 1990, Plate 11), but not enough to strongly rework the soil. Hence colluvial accumulation appears to have been rapid. Ard ploughing may only rework the upper 6 cm of the soil (Gebhardt 1990), so the presence of cultivation features throughout the 10 cm thick sample of the Beaker ploughsoil, indicates phases of accumulation at the same time as cultivation took place (Figure 18.8).

*Evidence of erosion*

Truncation of the early-to-mid-Holocene palaeosol

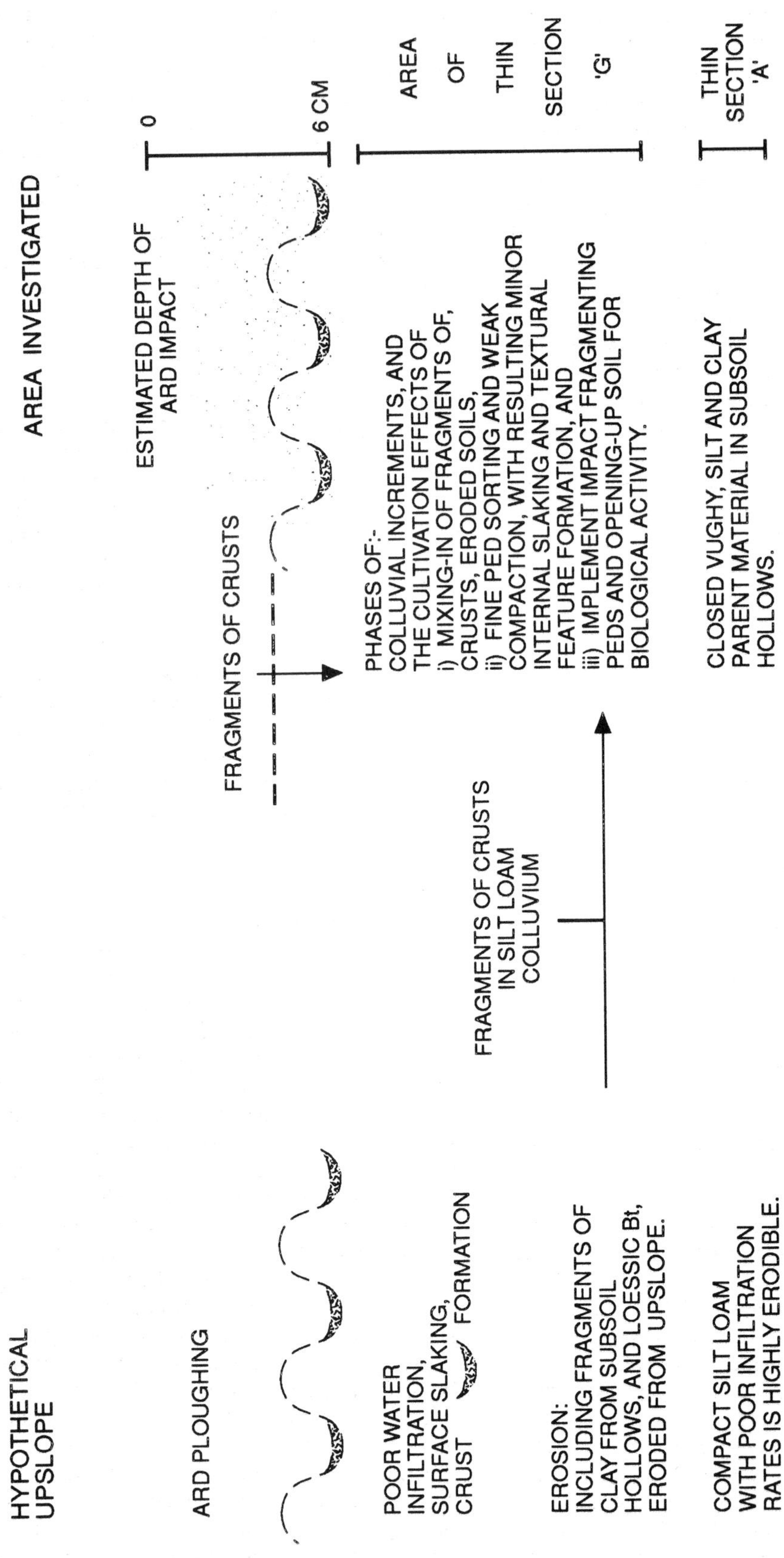


Figure 18.8 Possible mechanisms active during Beaker cultivation at Ashcombe Bottom, near Lewes, Sussex.

(argillic brown earth) probably resulted mainly from the late(?) Neolithic (as indicated from few pottery sherds; Allen 1984, and local archaeology; Drewett 1977; Thomas 1982; Allen pers. comm.) erosion that led to the deposition of the 30 cm thick lower colluvium. The palaeosol in the subsoil hollows (Figures 18.3 and 18.4), however, seems to have recorded both primary (inferred as early Neolithic, Allen 1984, pers. comm.) clearance and later (late Neolithic?) clearance of secondary woodland (Table 18.1). The presence of the Beaker colluvium itself testifies to further soil erosion, whereas the soil micromorphology indicates that cultivation may have induced both surface sealing and compaction of the plough soil, which in turn made the colluvium itself potentially erodible on a site where surface water flow could concentrate. Soil erosion, initiated in Beaker times, continued into the Bronze Age when a gravel fan was deposited. Colluviation with a generally higher magnetic susceptibility (Figure 18.1) continued into the Iron Age and Roman periods. Allen (1988, Table 6.1) calculated an average estimated depth of soil erosion for the valley at 5 cm (from Beaker to modern times), but soil micromorphology demonstrates that colluvium included both clay Bt and loessic Bt fragments (Figure 18.5), inferring both deeper erosion and probable soil loss down valley.

**Discussion: the effect of vegetation clearance and cultivation on soil erosion**

For convenience, sites additional to Ashcombe Bottom are tabulated according to the information they provide on the two major factors that activate ancient soil erosion. Thus, vegetation clearance and cultivation are dealt with separately (Tables 18.2 and 18.3). Sites also have been divided into four main landscape groups, namely; flatland, low slope, sloping and receiving sites, so that available data covers the whole landscape and allows information to be interpolated between site types. Soil erosion through occupation, animal trampling, the formation of drove-ways etc. is not discussed specifically.

*Neolithic and Beaker clearance (Table 18.2) in the UK Flatland and low slope sites*

On flatland sites (examples 1 and 2), much of the soil disturbed by clearance was probably washed into tree-throw hollows (Veneman *et al.* 1984; Schaetzl 1990). At Drayton Cursus (Macphail 1990a; Lambrick in prep.) and Irthlingborough (Macphail and Goldberg 1990) there are strong indications that large trees (the size of feature probably reflecting size of tree; recent studies, R. Bradley, pers. comm.) may have been killed by ring barking (G. Lambrick and M. Robinson, pers. comm.). Trees were toppled over, throwing-up the soil into hummocks, a surface morphology that would have lasted around 500 years (Stephens 1956). Soils from all horizons fell back into the hollows, and poorly stable subsoils slaked under rainfall impact and became washed into any soil fissures. Sometimes the *in situ* burning of the *dead* fallen tree preserved the disrupted soil by baking, as at Irthlingborough.

At Maiden Castle (Sharples 1991) (example 3) it appears that woodland clearance had disrupted the profile, and mobilised soil formed textural coatings and infills rapidly producing an argillic microfabric, whereas the original soil seems to have been more like a brown earth (decalcified Typical Calcaric Brown Soil; Avery 1990) in character (Macphail 1991a). Revegetation of the site is shown by the fine rooting of the disrupted soil fabric, as at Ashcombe Bottom.

Probably there was no net loss caused by the clearance of flatland sites, but on low angle slopes fragile subsoils could have been subject to some rainsplash and gravity movement, leading to erosion.

*Caternary studies*

When archaeological sites cover a slope or occupy a caternary position, data on the results of soil erosion may be recorded down slope (example 7). Some occupational effects (trampling, collapse of mud huts) may have come into play, for example at Uscio (see below). At Brean Down, however, the 'deforested soil' layer can be differentiated from an overlying cultivation colluvium (Macphail 1990b, Figure 128). Thus, it can be estimated, from the occurrence of eroded and transported fragments of ß B clay from the underlying breccia (Unit 8) within the 'deforestation colluvium' (lowest part of Unit 8a), that clearance upslope caused soil disturbance down to a depth of 40 cm. Resulting colluvium is still some 5 cm thick, notwithstanding soil loss from this slope induced by later cultivation (see Table 18.3, example 7).

*Secondary clearance*

The data presented in Table 18.2, suggests that many soil types are potentially erodible once disturbed by clearance. Revegetation of the site, as evidenced by fine rooting of disrupted soil fabrics, however, could rapidly stabilise the soil. So erosion by clearance may be regarded as a single phase event. Secondary clearance as inferred at Ashcombe Bottom, however, also has to be considered. At Carn Brea (example 6), original Neolithic clearance and occupation caused the acid brown earth to acidify, developing into a shallow humo-ferric podzol (Macphail 1990c). Secondary clearance associated with burning, prior to outer rampart construction, caused surface soil instability (alkali ash effect) at Carn Brea. Such a process (Courty *et al.* 1989) could cause increased acidification (soil beneath Bronze Age cairn at Chysauster, Cornwall; Macphail 1987; Macphail *et al.* 1990; Scaife, pers. comm.). This phenomenon, of secondary clearance (employing fire) taking place ahead of monument construction, has been recorded at a number of sites comparable to Chysauster and Carn Brea in Brittany, France (Gebhardt 1990; Gebhardt and Marguerie in press). Secondary clearance can therefore lead to an immediate erosion potential, whereas enhanced acidification over time (Dimbleby 1962) and ensuing deteriorating soil stability can result in protracted future erosion, as found on lowland

| *No/Site/date of clearance* | *Soil/parent material* | *Vegetation* | *Soil disturbance* | *Data source* |
|---|---|---|---|---|
| *Flatland sites (0-2°)* | | | | |
| 1. *Irthlingborough* Nene Valley Northamptonshire (Neolithic) | Argillic brown earth, loam over gravel | Large trees | 40-100 cm deep; hummocky ground. Upper and lower soil profile mixing; impure clay infills. | Field features Coarse burned soil fragments. Charcoal data. |
| 2. *Drayton Cursus* Thames valley Oxfordshire (Neolithic) | Argillic brown earth, loam over gravel | Large trees | 40-100 cm deep; hummocky ground. Upper and lower soil profile mixing; impure clay infills. | Field features Cursus buried soil. Charcoal data. |
| *Low slope sites (3-6°)* | | | | |
| 3. *Maiden Castle* plateau Dorset (Neolithic) | Typical brown earth, clay loam over chalk | Woodland | 30 cm deep; Horizon mixing, void infills. | Field features Bank buried soil. Charcoal and mollusc data. |
| 4. *Suvero* Sub-montane E. Liguria (Chalcolithic) | Argillic brown earth, loess over serpentinite | Woodland | 40 cm deep; Horizon mixing. | Colluvially and monument buried soil catena. |
| 5. *Uscio* Sub-montane terrace E. Liguria (1940s) | Argillic brown earth, impure limestone | Woodland | 60 cm deep; Horizon mixing, impure clay infills. | Modern soil and vegetation. Local history. |
| 6. *Carn Brea* Moorland Cornwall (Neolithic) | Podzol over acid brown earth, granitic Head | Shrubs | 5 cm deep; Disruption of shallow Bh horizon; infills of Ea horizon soil, fine charcoal and phytoliths. | Rampart buried soil. |

Table 18.2 Sites with soil micromorphological evidence of clearance (continued opposite).

| *No/Site/date of clearance* | *Soil/parent material* | *Vegetation* | *Soil disturbance* | *Data source* |
|---|---|---|---|---|
| *Sloping sites (8-20°)* | | | | |
| 7.<br>*Brean Down*<br>slope: 17-20°<br>coastal hillslope<br>Somerset<br>(Beaker) | Complex argillic brown earth, limestone Head and blown sand | Woodland | 40 cm deep; probably several metres horizontally. Rounding and mixing of various soil fragments, including the palaeosol and burned Ah horizon. | Colluvially buried soil catena. Charcoal data. |
| 8.<br>*Uscio*<br>Slope: 6-16°<br>Sub-montane<br>E. Liguria<br>(Chalcolithic) | Argillic brown earth, impure limestone | Woodland | 40 cm deep: possibly 20+ metres horizontally. *In situ* horizon mixing with void infills; rounded soil fragments down slope | Colluvially buried soil catena. Charcoal data |
| *Receiving sites* | | | | |
| 9.<br>*Prato Mollo*<br>Montane shallow peat basin<br>E. Liguria<br>(Chalcolithic) | Banded mineral-organic peats over serpentinite | Locally fir forest | Locally 40-50 cm deep. Horizontally tens of metres. 13 cm coarse (soil fragments?) and fine mineral accumulation, with burned peat at base of peat profile. | Peat buried sediment. Charcoal, pollen and local soil data. |
| 10.<br>*Lago Nero*<br>Montane shallow peat basin<br>E. Liguria<br>(Chalcolithic) | Silt band over serpentinite | Locally fir forest | Locally at least topsoil loss. Horizontally metres to hundreds of metres. 4 cm silt accumulation with fragments of burned soil and peat. | Peat buried sediment. Pollen data |

Table 18.2 Sites with soil micromorphological evidence of clearance (continued from opposite page).

heaths (Scaife, Macphail, in Drewett 1989).

In conclusion, differences in the stability of soil aggregates from different horizons and soil types (Burnham and Pitman 1986) is well known. For instance, A horizons are more stable than B horizons, with grassland A horizons being more stable than woodland A horizons (Greive 1980), whereas woodland soil A horizons are more stable than arable soil A horizons (Imeson and Jungerius 1976). The examples presented here show that soil profiles when disrupted by clearance permit the exposure of unstable subsoils to rainfall impact and potential downslope movement under gravity. These processes of slaking and soil movement seem to affect equally alluvial clay loams and decalcified chalk soils (although cultivated clay soils have been found to erode less frequently than sandy and light loamy soils as shown by Evans in this volume). Additionally, surface horizons can become coarsened by particle translocation (loess and brickearth soils) and by increased acidification through repeated clearance. This may lower the stability of *surface* horizons of soils formed on more coarse parent materials as occurred on lowland heaths, inducing erosion by both water (Macphail, in Drewett 1989) and wind (Radley and Simms 1967; Fisher and Macphail 1985), although some soils may become more stable because a larger particle is left (Boardman, pers. comm.). Subsoils, because of sesquioxidic cementation, however, become more stable (Grieve 1980).

### *Chalcolithic clearance in Eastern Liguria, north-west Italy*

The general eroded nature of Mediterranean soils is well documented (Thornes and Gilman 1983) as discussed by Bintliff in this volume, but sites investigated over the last decade permit the details of soil erosion on a number of soil types and climatic regimes in one specific Mediterranean region (Liguria, Italy) to be compared with the situations reported from the United Kingdom. Archaeologically, the sites are of similar age, Late Neolithic to Beaker (Chalcolithic), but in contrast to the United Kingdom situations, data from montane areas is presented. Also, Ligurian sites differ, because it was often coniferous forest that was cleared mainly for the purpose of montane pastoralism.

#### *Limestone terrain*

The sub–montane site of Castellaro di Uscio (example 8), at 728 metres altitude, is situated on impure limestones in the Sub–Mediterranean mixed oak forest vegetation belt (Maggi 1990). The site, which comprises a 50 m long caternary sequence of buried soils and colluvium is comparable to Brean Down, although the latter is at sea level. At Uscio, a long trench exposed a succession of colluvial deposits and a buried tree hollow upslope. The latter had microfeatures comparable to those of the local recently deforested soil (example 5) (Macphail 1990d). The tree hollow and most of the colluvium probably date to Chalcolithic clearance, after which the site was in continuous use for some 1,000 radiocarbon years, leading to further erosion and colluviation. Upslope, the Chalcolithic palaeosol is only 10–20 cm thick, whereas downslope soils are over a metre thick (not including the overlying archaeological layers), with micromorphologically-identified colluvial soil actually underlying the major Chalcolithic domestic occupation level. In both the tree hollow and the colluvium, features of disturbance (fissure infills), erosion, transportation (soil fragments), tend to be more diffused into the microfabric in comparison with the more 'fresh' features in the local recently deforested soil. This seems to be mainly because of such long term effects as wetting and drying on the buried soils, rather than through the more vigorous reworking by biological activity. Archaeological and environmental data strongly point to the site being cleared for occupation as part of a pastoral economy, rather than for cultivation during the Chalcolithic, whereas more permanent occupation and on–site cultivation took place in the Late Bronze Age when terracing was introduced to control soil erosion.

#### *Montane serpentinite landscapes*

The peat sites of Prato Mollo (1492 m) and Lago Nero (1475 m) (examples 9 and 10) are formed in shallow depressions situated in the present day montane beech woodland belt of the Ligurian Appenines. During the Chalcolithic the vegetation cover was primarily fir (*Abies*) forest (Cruise 1990), which is rarely found in the region today.

At Prato Mollo much of the serpentinite bedrock is exposed. The study of rare, telescoped (20–30 cm thick) residual soils, however, indicate that the original soil at this altitude is polyphased (Macphail 1988). One result of this telescoping is that Middle Palaeolithic to Chalcolithic stone tools can often be found concentrated in the same upper soil horizon in the montane region. At Prato Mollo the remains of a clay-rich pre-Holocene palaeosol (cf. palaeoargillic horizon) are found at the base of the profile, above which the soil is formed in a silt containing very abundant relic sand to gravel-size nodules. This part of the soil has a fragmented compact link capping fabric (Bullock *et al.* 1985, p. 100; sands coated by silts and clays resulting from freeze-thaw), residual of Late Glacial pedogenesis. Superimposed upon this, is an argillic microfabric. Most of the upper 10–15 cm is completely biologically worked and stone- (nodule) free. In this area of serpentinite, Late Glacial and early Holocene soil formation therefore mainly took place on superficial deposits, including loess, developing a polyphased soil. The nature of the inwashed material found at the base of the peat at Prato Mollo, suggests that it was mainly this upper argillic profile formed in loess, that was eroded during the Chalcolithic, probably through the clearance of the fir forest cover for increased pasturage (Maggi and Nisbet in press). The two radiocarbon dates (4300±60

BP Bln-3132, 4230±60 BP Bln-3131; Cruise 1990) through the 13 cm thick mineral-rich, wood charcoal and *in situ* burned peat layers may infer a period over some 200 radiocarbon years for this erosion. At Lago Nero, the basal silt band may have similar origins, i.e. Chalcolithic deforested and destabilised loessic soils, but the resulting sediment has been fluvially sorted. Even here, however, the junction between this basal silt and the overlying peat (4610±40 GrN-1443; Cruise 1990), contains probable local evidence of disturbance, in the form of burned soil and peat fragments.

At these sites, there is an inferred link between Chalcolithic montane activity (hunting and pastoralism) and impact through fire on the fir forest and soil erosion (Baffico *et al.* 1987; Courty *et al.* 1989; Cruise 1990). In these shallow basin sites, peat initiation at around 4,600–4,000 BP seems to have followed the sealing of the formerly *freely-draining* serpentinite substrate by silt and soil inwashes, the root traces of plants which decayed into peat, being preserved in the silts at Lago Nero, for example (Courty *et al.* 1989, Figure 18.1). At these two sites therefore, the weighting is on human factors for the causation of peat formation, rather than climatic or tectonic effects, although all three factors have been identified as being important for peat initiation in the western Mediterranean (Cruise 1990; pers. comm.).

*Submontane serpentinite landscape*

The site of Pianaccia di Suvero (example 4) at 580 m altitude, also occurs on serpentinite. The main occupation, which is Chalcolithic through to Bell Beaker (Maggi 1987), is present on a gently sloping (2–3°) natural terrace. Soil micromorphology shows that the original pedological situation is similar to that at Prato Mollo, some 30 km to the north west. At Suvero, the site buries a colluvium of inferred Chalcolithic date, which itself overlies a 50 cm thick disturbed (deforested?) loessic argillic brown earth developed over a relic nodular and clay-rich serpentinite soil (Macphail 1991b). Unlike the serpentinite soil and the colluvium that is in part derived from it (presumably from areas exposed upslope in the Chalcolithic), the loessic argillic brown earth is practically nodule-free. Here again, at this lower altitude, major soil destabilisation on the serpentinite landscape seems to have commenced in the Chalcolithic.

**Cultivation** (Table 18.3)

*Organic matter and soil stability*

Table 18.3 indicates that many Neolithic and Beaker cultivated soils suffered from surface soil instability, which induced particle translocation (examples 2, 3, 4 and 7). There is some justification for the suggestion that this is related to the low organic matter status of ancient cultivated soils (Romans and Robertson 1983; Macphail 1990e). Trees do not necessarily need to be removed to allow tillage (Lüning and Meurers-Balke 1980; Meurers-Balke 1985), but if cultivation immediately followed on soils already disrupted by clearance, then much of the soil that was tilled would be poorly organic, poorly stable subsoils. On sloping sites such as Brean Down (example 7), cultivation-induced erosion could constantly expose poorly humic subsoils, resulting in total soil loss upslope (Bell, this volume). Cultivation itself may produce a reduction in soil organic matter, increasing the erodibility of soils (Voroney *et al.* 1981; as reviewed by Allen 1991, p.43). But in the study of ancient soils how is their organic status to be estimated? Although data on carbon and nitrogen is available, how does this reflect original amounts after thousands of years of microbiological oxidation, especially in brown soils?

Generally, in biologically active brown soils, biological activity especially by meso- and macro-fauna (insects, arthropods, Enchytraeid worms and earthworms) commonly reflects amounts of available organic matter, and this activity is generally recorded by the microfabric. In addition, some organic features can be reconstructed from iron and manganese oxides, and calcite replaced pseudomorphs. The soil microfabrics (example 5; Gebhardt 1990; Macphail *et al.* 1990) examined from arable fields experimentally ard-ploughed at the Experimental Iron Age Farm at Butser, Hampshire (Reynolds 1981), show a strong relationship between aggregate stability, biological activity and organic matter status, and such soils have a high potential infiltration rate. In contrast, the examples of Strathallan (example 2; Romans and Robertson 1983), Hazleton (example 3; Macphail 1990f) and Maiden Castle (example 4; Macphail 1991a) are characterised by textural features (Macphail *et al.* 1987, Figures 4, 7 and 8) being more dominant than biological ones, and this could reflect diminishing organic matter status brought about by Neolithic cultivation practices. Occasionally, as in the cultivated soil at Kilham, Yorkshire (Dimbleby and Evans 1974), biological activity, such as rooting, was coeval with surface slaking (Macphail *et al.* 1990). Possible associated Neolithic soil erosion on low slope sites is recorded by probable on-site colluvium (of 4–8 cm thickness) and ditch infilling at Hazleton (Macphail 1990f) and ditchfills (Neolithic to Beaker, with soil stability during the Bronze Age) and local colluvium at Maiden Castle (Macphail 1991a; Staines 1991).

*Compaction*

Loss of soil structure can cause compaction (Jongerius 1970; 1983), whereas the concentration of textural coatings, infills and intercalations near the soil surface (example 2) also reduces soil water infiltration rates. The colluvium at Seaford Head was compact (example 8; Allen and Macphail 1987) and contained textural features, again through soil instability that may relate to low organic matter and biological activity. At Ashcombe Bottom, down-profile ped sorting (cultivation mixing sorting finer aggregates downwards; Dexter 1976; 1979) through tillage seems also to have induced compaction, and poor drainage

| *No/Site/period* | *Soil/parent material* | *Cultivation effect* | *Erosion* | *Comment* |
|---|---|---|---|---|
| *Flatland sites (0-2°)* | | | | |
| 1. *Phoenix Wharf* Thames valley London (Early Bronze Age) | Groundwater gley, Alluvial silt over sand | Mixing of organic matter, silts and phytoliths with sand. | (Accretion) | Manured? site. Wetness aiding organic matter preservation. |
| 2. *Strathallan* River terrace Perthshire (Neolithic) | Argillic brown earth, alluvial silty clay loam. | Concentration of dusty textural features at 0-5 cm (mound) and 12 cm (henge) and in subsoil (40-60 cm). Few faunal features. | Negligible | Little organic matter? and biological activity. Very fragile soil potentially highly erodible. |
| *Low slope sites (3-6°)* | | | | |
| 3. *Hazleton* Plateau Gloucestershire (Neolithic) | Argillic brown earth, clay on oolitic limestone | Surface soil concentration of dusty textural features; few faunal features. | Soil loss amounting to on-site 4-8 cm thick colluvium? Total amount? Ditch infills. | Little organic matter? and biological activity. Probably rapid erosion during short use of site. |
| 4. *Maiden Castle* Plateau Dorset (Neolithic) | Typical brown earth, clay loam over chalk. | Dusty clay coated biological fabric. Soil mixing. | Telescoping of soil profile. Ditch infills, colluvium. | Little organic matter? moderately low biological activity. Erodible soil. |
| 5. *Butser* Dry valley Hampshire (Experimental Iron Age) | Calcareous brown earth, on chalky colluvium. | Fragmentation of peds; well developed biological fabric. | Negligible erosion? Some fields are dish-shaped. Up to 30-35 cm thick Ap. | High organic matter, whether manured or not, with high biological activity; Stable soil. |

Table 18.3 Sites with soil micromorphological evidence of cultivation (continued opposite).

| *No/Site/period* | *Soil/parent material* | *Cultivation effect* | *Erosion* | *Comment* |
|---|---|---|---|---|
| *Sloping sites (7–20°)* | | | | |
| 6. *Chysauster* Slope: 4-7° Moorland Cornwall (Romano-British) | Humo-ferric podzol, on granitic Head. | Dusty clay coatings, crust? fragments, mixing-in of deep subsoil material. | 50-70 cm loss at negative lynchet; tens of metres horizontally; up to 50 cm accumulation. | Strongly erodible acid soil. Romano-British construction of walls on lynchets to stop soil loss? |
| 7. *Brean Down* Slope: 17-20° Coastal hillslope Somerset (Beaker) | Complex argillic brown earth, Limestone Head and blown sand. | Dusty clay coatings, infills and intercalations; soil mixing. | 20+ cm loss; 20 cm accumulation over 15 metres. | Little organic matter? and biological activity. Fragile soil and steep slope. |
| *Receiving sites* | | | | |
| 8. *Seaford Head* Plateau Sussex (Iron Age) | Argillic brown earth, over Clay-with-Flints. | Compact with dusty clay coatings and infills; Ap burial by colluvial Ap? | Upslope loss? 20 cm accumulation; 10-100 metres horizontally. | Little organic matter? and biological activity; erodible more coarse topsoil. |
| 9. *Suvero* Sub-montane E. Liguria (Chalcolithic) | Argillic brown earth, on loess over serpentinite. | Dusty clay coatings intercalations and faunal fabrics; mixing-in of nodule-rich subsoils. | Upslope 40-50 cm loss; 10+ cm accumulation in Chalcolithic; up to 70 cm accumulation since Chalcolithic over 80 metres. | Little organic matter? but periods of biological activity; Fragile, erodible soil. Site and soil preserved by terrace construction. |
| 10. *Drayton Cursus* Thames Valley Oxfordshire (Roman/Saxon) | Calcareous alluvium over clay loam over gravel. | Locally water meadows. | Locally and distant (kilometres). Rounded chalk and decalcified clay loam fragments. | Alluvially buried soil/sediment Mollusc data. |

Table 18.3 Sites with soil micromorphological evidence of cultivation (continued from opposite page).

resulted in internal slaking. The soil also suffered from surface crusting, because of the unstable ploughsoil. In either case, compaction and surface sealing made soils susceptible to erosion because infiltration rates had been strongly reduced allowing surface water to concentrate and cause rill erosion. On the other hand, the Butser experiments (example 5; Reynolds 1981; Gebhardt 1990; Macphail *et al.* 1990) show that when high organic matter contents are maintained, soil pore continuity is enhanced by biological activity and peds are basically stable. Little evidence of erosion is expected on these soils, although research is continuing on this site.

### *Manuring*

There is an inference that the last phase of Neolithic cultivation at Hazleton (example 3) shifted to a more organic midden-like area of the site, whereas at the Beaker period site of Ashcombe Bottom charred organic matter was present and biological activity was certainly noted in the arded horizons. As pointed out by Romans and Robertson (1983) 'subsistence' agriculture (i.e. without manuring) in the Neolithic may have led to highly unstable soils, for the reasons described above. At Phoenix Wharf (example 1; Macphail *et al.* 1990) there was an accumulation of organic matter and artifacts (J. Rackham, pers. comm.) to produce an Early Bronze Age Ap horizon some 17 cm thick, when ard ploughing on similarly light soils may only affect the soil down to 6 cm. The earliest examples of plaggen-like soils include a Bronze Age 58 cm thick humic epipedon from Germany (Blume and Kalk 1986), whereas manuring with seaweed has also been suggested from as early as the Bronze Age (Bell 1981; 1990). It is possible, all things being equal (soil type, cultivation methods etc.), that if Bronze Age soils were manured, they may have been (as a side effect) relatively more stable and resistant to erosion than Neolithic cultivated soils. The overall impact on soil erosion by Bronze Age peoples, however, was probably greater (Evans 1990), and so the above suggestion may be very difficult to substantiate.

### *Additional soil types and later periods*

In addition to the soil types already discussed, evidence of cultivation-induced textural features and deep soil erosion can be found in upland podzols (example 6; Macphail 1987) and in Italian sub-montane loessic argillic brown earths (example 9). Cultivation higher up in river catchments can also be recorded down river (example 10; Robinson and Lambrick 1984; Macphail 1990a).

## Conclusions

Vegetation clearance, whether for cultivation or pastoralism, could potentially initiate soil erosion. The amount of resulting soil erosion probably related to slope angle, and is best studied on soil catenas with both features of deforestation and ensuing downslope colluviation. Some soils could be disrupted down to depths of 40–100 cm, and regularly more than 10 cm of resulting colluvium has been recorded. All soil types investigated from the United Kingdom and north west Italy appear to be potentially erodible, because poorly stable subsoils were often exposed to rainsplash and gravity movement. In Eastern Liguria, Italy, major soil instability seems to have been initiated across the sub-montane to montane landscape by Chalcolithic clearances, often because of pastoral activity.

Vegetation seems to have rapidly regenerated on deforested soils, restabilising them, and so soil erosion from clearance may be considered as a single phase. The effect of secondary clearance(s) (with fire) on some coarse acid soils, was to further acidify them, producing unstable surface horizons, that led to protracted periods of erosion.

Neolithic cultivation seems to have produced highly erodible soils, possibly because levels of soil organic matter were reduced. Soil instability and diminishing faunal activity ensued. The resulting reduction in infiltration rates through surface sealing, compaction and pore infilling by textural features, may have accelerated erosion. There is evidence that Beaker cultivation also strongly eroded forest soils, sometimes removing them completely. The resulting colluvium itself could be equally erodible, because of surface sealing and compaction, though on some sites there may have been instances of manuring that enhanced biological activity and soil stability. In the long term, it is possible that manuring, which may have begun in the Early Bronze Age in Europe, could have made soils less erodible because, as shown experimentally, such soils have higher infiltration rates. This is because peds are more stable and soils are strongly perforated (interconnected void space) by biological activity. In sharp contrast, many of the Neolithic soils investigated were compact with textural features infilling pores.

Only a limited number of ancient soil sites have been studied by soil micromorphology. Yet if the above contentions concerning the mechanisms and factors affecting rates of past soil erosion are to be tested, this technique needs to be employed routinely. Modern studies (Jongerius 1970; 1983; Kooistra 1987) as noted by Farres *et al.* in this volume, are also important, whereas additional experimental data on clearance and ancient cultivation methods should be sought.

## Acknowledgements

The author wishes to acknowledge the long term funding by English Heritage and the Soprintendenza Archeologica della Liguria, and the close collaboration of Mike Allen (Ashcombe Bottom) and Anne Gebhardt (experiments). He also warmly thanks M.A. Courty and N. Fedoroff for the collaborative discussion of soil microfabrics, G.M. Cruise for her comments on the Ligurian data and the text in general, and Stewart Laidlaw for photographic assistance.

## References


Allen, M. J. 1984 *Ashcombe Bottom Excavation AB84*. Lewes Archaeological Group, Lewes.

Allen, M. J. 1988 Archaeological and environmental aspects of colluviation in south-east England. In: Groenman-van Waateringe, W. and Robinson, M. (eds.) *Man-Made Soils*. British Archaeological Reports, International Series, 410, Oxford, 67–92.

Allen, M. J. 1991 Analysing the landscape: a geographical approach to archaeological problems. In: Schofield, A. J. (ed.) *Interpreting Artefact Scatters: Contributions to Ploughzone Archaeology*. Oxbow Monograph 4, Oxford, 39–57.

Allen, M. J. and Macphail, R. I. 1987 Micromorphology and magnetic susceptibility studies: their combined role in interpreting archaeological soils and sediments. In: Fedoroff, N., Bresson, L. M. and Courty, M. A. (eds.) *Soil Micromorphology*. Association Française pour Etude du Sol, Plaisir, 669–676.

Avery, B. W. 1980 *Soil Classification for England and Wales*. Soil Survey Monograph 14, Harpenden.

Avery, B. W. 1990 *Soils of the British Isles*. CAB International, Wallingford.

Avery, B. W. and Bascomb, C. L. (eds.) 1974 *Soil Survey Laboratory Methods*. Soil Survey Monograph No. 6, Harpenden.

Baffico, O., Cruise, G. M., Macphail, R. I., Maggi, R. and Nisbet, R. 1987 Monte Aiona – Prato Mollo. In: Melli, P. (ed.) *Archeologia in Liguria. III*. 1, Soprintendenza Archeologica della Liguria, Genoa, 57–66.

Bell, M. 1981 Seaweed as a prehistoric resource. In: Brothwell, D. and Dimbleby, G. (eds.) *Environmental Aspects of Coasts and Islands*. British Archaeological Reports, International Series, 94, Oxford, 117–126.

Bell, M. 1990 *Brean Down Excavations 1983–1987*. English Heritage Archaeological Report 15, London.

Blume, H. P. and Kalk, E. 1986 Bronzezeitlicher auftragsboden bei Rantum auf Sylt. *Zeitschrift Pflanzenernaehr. Bodenkunde* 149, 608–613.

Boiffin, J. and Bresson, L. M. 1987 Dynamiques de formation des croutes superficielles: apport de l'analyse microscopique. In: Fedoroff, N., Bresson, L. M. and Courty, M. A. (eds.) *Soil Micromorphology*. Association Française pour l'Etude du Sol, Plaisir, 393–399.

Bullock, P., Fedoroff, N., Jongerius, A., Stoops, G. and Tursina, T. 1985 *Handbook for soil thin section description*. Waine Research Publications, Wolverhampton.

Bullock, P. and Murphy, C. P. 1979 Evolution of a palaeo-argillic brown earth (Paleudalf) from Oxfordshire, England. *Geoderma* 22, 225–53.

Burnham, C. P. and Pitman, J. I. (eds.) 1986 *Soil Erosion*. SEESOIL 3, South East Soils Discussion Group, Ashford.

Courty, M. A., Goldberg, P. and Macphail, R. I. 1989 *Soils and micromorphology in archaeology*. Cambridge University Press, Cambridge.

Cruise, G. M. 1990 Holocene peat initiation in the Ligurian Appenines, northern Italy. *Review of Palaeobotany and Palynology* 63, 173–182.

Dexter, A. R. 1976 Internal structure of tilled soil. *Journal of Soil Science* 27, 267–78.

Dexter, A. R. 1979 Prediction of soil structure produced by tillage. *Journal of Terramechanics* 16(3), 117–27.

Dimbleby, G. W. 1962 *The development of British Heathlands and their soils*. Clarendon Press, Oxford.

Dimbleby, G. W. and Evans, J. G. 1974 Pollen and land snail analysis of calcareous soils. *Journal of Archaeological Science* 1, 117–33.

Drewett, P. L. 1977 The excavation of a Neolithic causewayed enclosure on Offham Hill, East Sussex, 1976. *Proceedings of the Prehistoric Society* 43, 201–241.

Drewett, P. L. 1989 Anthropogenic soil erosion in prehistoric Sussex: excavations at West Heath and Ferring, 1984. *Sussex Archaeological Collections* 127, 11–29.

Evans, J. G. 1990 Notes on some Late Neolithic and Bronze Age events in long barrow ditches in southern and eastern England. *Proceedings of the Prehistoric Society* 54, 111–116.

Fisher, P. J. and Macphail, R. I. 1985 Studies of archaeological soils and deposits by micromorphological techniques. In: Feiller, N., Gilbertson, D. D. and Ralph, N. G. A. (eds.) *Palaeoenvironmental Investigations: Research Design, Methods and Data Analysis*. British Archaeological Reports, International Series 258, Oxford, 93–112.

Gebhardt, A. 1990 *Evolution du paléopaysage agricole dans le nord-ouèst de la France: apport de la micromorphologie*. Unpub. thèse, Université de Rennes 1.

Gebhardt, A. and Marguerie, D. In Press Etude paléo-environmentale de la transformation du paysage Armorican sous l'influence de l'homme néolithique. Actes du XVIe Colloque Interrégional sur le Néolithique. *Documents Archéologique Français*.

Grieve, I. C. 1980 Some contrasts in soil development between grassland and deciduous woodland sites. *Journal of Soil Science* 31, 137–45.

Guilloré, P. 1985 *Méthode de Fabrication Méchanique et en Série des Lames Minces*. Institut National Agronomique, Paris.

Hodgson, J. M. 1974 *Soil Survey Field Handbook*. Soil Survey Technical Monograph No. 5, Harpenden.

Imeson, A. C. and Jungerius, P. D. 1976 Aggregate stability and colluviation in the Luxembourg Ardennes: an experimental and micromorphological study. *Earth Surface Processes* 1, 259–271.

Jongerius, A. 1970 Some morphological aspects of regrouping phenomena in Dutch soils. *Geoderma*

4, 311–31.
Jongerius, A. 1983 Micromorphology in agriculture. In: Bullock, P. and Murphy, C. P. (eds.) *Soil Micromorphology*. A B Academic Publishers, Berkhamsted, 111-38.
Kemp, R. A. 1986 Pre-Flandrian Quaternary soils and pedogenic processes in Britain. In: Wright, V. P. (ed.) *Palaeosols: their Recognition and Interpretation*. Blackwell Scientific Publications, Oxford, 242–262.
Kerney, M. P. 1977 A proposed zonation scheme for late-glacial and post-glacial deposits using land mollusca. *Journal Archaeological Science* 4, 387-390.
Kooistra, M. J. 1987 The effects on compaction and deep tillage on soil structure in a Dutch sandy loam soil. In: Fedoroff, N., Bresson, L. M. and Courty, M. A. (eds.) *Soil Micromorphology*. Association Française pour l'Etude du sol, Plaisir, 445–450.
Kwaad, F. J. P. M. and Mücher, H. J. 1977 The evolution of soils and slope deposits in the Luxembourg Ardennes near Wilts. *Geoderma* 17, 1–37.
Kwaad, F. J. P. M. and Mücher, H. J. 1979 The formation and evolution of colluvium on arable land in northern Luxembourg. *Geoderma* 22(2), 173–92.
Longworth, G. and Tite, M. S. 1977 Mossbauer and magnetic susceptibility studies of iron oxides in soils from archaeological sites. *Archaeometry* 14, 1–14.
Lüning, J. and Meures-Balke, J. 1980 Experimentaller Getreideanbau im Hambacher Forst, Gemeinde Elsdorf, Kr. Bergheim/ Rheinland. *Sonderdruck aus Bonner Jahrbücher* (Rheinischen Landesmuseums in Bonn) 180, 305–44.
Macphail, R. I. 1987 *Soil Report on the cairn and field system at Chysauster, Penzance, Cornwall*. Ancient Monument Laboratory Report, 111/87, English Heritage, London.
Macphail, R. I. 1988 *Soil report on the soil and basal peat at Prato Mollo, eastern Ligurian Apennines*. Unpub. report, Soprintendenza Archeologica della Liguria, Genoa.
Macphail, R. I. 1990a *Soil Report on Drayton Cursus, near Abingdon, Oxfordshire*. Ancient Monument Report, 57/90, English Heritage, London.
Macphail, R. I. 1990b Soil history and micromorphology. In: Bell, M. (ed.) *Brean Down excavations 1983–1987*. Archaeological Report No. 15, English Heritage, London, 187-196.
Macphail, R. I. 1990c *Soil report on Carn Brea, Redruth, Cornwall; with some reference to similar sites in Brittany, France*. Ancient Monument Report, 55/90, English Heritage, London.
Macphail, R. I. 1990d Micromorphological investigation of soils and sediments. In: Maggi, R. (ed.) *Archeologia dell'Appennino Ligure. Gli Scavi del Castellaro di Uscio; un insediamento di crinale occupato dal Neolitico alla conqista Romana*. Instituto Internazionale di Studi Liguri, Bordighera, 179–196.
Macphail, R. I. 1990e Soil micromorphological evidence of the impact of ancient agriculture. *Transactions 14th International Congress of Soil Science*, VII, Kyoto, Japan, 264–269.
Macphail, R. I. 1990f The soils. In: Saville, A. (ed.) *Hazleton North: The excavation of a Neolithic long cairn of the Cotswold-Severn group*. Archaeological Report No. 13, English Heritage, London, 223–226.
Macphail, R. I. 1991a The archaeological soils and sediments. In: Sharples, N. M. (ed.) *Maiden Castle Excavations and Field Surveys 1985-86*. Archaeological Report No. 19, English Heritage, London, 106–118.
Macphail, R. I. 1991b *Third soil report on Pianaccia di Suvero, near Rochetta, GE., Liguria*. Unpub. report, Soprintendenza Archeologia della Liguria, Genoa.
Macphail, R. I., Courty, M. A. and Gebhardt, A. 1990 Soil micromorphological evidence of early agriculture in NW Europe. *World Archaeology* 22(1), 53–69.
Macphail, R. I. and Goldberg, P. 1990 The micromorphology of tree subsoil hollows: their significance to soil science and archaeology. In: Douglas, L. (ed.) *Soil Micromorphology: A basic and applied science*. Elsevier, Amsterdam, 431–440.
Macphail, R. I., Romans, J. C. C. and Robertson, L. 1987 The application of micromorphology to the understanding of Holocene soil development in the British Isles: with special reference to early cultivation. In: Fedoroff, N., Bresson, L. M. and Courty, M. A. (eds.) *Soil Micromorphology*, Association Française pour l'Etude du Sol, Plaisir, 647–565.
Maggi, R. 1987 Pianaccia di Suvero. In: Melli, P. (ed.) *Archeologia in Liguria III*. 1, Soprintendenza Archeologia della Liguria, Genoa, 23–32.
Maggi, R. 1990 *Archeologia dell'Appennino Ligure. Gli Scavi del Castellaro di Uscio: un insediamento di crinale occupato dal Neolitico alla conquista Romana*. Instituto Internazionale di Studi Ligure, Bordighera.
Maggi, R. and Nisbet, R. In press 1990 Prehistoric pastoralism in Liguria. In: Barker, G., Maggi, R. and Nisbet, R. (eds.) *The Archaeology of Pastoralism in Southern Europe*. Instituto Internazionale di Studi Ligure, Bordighera, 56.
Meures-Balke, J. 1985 Experimentaszum Anbau und zur Verarbeiterung prähistorischer Getreidearten. *Archäologische Information* 8(1), 8–17.
Mücher, H. J. 1974 Micromorphology of slope deposits: the necessity of a classification. In: Rutherford, G. K. (ed.) *Soil Microscopy*. Limestone Press, Kingston, Ontario, 553–556.
Murphy, C. P. 1986 *Thin Section Preparation of Soils and Sediments*. A B Academic Publishers,

Berkhamsted.
Radley, J. and Simms, C. 1967 Wind erosion in East Yorkshire. *Nature* 216, 20–22.
Reynolds, P. J. 1981 Deadstock and livestock. In: Mercer, R. (ed.) *Farming practice in British Prehistory*. Edinburgh University Press, Edinburgh, 97–122.
Robinson, M. and Lambrick, G. 1984 Holocene alluviation and hydrology in the upper Thames basin. *Nature* 308, 809–14.
Romans, J. C. C. and Robertson, L. 1983 The general effects of early agriculture on the soil. In: Maxwell, G. S. (ed.) *The Impact of Aerial Reconnaissance on Archaeology*. CBA Research Report No. 49, London, 136–41.
Schaetzl, R. J. 1990 Effects of treethrow microtopography on the characteristics and genesis of spodosols, Michigan, USA. *Catena* 17, 111–126.
Sharples, N. M. 1991 *Maiden Castle Excavations and Field Surveys 1985-6*. Archaeological Report No. 19, English Heritage, London.
Staines, S. 1991 The soils. In: Sharples, N. M. (ed.) *Maiden Castle Excavations and Field Surveys 1985-6*. Archaeological Report No. 19, English Heritage, London.
Stephens, E. P. 1956 The uprooting of trees: a forest process. *Soil Science Society of America Proceedings* 20, 113–116.
Thomas, K. D. 1982 Neolithic enclosures and woodland habitats on the South Downs in Sussex, England. In: Bell, M. and Limbrey, S. (eds.) *Archaeological Aspects of Woodland Ecology*. British Archaeological Reports, International Series, 146, Oxford, 147–170.
Thornes, J. B. and Gilman, A. 1983 Potential and actual erosion around archaeological sites in south east Spain. In: de Ploey, J. (ed.) Rainfall Simulation, Runoff and Soil Erosion. *Catena Supplement* 4, 91–114.
Tite, M. S. and Mullins, C. E. 1971 Enhancement of the magnetic susceptibility of soils on archaeological sites. *Archaeometry* 13, 209–219.
Veneman, P. L. M., Jacke, P. V. and Bodine, S. M. 1984 Soil formation as affected by pit and mound relief in Massachusetts, USA. *Geoderma* 33, 89–99.
Voroney, R. P., Van Veen, J. A. and Paul, E. A. 1981 Organic C dynamics in grassland soils 2. Model validation and simulation of the long term effects of cultivation and rainfall erosion. *Canadian Journal of Soil Science* 61, 211–224.

# CHAPTER 19

# A CONCEPTUAL MODEL OF SOIL DEPOSITION AND ITS IMPLICATIONS FOR ENVIRONMENTAL RECONSTRUCTION

P. J. FARRES, S. J. WOOD AND S. SEELIGER

*A general conceptual model of the erosional processes and landscape conditions is initially considered. The model shows how at various parts in the landscape the sequential deposition of eroded material can take place. At these positions the development of 'colluvial soils' takes place where deposited material is integrated with the underlying soil by the mechanical process of ploughing and by faunal mixing. The paper formalizes this process by developing a mathematical model which considers the mixing and survival of deposited material through a sequence of additions. The model is used to simulate the development and characteristics of a colluvial soil.*

## Introduction

Soil material, forming as it does in the surface and near surface environment, reacts and responds to any changes in the characteristics of the external control environment. The characteristics of surface soils and buried soil units have been successfully used to say something about the nature of environmental change (Langhor and Sanders 1985). However, to make such statements requires a conceptual framework in which the relationships between the control environment and measurable soil responses can be directly compared. One such framework has been stated generally by Yaalon (1971), who considers the stability of soil properties and horizons as external environmental conditions change. Soils developed in colluvial material (colluvial soils) provide a potentially important source of information on changes in environmental conditions. To understand such material requires not only a Yaalon-type framework, but also an appreciation of the nature and origins of the colluvial material in addition to the possible effects of post-depositional landscape processes. This paper seeks to provide a conceptual framework in which the characteristics of agricultural colluvial soils may be more realistically evaluated.

## General context of the model

Research into soil erosion within North West Europe readily demonstrates that much of Belgium, Southern Netherlands, Northern France, and many areas in the UK show extensive evidence of soil erosion and deposition (Kwaad and Mücher 1979; Imeson *et al.* 1980; Vreeken and Mücher 1981; Auzet 1987; Boardman 1990). Many of these areas have a number of morphological, pedological and land-use attributes in common (Farres *et al.* 1991). The soils are of loam/fine sand textures associated with loess, its reworked derivations, and Tertiary sand materials. The profiles of the soils show the characteristics of Cambisols and Luvisols (FAO 1988). Such soil materials have in general a low structural stability to raindrop impact forces and mechanical compaction, they therefore readily crust and seal which may enhance runoff generation. In addition, the individual mineral grains are of sizes easily transported by flowing water. The soils are generally well drained, easily worked, relatively fertile and have been used, more or less continually, for agricultural production from the Roman period to the present day.

The physiography of these old arable landscapes are also remarkably similar. They tend to be associated with low amplitude rolling relief with high drainage densities. Significantly most of the valley forms are dry. The implications of this configuration is that most of the removed material is redistributed and deposited within the landscape system, and only on rare occasions finds its way to permanent drainage lines for complete removal from the local agricultural catchment.

The soil erosion processes operative in these landscapes are well documented (Boffin 1984; Poesen 1989; Govers 1987; Poesen and Govers 1990). Runoff tends to be generated on the side slopes and concentrated in the thalwegs of the dry valleys. It is here the major soil movement takes place. Landscape linearities created by tractor wheeling, drilling lines etc. often play a major role in defining the spatial and temporal dynamics of this phase of the erosion at a local scale. Barriers across the primary flow lines, created typically by hedgerows or old sunken route ways result in the deposition of colluvial materials as ponding of runoff occurs and the transport power is reduced. In fields high in the catchments, barriers often serve as reservoirs which funnel water into the next field down the catchment cascade. This concentration of water at particular critical points in the landscapes is achieved by water flowing over the barrier, or more typically through enlarged faunal burrows and root channels. The result of such localised water concentrations is the initiation and development of gully forms, even in fields not themselves generating significant side slope runoff. The implication of such erosional responses is that erosion in any one field cannot be considered in

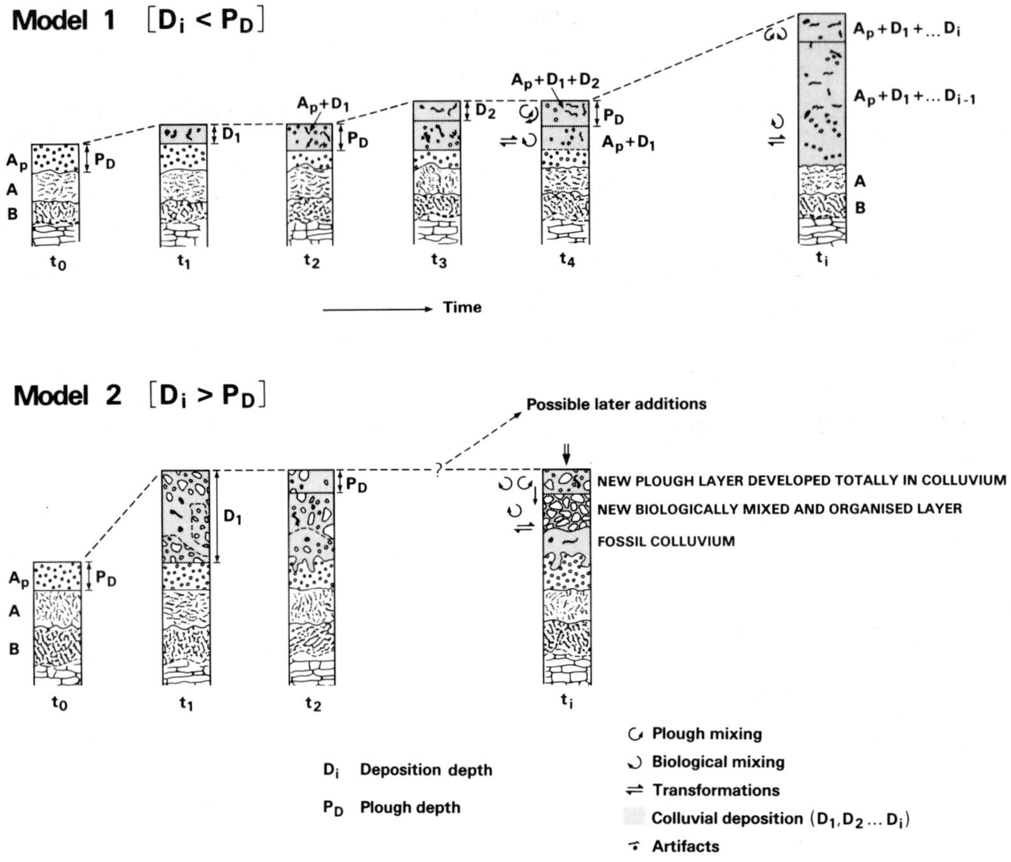


Figure 19.1 Schematic model of the development of colluvial foot slope soil profiles

isolation, its relationship to all connecting fields in a landscape context is vital.

However, many of these natural barriers, especially those in lower physiographic positions, give rise to sites where sequential deposition can take place. It is in such landscape positions colluvial deposition will be best preserved and on which the colluvial soil development model outlined here is focused.

### Specific context of the model

The activation of the erosion processes must result from a disturbance to the natural soil slope evolution. This disturbance creates upslope instability, soil mobilisation, transportation and finally deposition in footslope facets of the landscape. In the areas of North West Europe central to the model these disturbances will essentially be a result of the coupling of storm events with anthropogenic effects. Not all land-uses bring about such radical soil slope instability, some may be in sympathy with the natural soilscape genesis (Hoel 1978), while others may even enhance the stability. However, it is in these old agricultural landscapes where alternating states of stability and instability may be found, resulting from different agricultural practices. These changes will then be reflected in the resulting colluvial deposits. This concept of alternating phases are to be found embodied in the 'K' cycle concept of landscape soil evolution developed by Butler (1982). He writes:

> 'Only in the depositional part of the landscape should one expect to find the regular stack of soil-on-soil and sediment-on-sediment such as the formal concept of stratigraphy conjures up'.

Unfortunately, for the soils and their associated colluvial deposits considered here, few, if any, are likely to be so stratigraphically simple. This is because these depositional areas also form part of the agricultural system itself and are therefore subject to ploughing, arable crop growth, and periodic ley grassland management. The deposits are themselves composed of soil material, which has been transported to a new pedoenvironment. Consequently apart from all the man-induced disturbances natural pedogenic processes will also play a part in developing the material characteristics.

Soils at the bases of such slopes are normally recognised (at least at profile scale) as over deep soil units. The model developed here seeks to provide a conceptual framework for any area where this type of erosion and deposition occurs: a framework in which the potential of such soils as a source of evidence for past erosional conditions might be realistically considered.

### The conceptual model

Consider a site in the landscape where deposition of soil material removed by erosion from up slope locations may take place. As a starting condition let us assume the soil at the site is a brown earth, also, that the soil is under arable cultivation and has an open ploughed surface structure. Colluvial material is now deposited on the surface of the soil at this site. The fabric of the material will reflect its mode of fluvial deposition, contain organic matter and may contain artifacts of agricultural production. The material was once part of an agricultural soil removed from higher up in the catchment and although significantly altered during transportation and deposition, some elements of the original pedofabric may still remain contained in the colluvial material.

The next stage depends on the landscape history of the site but certainly in this type of environment mechanical disturbance by ploughing will take place. As a consequence the deposited material will be mixed in with the original A horizon, so partially disturbing the original depositional structure. As well as mechanical mixing normal processes of pedogenesis will also be operative, in particular faunal mixing will degrade the original colluvial structure and help create a well integrated soil material. It is important to note that typically the plough depth will exceed the deposition depth of the colluvial addition: this we call the *normative* situation. Because of the nature of these landscapes this process will be repeated many times ultimately producing the over deep foot slope colluvial soil as shown schematically in Figure 19.1. As the depth of deposition is always less than the plough depth the intense mechanical and natural mixing consequently means almost all of the latent colluvial structures are destroyed and contorted. This makes the recognition of individual colluvial episodes by traditional field observations from a pit face almost impossible.

A second variation on this model can be identified (non-normative model). In this model the added colluvial material exceeds the normal plough depth. It is only under such extreme circumstances, in the context of a pedologically active mixing environment, the true buried soil colluvial layer can be preserved. This variation is also shown in Figure 19.1. There must also exist the possibility of a hybrid version combining both sets of conditions in one colluvial soil profile.

### A mathematical model and some implications

#### *Distributions of deposited material*

One implication of the conceptual model is that material from the very first depositional event may be found distributed at all depths throughout the colluvial profile, even on the present soil surface. This and other aspects of material distribution can best be considered with the aid of the following mathematical model.

Consider just one initial colluvial event deposited on the soil surface. If the deposit is $D_i$ thick, and the plough depth is $P_D$. Then, when mixing takes place by ploughing in subsequent years the material in the new surface plough layer will be made essentially of the two materials:

(1) The colluvial material (deposit 1)
(2) The original soil surface material.

The proportion of each material in the new plough layer will be:

$$\left(\frac{D_i}{P_D}\right)^1 + \left(\frac{P_D - D_i}{P_D}\right)^o \qquad (1)$$

If we let *a* be the ratio between deposition and plough depth:

$$a = \frac{D_i}{P_D} \qquad (2)$$

Then we have:

$$a_{d_1} + (1-a)_{o_s} \qquad (3)$$

Where the subscript $d_1$ indicates deposit 1 and the $o_s$ the original soil material.

If a second deposit is now added to the surface, again say $D_i$, thickness, by rearranging the terms a little we have:

$$a(1-a)_{d_1} + a_{d_2} + (1-a)^2_{o_s} \qquad (4)$$

This can now be written generally to give the proportions of each depositional event found at a point in time at the soil surface by the series:

$$a(1-a)^{k-1}_{d_1} + a(1-a)^{k-2}_{d_2} \ldots + a(1-a)^{k-k}_{d_k} + (1-a)^k_{o_s} \qquad (5)$$

Where K = total number of depositional events, and subscript $d_i$ indicates proportion of deposit i and $O_s$ the original soil material.

A specific version of the model can be shown diagrammatically (Figure 19.2). Here 60 cm of colluvial material has developed formed from 5 equal 12 cm depositional additions mixed by a constant plough depth of 30 cm (i.e. a = 0.4). The proportion of each deposit at various depths down the profile can be shown.

Specifically for a bulk sample taken at a depth x from the soil surface the proportion of the sampled material derived from deposit i (i.e. $P_{ix}$) will be given by:

$$P_{iX} = a(1-a)^{N-i} \qquad (6)$$

Because the model assumes uniform mixing within the plough depth and a series of distinct depositional events the model operates as a series of discrete steps. We can define N as the integer part of the function below, where $X_T$ is the total depth of the colluvial build up for the hypothetical section.

$$N = \frac{X_T - X}{D_i} \qquad (7)$$

Continuing initially with the assumption that each depositional event adds the same depth of material ($D_i$), and in addition the plough depth remains constant, it is possible to investigate the effects of differing values of $D_i$ on the proportion of a specific deposit with depth by varying the parameter a. If for example we take a 1 m depth profile of mixed colluvial material (i.e. the first deposit on top of a surface now 1 m below the present surface) and in addition we also assume a reasonable constant plough depth of say 30 cm, we derive a set of curves shown in Figure 19.3 for the proportion of the initial deposit.

Consider the material derived from just one specific deposit. As sequential depositional layers are added to the surface of this layer and mixed by ploughing the proportion of the initial deposited material decreases up the colluvial unit generated by each accretionary event. In terms of investigation of colluvial units a significant factor to consider would be the number (hence depth) of subsequent depositional events required to be added to the top of a reference deposit before the proportion of that deposited material falls below some critical detectable limit. From this type of argument it is then possible to develop some sort of *critical detectable survival depth*. For example if we say a critical threshold percentage of 5% i.e. 5 gm in a 100 gm bulk sample, then the reference deposit will require a minimum of M cycles of additional deposits $D_i$ in thickness to be added before a reduction to the critical proportion of detection occurs e.g.:

$$a(1-a)^M = 0{\cdot}05 \qquad (8)$$

Taking the integer part of the following expression we can solve for M:

$$M = \left[\frac{Log\left(\frac{0{\cdot}05}{a}\right)}{Log(1-a)}\right] \qquad (9)$$

The product of M and the depositional depth $D_i$ will give the minimum depth from the original depositional level before the threshold proportion is reached. The addition of $D_i$ to this produces the maximum survival depth. If $P_D$ is the plough depth and $P_{CRIT}$ a defined critical detectable proportion, then, the survival depth range ($S_i$) is given by the general expression:

$$D_i\left[\frac{Log\left(\frac{P_{CRIT}\cdot P_D}{D_i}\right)}{Log\left(1-\frac{D_i}{P_D}\right)}\right] < S_i < \left(D_i\left[\frac{Log\left(\frac{P_{CRIT}\cdot P_D}{D_i}\right)}{Log\left(1-\frac{D_i}{P_D}\right)}\right] + D_i\right) \qquad (10)$$

The effects of different values for the parameter $P_{CRIT}$ on the relationship between $S_i$ and $D_i$ can now be investigated. Some of the generated curves are shown in Figure 19.4. The inherent features of the model are that for small depositional thicknesses, compared with the plough depths, vertical survival depths are small and the units are highly homogenised. Where depositional depths are large the critical proportion of recognition is achieved after the addition of very few layers. So again the survival depth is small, though depositional layers are far less integrated. Between these two extremes survival depths are greatest varying with the degree of mixing and diffusion. The curves generated are of course not smooth because of the step nature of the model. The critical proportion is reached for all depositional thicknesses between two limits, i.e. the top and base of each individual buried deposit. This can be graphically illustrated (Figure 19.5). Here the depth envelope for the critical proportion of 0.03 is shown.

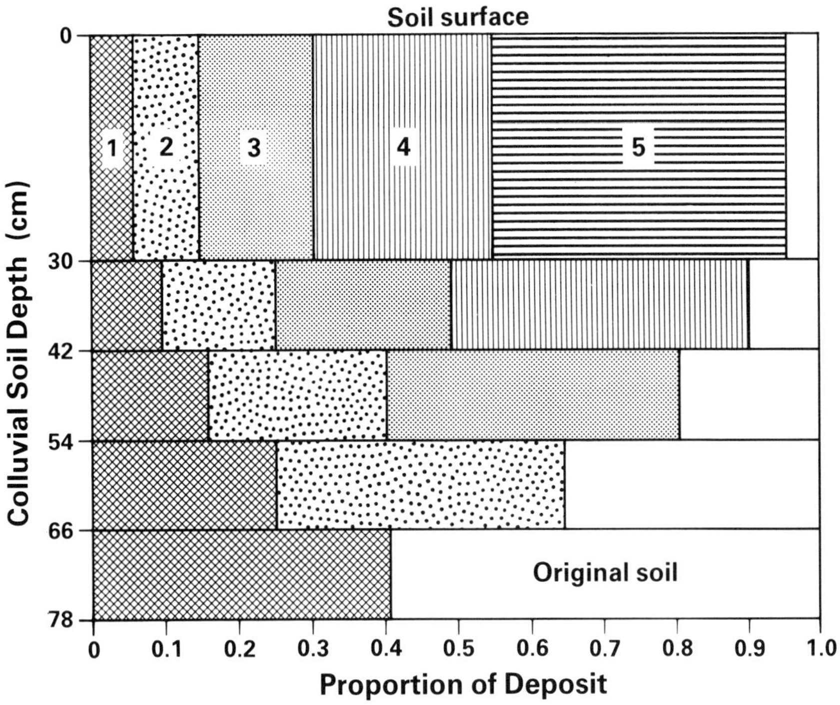


Figure 19.2 Proportion of each depositional event at various depths.

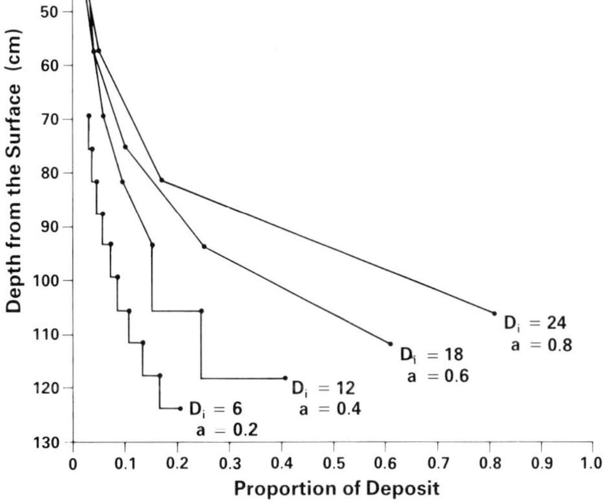


Figure 19.3 Changes in the proportions of the depositional events for different values of a.

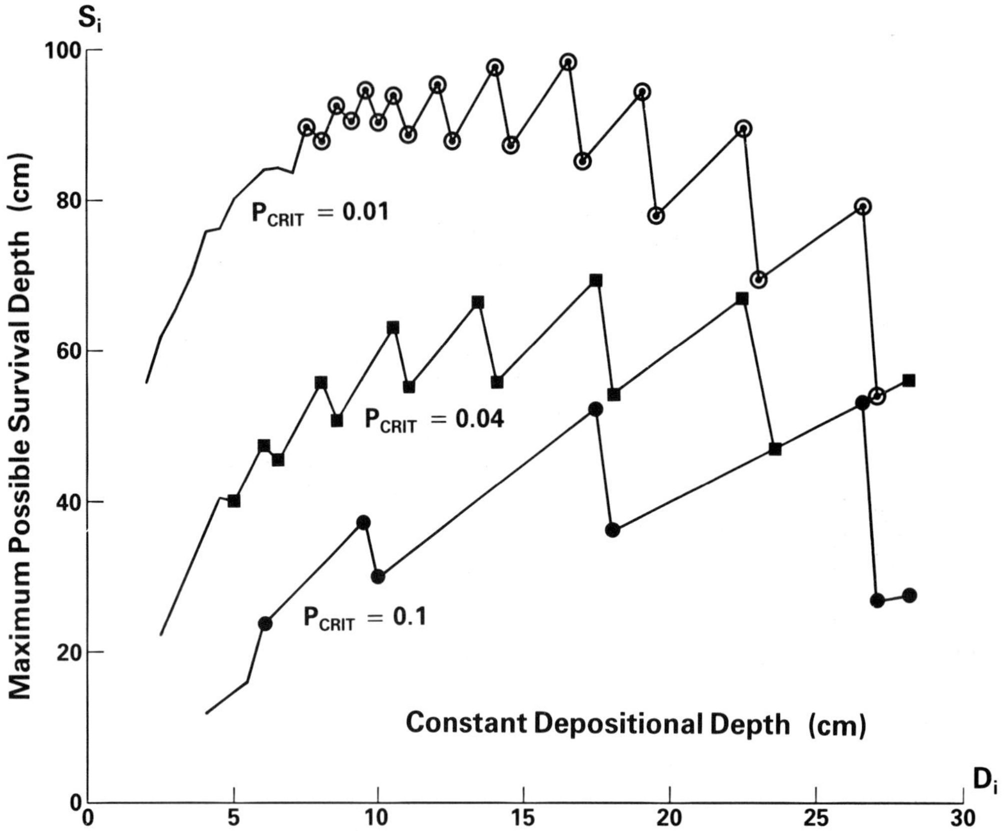


Figure 19.4 Survival as a function of depositional depths for different values of $P_{CRIT}$.

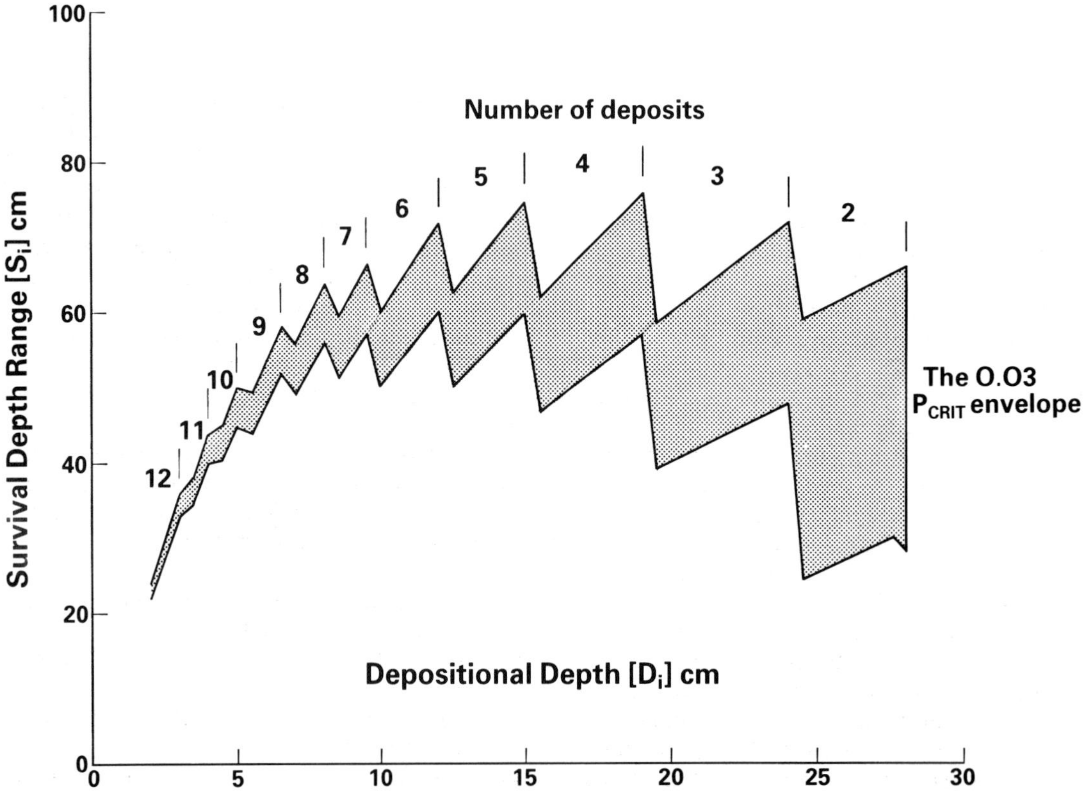


Figure 19.5 Survival depth range for $P_{CRIT}$ = 0.03 for different depositional depths.

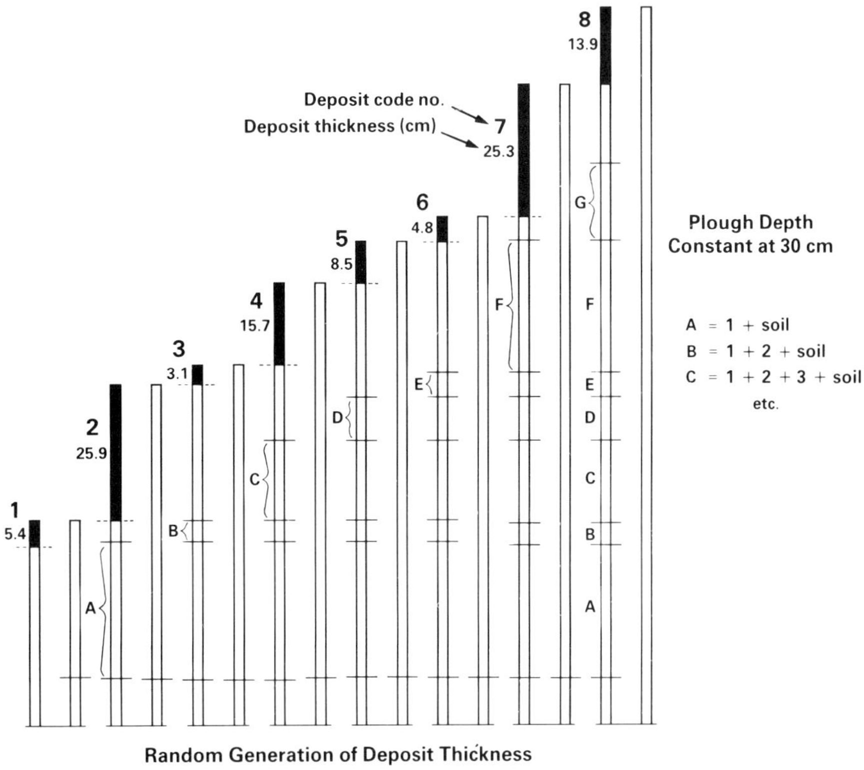


Figure 19.6 Schematic diagram showing the burial of mixed colluvial deposits.

*Distributions of colluvial features*

The implications of the model so far has concerned itself only with the presence and distribution of specific depositional events in terms of proportions of deposited material at various depths. The question to be addressed now is, can the material from any one specific depositional event be isolated from that of another also represented at a particular level in the colluvial sequence? To consider this aspect the conceptual ideas developed by Yaalon (1971) also need to be taken into account.

This framework concerns itself with the relative stability through time of specific observable and measurable characteristics of soil material. Each individual deposit contributing to the total colluvial soil material may contain some characteristics that can be said to be irreversible (i.e. not significantly degraded by post-depositional pedological processes) e.g. charcoal fragments, artifacts like farm nails, pieces of brick. The distribution of properties such as these will simply mirror the pattern of variations in proportions of particular deposits already outlined. However, it must be appreciated that some characteristics of the original depositional material are reversible, and as such they adjust and degrade through time. For example, the original depositional fabric of the deposit, or its initial organic matter content are such properties. The recognition, and significance of these during the analysis of the colluvium are therefore not simply a function of the proportion of material from the deposit at a particular depth, but in addition, the inherent stability of the original feature to the processes of faunal and mechanical mixing. Once this dimension of the model is examined the relative ages, and time between depositional events, becomes a significant consideration.

To interpret the possible depositional history by looking at the vertical distribution of features and properties represented within the over deep colluvial soil sequence one is constrained by:

(i) The frequency per unit volume of features originally present in the deposit.
(ii) The stability of the features to pedogenic processes.
(iii) The relative mobility of the features to pedogenic processes.
(iv) The uniqueness of the features to specific deposits.

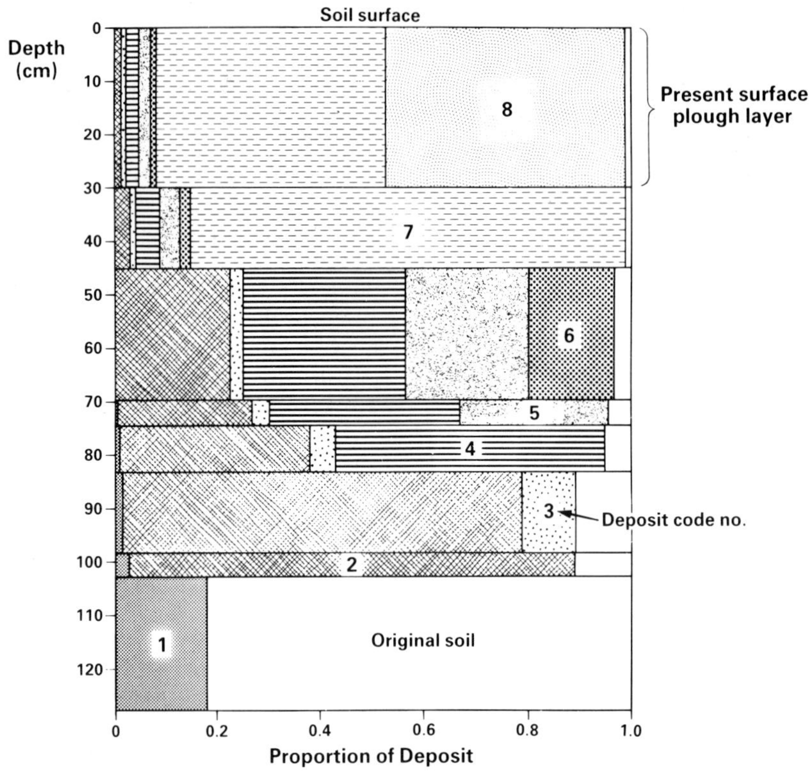


Figure 19.7 Proportion of each deposit in the buried layers.

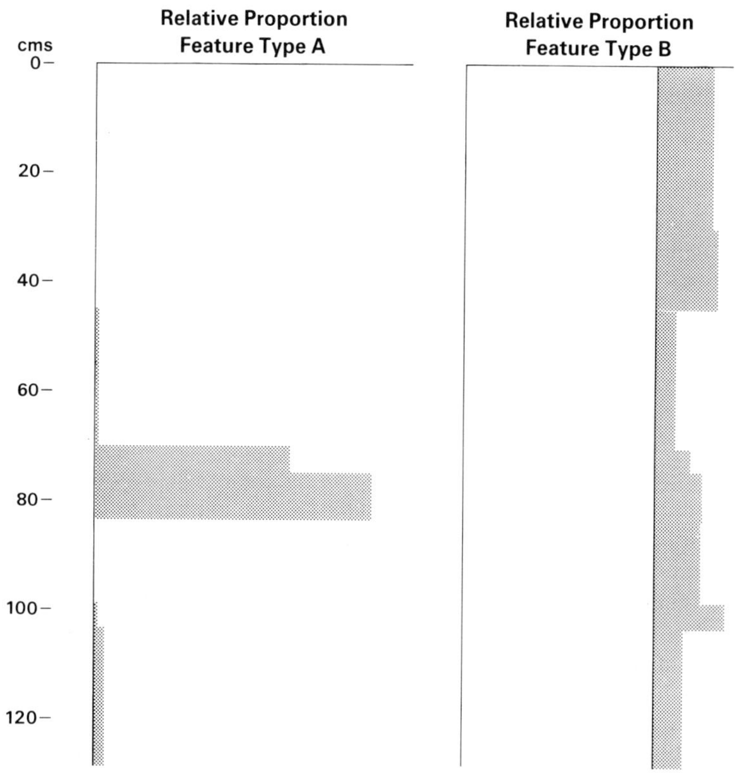


Figure 19.8 Vertical distribution of the mixed buried layers from the random depth additions of material. Profile A assumes an input of an irreversible feature found in deposit 1 and 4. Profile B shows a feature present in each deposit proportional to the depth of the deposit added and decaying at a linear rate.

The final complicating factor results from the formation of *in situ* features independent of the characteristics of the individual deposits. The frequency of the depositional phases, and the land-use history of the depositional site control this additional complicating factor. The lower the frequency of events the longer soil forming processes have to operate and therefore the greater the possibility for post-depositional pedofeatures to develop reflecting the soil forming processes at the site, e.g. agricutans. The presence of such features is of great significance in reconstruction because they can give both an indication of the temporal depositional dynamics of the site as well as the possible external pedogenic environment.

### Using the model as a basis for simulation

With the ideas developed so far it is possible to form a simulation-type model of possible depositional sequences and by implication some theoretical depth curves for features of various types.

To produce the simulation model initially as simple as possible let us assume the probability density function of depositional depths to be rectangular, i.e. all thicknesses between 0 and 30 cm are equally possible. Additionally the assumption that the plough depth is constant through time will be used, though it is appreciated that plough depths in the past may well have been less than those found today (Van Vliet-Lanoë *et al.* 1992). Random numbers were used to generate a typical sequence of depositional thicknesses: 5.4. 25.9. 3.1. 15.7. 8.8. 4.8. 25.3. 13.9 cms respectively. The previously developed model was adjusted for constant depositional depths. The proportions of each deposit at the surface would be given by:

$$
\begin{aligned}
&[a_1(1-a_2)(1-a_3)\ldots(1-a_k)]_{d_1} \\
&+[a_2(1-a_3)\ldots(1-a_k)]_{d_2}\cdots \\
&+[a_{k-1}(1-a_k)]_{d_{k-1}}\cdots[a_k]_{d_k} \\
&+[(1-a_1)(1-a_2)\ldots(1-a_k)]_{o_s}
\end{aligned}
\qquad (11)
$$

The specific generation used can be shown diagrammatically as in Figures 19.6 and 19.7.

Once the proportions of material at each depth have been calculated, attention can be turned to the observable features of the material. If the property under consideration is an irreversible one then the distribution of the property simply reflects the proportion of the deposit at each depth and the initial proportion of the particular feature in the deposit. For example let us say deposits 1 and 4 are characterised by the irreversible property of charcoal fragments. In addition assume that 10% of the original depositional material was made up of such features. With these constraints the distribution of such a feature with depth is shown in Figure 19.8. If we also take another type of feature, for example organic matter content, we could assume each deposit has an organic content proportional to its thickness and that this feature is destroyed in a linear way with respect to time. With these simple conditions imposed on the model the generated profile is also shown in Figure 19.8.

A simulation approach such as this has many possibilities of exploration; these include variations in the plough depth for different deposition events, different stabilities of features, and chemical properties with known decay rates like radiocaesium. It may even be possible to predict the most probable distribution of depositional events from measured profiles by running the model in reverse. These aspects of the model are not considered in this paper.

### Conclusions and observations

The model proposed shows the complexity possible in depositional sequences of soil eroded material in these landscapes of North West Europe. It also clearly demonstrates that the interpretation of vertical profiles of soil features must be treated with great care, particularly with regard to post depositional mixing problems. One method of analysis of these colluvial soils which may prove to be of significance is the use of soil micromorphology. Although many of the latent structures of deposited material may have been removed from view at the pit scale of description they may still be observable at the microscale in thin section. The application of such an approach to these erosion deposits will be the subject of a future paper. Finally, to make conclusions about the relationships between the characteristics of the deposits and the up slope land-use one needs increased understanding of the coupling of processes of soil erosion and deposition and the nature of the agricultural activities.

### References


Auzet, A. V. 1987 L'érosion des sols par l'eau dans les régions de grande culture, aspects agronomiques. *Min Env/Min Agric. GEREG-UA 95 CNRS.*

Boardman, J. 1990 Soil erosion on the South Downs: a review. In: Boardman, J., Foster, I. D. L. and Dearing, J. A. (eds.) *Soil Erosion on Agricultural Land.* J. Wiley, Chichester, 87–106.

Boffin, J. 1984 *La dégradation structurale des couches superficelles du sols sons l'action des pluies,* Thésis, Paris.

Butler, B. E. 1982 A new system for soil studies. *Journal of Soil Science* 33, 581–595.

FAO 1988 *Legend for the Soil Map of the World.* UNESCO, Rome.

Farres, P. J., Poesen, J. and Wood, S. J. 1991 Some characteristic soil erosion landscapes of N.W. Europe. *Geography Review* (in press).

Govers, G. 1987 Spatial and temporal variability in rill development process at the Huldenberg experimental site. In: Bryan, R. (ed.) *Rill Erosion*. Catena Supplement 8, 17–34.

Hole, F. D. 1978 An approach to landscape analysis with the emphasis on soils. *Geoderma* 12, 17–34.

Imeson, A. C., Kwaad, F. J. P. M. and Mücher, H. J. 1980 Hillslope processes and deposits in forested areas of Luxembourg. In: Cullingford, R. A., Davidson, D. A. and Lewin, J. (eds.) *Timescales in Geomorphology*. J. Wiley, Chichester, 30–42.

Kwaad, F. J. P. M. and Mücher, H. J. 1979 The formation and evolution of colluvium on arable land in Northern Luxembourg. *Geoderma* 22, 173–192.

Langhor, R. and Sanders, J. 1985 The Belgium Loess belt in the last 20,000 years: Evolution of soils and relief in the Zonian Forest. In: Boardman, J. (ed.) *Soils and Quaternary Landscape Evolution*. J. Wiley, Chichester, 359–371.

Poesen, J. 1989 Conditions for gully formation in the Belgium loam belt and some ways to control them. In: Schwertman, U., Rickson, R. J. and Auerswald, K. (eds.) *Soil Erosion Protection Measures in Europe. Soil Technology*, 39–52.

Poesen, J. and Govers, G. 1990 Gully erosion in the loam belt: typology and control measures. In: Boardman, J., Foster, I. D. L. and Dearing, J. A. (eds.) *Soil Erosion on Agricultural Land*. J. Wiley, Chichester, 87–106.

Van Vliet-Lanoë, B., Helluin, M., Pellerin, J. and Valadas, B. 1992 Soil erosion in western Europe: from the Last Interglacial to the present. In: Bell, M. and Boardman, J. (eds.) *Past and Present Soil Erosion*. Oxbow Books, Oxford, 101–114.

Yaalon, P. A. 1971 Soil-forming processes in time and space. In: Yaalon, D. H. (ed.) *Paleopedology. Origin nature and dating paleosols*. International Society Soil Science and Israel Universities Press, Jerusalem, 29–39.

# CHAPTER 20

# ASSESSING THE IMPACT OF EROSION ON SEMI-ARID ARCHAEOLOGICAL SITES

JOHN WAINWRIGHT

*Erosion, especially in semi-arid regions, is of vital significance to archaeology. Many aspects of the discipline are affected, notably the study of site formation and the limits of cultural interpretation from micro to regional scales; the survey and discovery of sites; and cultural resource management. A process-based approach to the study of erosion on archaeological sites allows the use of powerful analytical techniques. Foremost amongst these is the use of numerical models which may be used to define conditions of erosion in the past, as well as predict future patterns. A new computer model has been developed for the erosion of sites by rainstorms. This couples hydrological and sedimentological components, and takes various interactions into account. This model is unique in examining hydrology in relation to its archaeological context. To parameterize this model, a series of field experiments has been carried out in France and Spain. Together with a sensitivity analysis of the model to the spatial distribution of parameters, these further confirm that hydrology is strongly related to site history Sediment transport experiments have been undertaken, and provide the first quantitative data on the transport of archaeological materials under laboratory conditions. The numerical model has been applied to sites in semi-arid environments in Southwest France. Preliminary results suggest that erosional processes are important, but their effects are spatially and temporally complex. This has important implications both in the study of specific sites, and also in wider issues of site formation, cultural resource management and landscape development.*

### Introduction

Semi-arid regions are characterized by severe erosion by water occurring as runoff from rainfall events. In such active environments, understanding the role played by erosion is fundamental to archaeology for several reasons. First, erosion is an important factor in site formation, so that on a site and landscape scale, it controls what can be known about the archaeological record. Secondly, understanding the movement of artifacts by hydrological processes can show the limits of cultural interpretation on the intra-site level. This can be expanded into the subsequent displacement of artifacts, which is of vital relevance to field survey work, and can show the limits of the potential recovery of the archaeological record. Off-site movement of artifacts is also geomorphologically important, where they are used as dating evidence for sediments. Understanding the processes can lead to greater precision in such techniques. Finally, erosion also controls what will remain in the archaeological record for the future, and what will be destroyed. Erosion studies must form a part of the effective management of our cultural resources.

The erosion of archaeological sites in semi-arid environments can be understood in terms of the hydrological and sedimentological processes causing and resulting from the generation of overland flow. The principal advantage of a process-based approach is the ability to use powerful modelling techniques. These may be applied either to produce predictions on the nature of the past or future archaeological record, or as an interpretative aid.

The development of a relevant model requires various interactive components. The first of these is the hydrology of archaeological sites, which provides the essential interface between the rainfall event and the generation of runoff. This has been developed in theory, and verified by a series of field experiments. Secondly, quantitative data are required on the transport of archaeological materials by overland flow and rainsplash. Finally, the model may be assembled by making certain assumptions taken from the literature on the nature of various interactions between erosion, water flow and vegetation.

### Hydrology of archaeological sites

Within a site, the structures may be divided into five overlapping groups or contexts (Figure 20.1). The designation of these contexts is based on morphology; the relationship between these and other notions of 'context' as used in archaeology will not be entered into here. The hydrology and sedimentology of these contexts may be understood in terms of five phases of site development – clearance, construction, occupation, abandonment and regeneration (Figure 20.2). These spatial and temporal aspects are related to five factors:

(i) the input of new materials and their textures;
(ii) the removal of materials;
(iii) the amounts and spatial patterns of surface compaction;
(iv) the input of organic materials; and
(v) direct modification of the physiochemical composition of the soil.

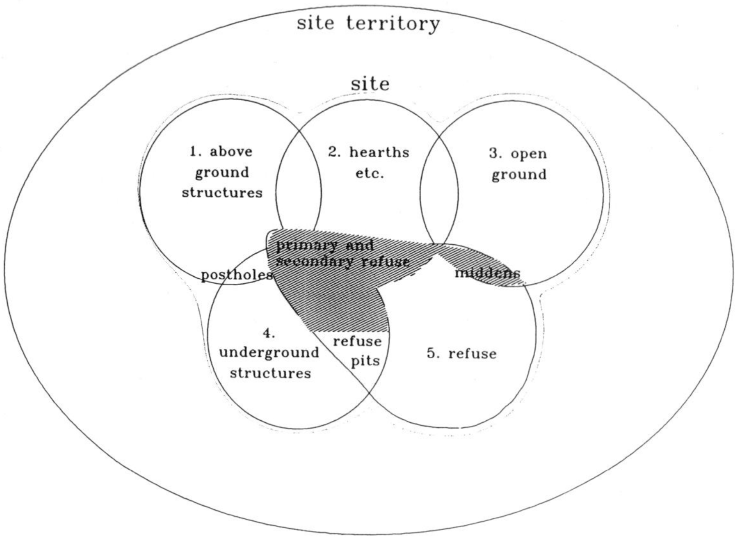


Figure 20.1 Morphological definitions of site context.

For each of the context groups, the relationships between these variables will be described, together with empirical data from archaeological or modern comparative contexts. These data are the result of field experiments into the form of the infiltration curve using cylinder infiltrometry, the practical use of which is described in the literature (Hills 1970; Bouwer 1986). The results were then used to derive two hydrological parameters – namely the saturated hydraulic conductivity ($K_{sat}$), and the soil moisture storage capacity (cf. Scoging and Thornes 1979; Gilman and Thornes 1985). $K_{sat}$ is defined as the long-term infiltration rate, taken here as being the value after sixty minutes. The soil moisture storage is calculated as the integral of the infiltration rate between one and sixty minutes.

*(a) Above-ground structures*

Sediment may be provided in the constructional phase, in the form of organic debris both as matrix and clasts. This may decay into a soil humus or may be preserved, depending on the specific microclimate. Charcoal may also be derived from primary or secondary sources and be incorporated into the matrix. The digging of construction trenches may form sinks for sediment. Overland flow is likely to increase due to the clearance of the site, and this will significantly increase erosion rates, together with the higher detachment capacity of rain falling on bare ground. Otherwise, soil storage capacities are likely to be little affected.

Primary soil matrix development will be subdued in the occupation phase, consisting of a small amount of material weathered from the walls, or eroded by undercutting. This is part of the 'slow collapse' phase considered by McIntosh (1974; 1977; Figure 20.3). Soil water storage is likely to remain at equivalent levels to before, and may even be increased slightly due to capillary uptake into the walls, although the surface area allowing infiltration in a closed structure will be reduced due to the roofing of the structure. Rain dripping off the roof tends to form small rain gullies in the surface around the base of the structure. Increases in the amount of water falling on these

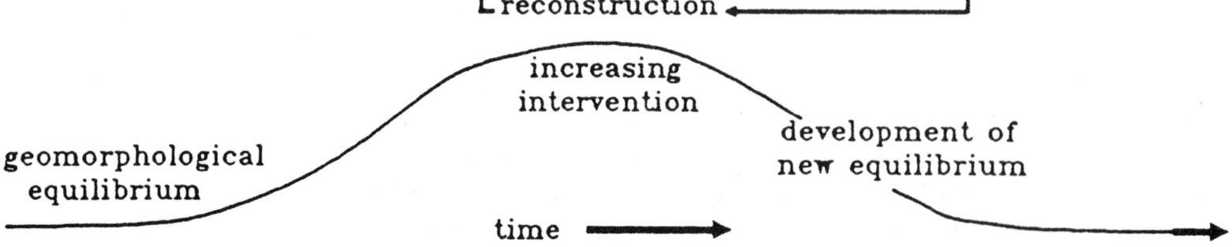


Figure 20.2 Phases in archaeological site development.

relatively small areas may lead to the preferential initiation of overland flow, starting from these channels and either continuing or being reabsorbed, depending on the conditions of rainfall and the soils in the immediate vicinity. Any flow from upslope will (hopefully for the occupants) be diverted around the structure, again altering the spatial configuration of runoff.

In the abandonment phase, some artifacts may be removed (Schiffer 1985; Stevenson 1985). Any burning may result in the input of large amounts of charcoal into the matrix. For wattle and daub or adobe structures, varying amounts of baked clay materials will also be added from the walls and also possibly the roof, depending on the temperatures reached. Otherwise, the slow decay of the walls will continue, perhaps with greater sediment accumulation due to the absence of removal by deliberate cleaning, together with the decay of other structural elements. Thresholds will be reached when the collapse of the roof allows accelerated erosion to occur within and at the top of the structure, and also when the amount of undercutting allows toppling to occur. These materials are deposited about the stumps of the former walls in a form of talus deposit, with coarser materials nearer the stump, and a progressive outward-fining in both directions, as sorting and breakdown of the wall material gradually occurs.

Hydrologically, the abandonment and regeneration phases have the greatest effects. In the areas with coarser materials, storage capacities are likely to increase, due to the addition of loose sediments. The decay of organic materials used in the structure may also increase this by increasing permeability, and also by developing macropores, which will also alter the flow patterns. Where the clayey deposits are more common, storage values will be decreased. Infiltration rates will also generally reflect this pattern, and may also be reduced where the rain gullies were previously to be found, if there has been surface armouring (the removal of erodible sediments enabling the sealing of the soil surface). Compacted floor surfaces will also act initially to reduce infiltration, with the development of a perched water table, and allow preferential runoff. The erosion which sorts the materials about the wall stumps will initially be increased, due to the locally steepened slope.

Except where qualified, these hypotheses seem generally valid for walls constructed of wattle and daub, adobe (McIntosh *op. cit.*; Kirkby and Kirkby 1976; Schiffer *et al.* 1987) and stone after personal observations. In the latter case, the coarseness of the sediment is likely to increase storage capacities and infiltration rates. Purely wooden or organic structures will behave slightly differently due to the absence of a permanent sediment build up. In the latter two phases, infiltration and storage are likely to increase, at least in the short term, due to the protection afforded to the surface by the structure, even following collapse. Unless transported due to other factors, this will continue as the structural materials decay, as the open, humic soil structure produced allows high permeability.

Experiments were conducted on three field sites in Southwest France, providing data on the hydrology of this type of structure. At Le Laouret (Aude), a small Late Bronze Age IIIb hamlet situated on several limestone terraces, ten infiltration experiments were conducted around one hut which was subsequently excavated to give details of the structure. The results show that soil storage values are lowest around the immediate edges of the structure, increasing both within and outside it. This concurs with the model of house collapse with concentrations of clay particles at the base of the walls.

In the stratified Late Bronze–Early Iron Age site of Pech Egos (Tarn), three levels provided comparative evidence on two types of above-ground structure. The uppermost level tested, C3b, shows an oscillation in the soil water storage values running across the area analysed. By comparison with the excavated remains, the low points again correspond to the footings of walls. In level C4 (Figure 20.4), the large peak of soil storage values to the rear of the diagram is related to the underlying drystone wall remains. Here there is confirmation of the hypothesis that collapsed stone structures form more open sediment textures and thus develop into soils with higher soil moisture storage capacities.

Finally, at the Chalcolithic site of Roquemengarde (Hérault) a sample of eight experiments in Structure 35 show a similar pattern for clay buildings as seen at Le Laouret and the upper level tested at Pech Egos.

*(b) Hearths*

Much less sediment is derived from these, although some material may be imported in the form of a stone or clay lining during construction. During occupation this may be replenished or renovated, and further stone may be imported for heating purposes, together with the input of some artifactual material. Charcoal and ash will form a predominant part of the matrix, although successive cleaning operations may reduce their importance. Depending on the original form, these structures may provide sinks for natural erosion at any time, as well as having secondary usage as a depository for cultural material in back-fill.

Little hydrological importance will be apparent until the abandonment phase, when they are most likely to be exposed. The loose structure may increase the storage capacity, but the effects of this are more than likely to be outweighed by the shallowness of the structure (or non-compact sediment), but more so due to the effects of burning, which alter the chemical and physical composition of the subsurface layer, especially by inducing hydrophobicity. If a clay lining has been used, this is also likely to reduce the rates of infiltration and the storage capacity, whereas a stone lining would have the opposite effect.

No consistently marked archaeological hearths were available for testing on any of the test sites used. However, on the plateau of Pech Egos, experiments were carried out on two modern burnt areas. This allowed a direct comparison on the same soil type as

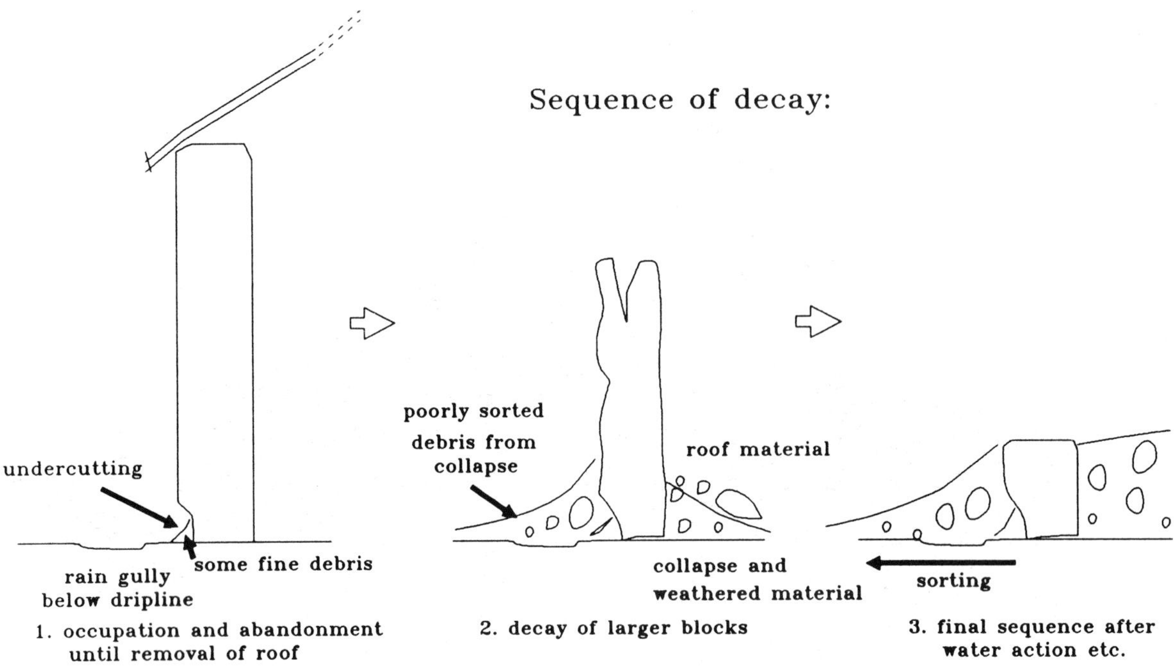


Figure 20.3 Models of collapse of above-ground structures.

archaeological remains. For both areas, samples were taken in the burnt areas and the surrounding unburnt soil. The results show in both cases a marked and statistically significant difference in the parameters (using Wilcoxon-Mann-Witney tests, the significance levels are 0.0011 and 0.0009 for $K_{sat}$ values in F1 and F2 respectively, and 0.0008 and 0.0022 for the soil storage values). The burnt areas tend to have lower $K_{sat}$ and storage values, due to the fact that below the loose ashy layer on the surface, the high temperature has baked the clay minerals in the soil, increasing the density. The ash layer is also very vulnerable to erosion by wind and water action, quickly removing any slight increased soil storage capacity. A corresponding feature is the lower degree of variability in the burnt parameters, probably relating to a homogenization of the soil structure. This hydrophobicity of a clayey soil after burning corresponds well with other examples in the literature (cf. Trimble 1988).

*(c) Open ground*

Some primary refuse may be generated, depending on the location of activity areas, but most sediment inputs will be secondary. These will depend on types of site occupation and cultural attitudes to refuse in these areas, which is equally the case in other contexts. Organic matter, phosphate and nitrate may be important matrix elements where animals are present. This may increase infiltration once again, for example by allowing the aggregation of soil particles, which will also tend to make them less erodible if stable. Other sediment factors will depend on secondary inputs, but more so the initial topography and composition, as well as any levelling or redeposition of material that takes place in the construction phase.

Vegetation removal will again increase erosion rates. Here this will be accentuated further as the compaction due to trampling of defined pathways will increase surface runoff. With abandonment and burial, these are likely to continue as spatially discontinuous impeding horizons at a depth determined by the depth of burial. Storage is thus likely to gradually increase. This position may be changed if the soil structure is disturbed, for example by faunal or floral activity, or by subsequent ploughing.

For this type of structure, it was again necessary to use a modern control. A modern path was studied just below and on the same soil type as Le Laouret. Eleven samples were taken on the path itself, with seven surrounding controls. The values of $K_{sat}$ for the path have a range from $8 \times 10^{-5}$ m $s^{-1}$ to $5 \times 10^{-3}$ m $s^{-1}$, and differ significantly at the five percent level from the controls, which have values between $6 \times 10^{-4}$ m $s^{-1}$ and $6 \times 10^{-3}$ m $s^{-1}$. This agrees with other data that compaction of the soil by trampling tends to decrease the value of $K_{sat}$, and the corresponding storage values.

*(d) Underground structures*

Primary materials may be derived from any lining to these features, as well as any object relating to their use. The latter may not always be preserved, however,

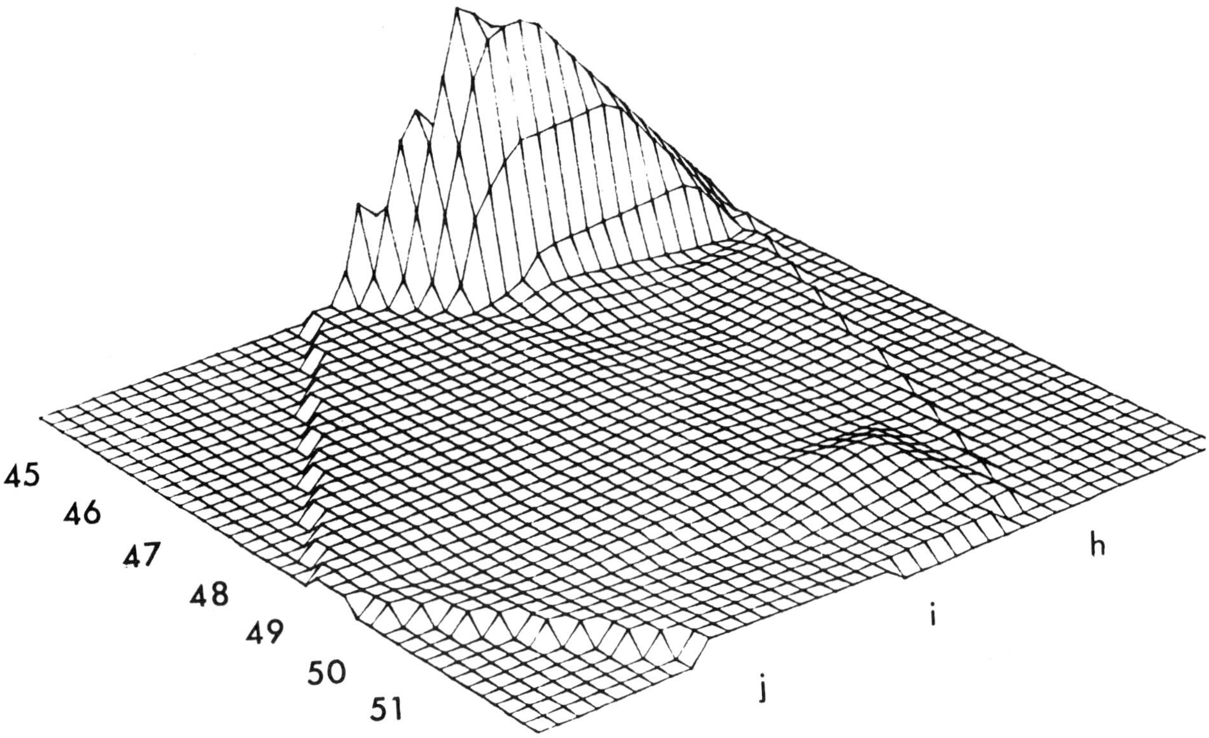


Figure 20.4 Spatial variation of soil storage values in level C4 at Pech Egos.

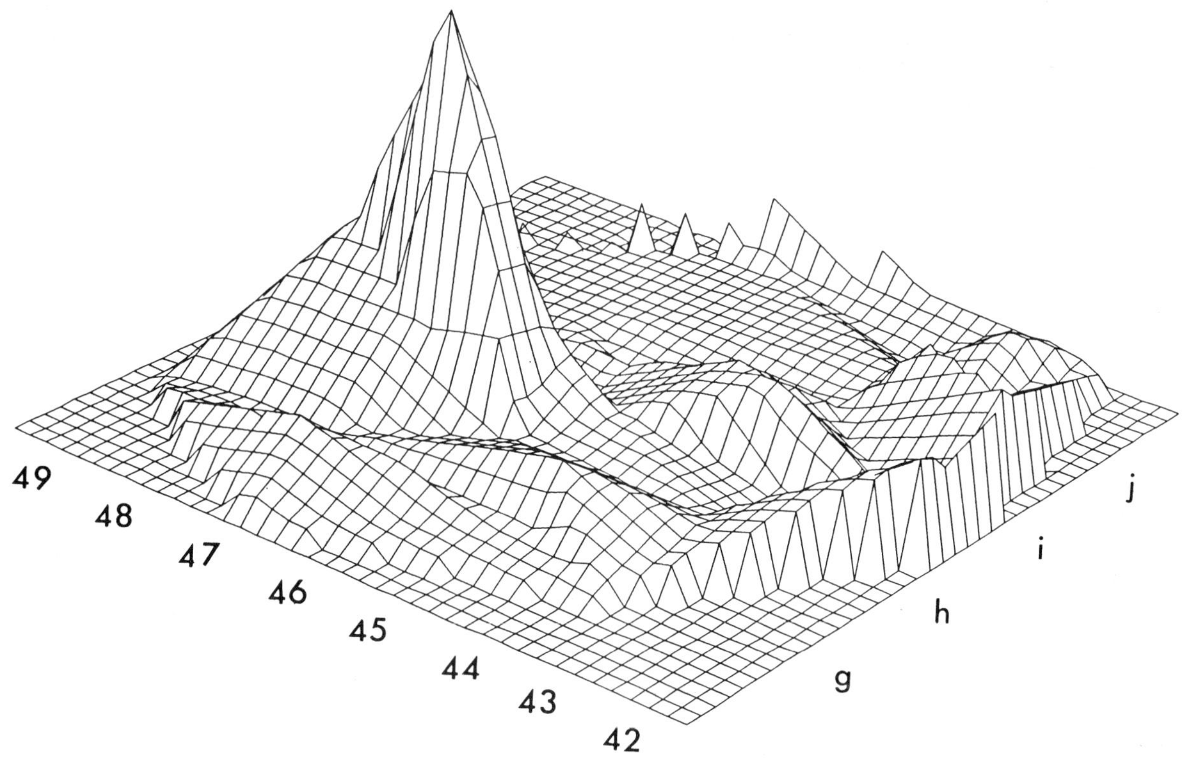


Figure 20.5 Spatial variation of $K_{sat}$ values in level C6 at Pech Egos.

unless there is the intervention of fire or other conditions out of cycle with the normal usage pattern, as for example with the case of grains. More commonly, added material will be of secondary type, after the primary use of the structure has ceased. Depending on the amount of deliberate filling, it will provide a sediment sink until the remaining capacity has been filled.

The hydrological effects of these structures are again more likely to become apparent in the abandonment phase, or in the main occupation phase if they fall into disuse. Subsurface storage may be significantly increased, due to the space left, or the open texture of the fill. If part of the structure is left open, it may be filled with a cap of fine, wind-blown or water-borne sediment which allows the development of surface runoff and/or an impeding horizon. The erosion due to this may explain the truncation of the upper part of the profile of many of these structures in the archaeological record. During the occupation (use) phase, the presence of an impermeable clay seal for example on storage pits may have similar results.

Testing of this type of structure was possible in three instances. In the distribution of values for the level C4 at Pech Egos seen above (Figure 20.4), there is a second small peak. This corresponds to the pit excavated in squares ij 46–47 in the underlying C5. The lesser apparent influence is in part due to the large relative range on the diagram with the drystone wall collapse sediments, the small size of the structure, and its depth at sampling. A much larger peak occurs in the $K_{sat}$ distribution in C6 (Figure 20.5). This corresponds to squares i46 and i47, where a large but shallow (probably truncated) pit was excavated. The values for the pit parameters were tested against those for the rest of the layer using Wilcoxon-Mann-Witney tests. These show significant differences in both distributions (significance levels of 0.0019 and 0.0058 for the $K_{sat}$ and storage distributions respectively).

On a much larger scale, at Roquemengarde thirty-one samples were made in the ditch and surrounding berms. The ditch survives up to one metre deep, and forms the enclosure of the barred spur. The spatial pattern of the soil storage values clearly demonstrates much higher values for the ditch samples, relating to the lower fill density. Again this difference is significant at the five percent level.

### *(e) Refuse*

All material deposited here will be of secondary derivation, and will be controlled by the rate and location of deposition of objects during the occupation and abandonment phases. Initial considerations imply that the spatial distribution and composition of refuse will depend on four major factors:

(i) the location of activity areas
(ii) ideological and power relations
(iii) gender relations
(iv) the interaction of (i), (ii), and (iii) with site topography and microtopography.

A fuller analysis of this problem is beyond the present aims of this study, but would draw on various ethnographic, ethnoarchaeological and sociological sources (cf. Moore 1982; Hodder 1982; Stevenson 1985; Yellen 1977).

Erosionally, this context is likely to be highly variable. Often, however, the open, uncompacted structure is likely to increase infiltration and storage rates. Again, high organic matter contents may accentuate this situation. Alternatively, the sediment is likely to be much more erodible where it has a loose texture and low density, for example if ash makes up a large proportion of the deposit.

Because of the heterogeneity of this type of context, it was not possible to perform any direct tests. Any parameters required would need to be estimated taking into account the texture of the soil and composition of the deposit.

### Transport of archaeological sediments

Very little is published in the literature on the transport of artifact materials. Hanson (1975) gives data for the movement of faunal remains. Frostick and Reid (1982; Reid and Frostick 1985) have published experimental data for flint and bone objects on thrity degree, straight slopes at Koobi Fora, East Africa over two to three years. Further, although there has been more concentration on stoney soils in recent years in the geomorphological literature (e.g. Nicols *et al.* 1984; Poesen 1987; Torri and Poesen 1988; Abrahams *et al.* 1988) very little of this is applicable due to the concentration on bulk soil properties (but cf. Torri and Poesen 1988). To counter this, two series of experiments were conducted in order to test the conditions of entrainment of pot and flint artifacts due to overland flow and rainsplash. These experiments were undertaken at the Bristol University flume facility. The flumes consist of two bed types – clay and loam – and allow the generation of variable overland flow and / or rainfall.

The theory and results from the overland flow series of experiments has been published previously (Wainwright and Thornes 1991), and so will only be summarized here. Significant amounts of movement were caused – up to 89% of artifacts less than 2 cm in largest dimension were displaced in a single large event. Conditions for the start of movement are complex (*ibid.* Figure 27.4). This seems to be related to irregularities in particle shape. Entrainment parameters are also not constant through time, decreasing as particles become stabilized in relation to the joint conditions of flow and surface form. Any further increase can only occur if higher flows occur, or the surface is transformed by erosion, biological activity, etc. Once entrained, the amount of movement is also highly variable (*ibid.* Figures 27.5 and 27.8). Considering these factors, the best description of the movement of artifacts is as a stochastic process, with gamma distributions describing the conditions and amount of movement, and a renewal process of exponential form describing the temporal variation in entrainment.

Two geomorphic studies (Kirkby and Kirkby 1974; Moeyersons and de Ploey 1976) suggest that the movement of gravels does occur due to rainsplash. It was therefore decided to consider the movement of artifacts by splash, in theory and experimentally.

The forces exerted on an artifact by a raindrop are shown in Figure 20.6. The raindrop force is split on impact into two vectors, one acting in a downslope direction:

(1) $F_{SD} = m'g' \cos(90 - \theta - \alpha)$

where
$F_{SD}$ is the downslope component of the splash force [N]
$m'$ is the mass of the raindrop [kg]
$g'$ is the gravity force acting on the raindrop [m s$^{-2}$]
$\theta$ is the slope angle [°]
$\alpha$ is the angle of raindrop incidence on the particle measured from the vertical (positive upslope, negative downslope) [°]

and one acting normal to the slope:

(2) $F_{SN} = m'\, g' \cos \alpha$

The gravity forces acting on the artifact are also split into two vectors. The component acting to move the particle in the downslope direction is:

(3) $F_{GD} = m\, g \sin \theta$

The normal force, which acts to keep the artifact on the surface is:

(4) $F_{GN} = m\, g \cos \theta$

Taking into account the friction force acting on the artifact by contact with the slope, the full conditions for incipient motion by splash are:

(5) $$FS = \frac{F_{SD} + F_{GD}}{F_{SN} + F_{GN} \tan \beta}$$

where $\beta$ is the angle of repose of the particle [°]

If the downslope components are dominant, FS is greater than unity and downslope motion occurs. Because of the cosine term in equation 1, negative values may be obtained, predicting upslope movement if FS is less than minus unity.

Figure 20.7a shows the incipient motion conditions for large artifacts hit by a single raindrop falling vertically. A single raindrop is considered because of the extremely low probability ($3 \times 10^{-7}$ for a particle with an exposed surface area of 1 cm$^2$ and less than $3 \times 10^{-5}$ for a 10 cm$^2$ particle; cf. Marshall and Palmer 1948) of encountering more than one raindrop in the 7 milliseconds required for the dispersion of the splash shear forces (Al-Durrah and Bradford 1982). The slope angle increases along the x-axis, reaching a value for FS of one where it equals $\beta$, the angle of repose of the particle, which has a value of 35° here. These results show that only where the artifact is at rest very close to its angle of repose would raindrop impact cause movement. Entrainment is highly insensitive both to the raindrop diameter and the artifact size. In the case of the former, the effect of doubling the diameter of the raindrop from two to four millimetres only raises the value of FS by $6.35 \times 10^{-9}$ for a 0.25 cm$^2$ artifact, $9.98 \times 10^{-10}$ for a 2 cm$^2$ artifact, $2.28 \times 10^{-10}$ for an 8 cm$^2$ artifact and $1.91 \times 10^{-11}$ for a 96 cm$^2$ artifact. Similarly, for a constant drop diameter of two millimetres, the respective increases in FS are $8.98 \times 10^{-10}$, $9.78 \times 10^{-11}$ and $2.99 \times 10^{-11}$ with increasing artifact size.

The case for angular incident raindrops is presented in Figure 20.7b for artifacts on a 20° slope. Again, there is no movement suggested. For the 0.5 cm$^2$ artifacts, the FS=1 threshold is crossed with highly oblique raindrops (greater than 64°) of 6 mm in diameter. However, in this case, the raindrop diameter is greater than the largest dimension of the artifact, so that the consideration of the transfer of momentum to the artifact is more complex. As a relatively large proportion of this energy is likely to be dissipated in the surrounding soil, it may effectively be considered that no movement occurs. No thresholds are crossed on shallower slopes.

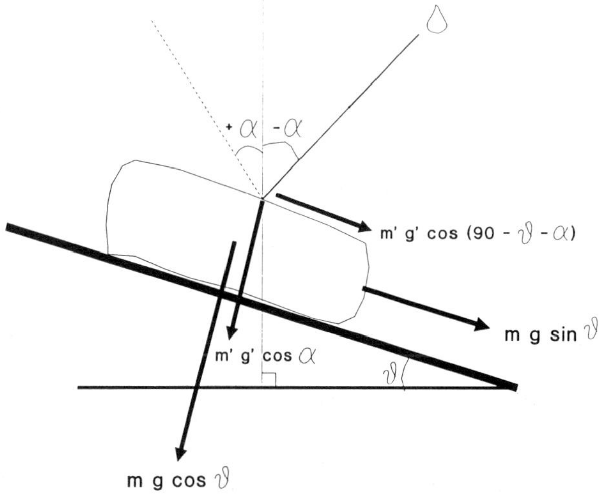


Figure 20.6 Forces incident on a particle at rest on a slope upon impact by rain drop.

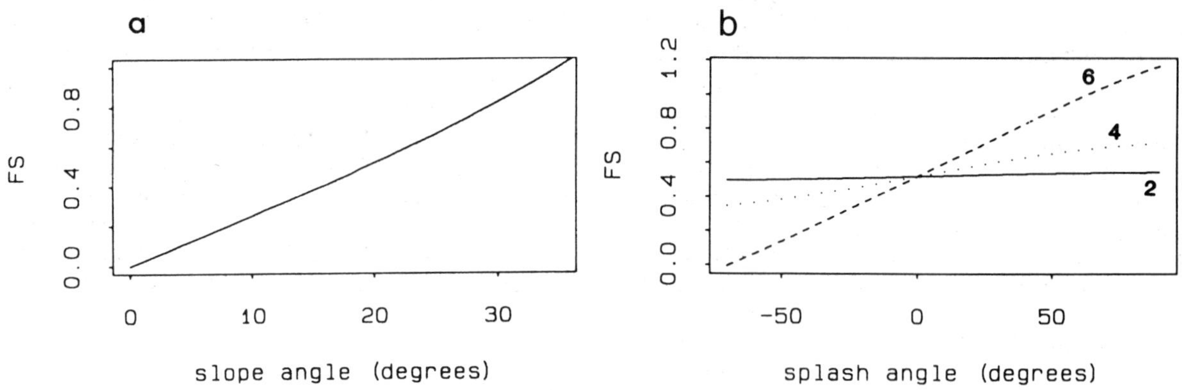


Figure 20.7a Balance of forces on impact of vertically falling raindrop on slopes up to the angle of internal friction of an artifact of 0.25 cm$^2$; b. Balance of forces on impact on the same artifact of an angularly incident raindrop on a slope of 20°. Numbers are raindrop diameters in millimetres.

Splash experiments on both slopes, with rainfall intensities of 45 and 90 mm hr$^{-1}$ accord with the theory. No movement was observed, except where the rainfall generated runoff which led to movement. These results confirm the hypothesis generated by the theoretical consideration of rainsplash impact on large artifacts, that insufficient energy is available to cause saltation or creep. The cases for gravel noted above seem to contradict this. Kirkby and Kirkby (1974) show movement for stones of four to eight millimetres on a seventeen degree slope. However, these measurements were carried out in field conditions where runoff was possibly the cause of movement. Moeyersons and de Ploey (1976) observe in simulated storm events in the field that larger gravels (1–2 cm) were moved by 'splash creep' due to lateral drop strikes. While this may in part be due to interactions between the drop and the surface – in this case a gravel lag pavement – perhaps taking the form of lateral jetting (cf. Huang *et al.* 1982), their own laboratory experiments suggest that the majority of this movement is related to surface flow.

To summarize, it is possible using laboratory experiments to show the importance of the movement of artifacts by overland flow, and define the parameters useful in describing this. Conversely, rainsplash is not an important process in the movement of large particles of the density of interest here. This has been demonstrated both in theory and practice.

**Computer model**

The hydrological and sedimentological data are developed into a numerical model as outlined in Wainwright and Thornes (1991: Figure 27.10). Overland flow is generated using a simple storage-type model (cf. Gilman and Thornes 1985). This is parameterized by means of the infiltration experiments described above, with parameters for the form of the infiltration curve, and saturated hydraulic conductivity, and total soil storage. These parameters are fully distributed because of the variability shown in the above section on hydrology (this proposition is also tested in the section below on hydrological model results. The rainfall input to this is defined using a Markov chain model (cf. Roldán and Woolhiser 1982). Occurrence probabilities are defined from modern long rainfall series. For the Western Mediterranean, it can be shown by statistical analysis of historical records and from palaeoenvironmental data that the climatic system has been stable for at least the last six millennia (cf. Gilman and Thornes 1985; Vernet 1988; Wainwright 1991). It is also possible to derive scenarios for particularly wet or dry conditions by altering the parameters.

Sediment transport is related to a pair of probability distributions for each grain size. These represent the rest period – the length of time under flow conditions during which a single particle does not move – and the step length, or the distance moved by any particle once entrained. The results of the experiments described above, and published data (cf. Einstein 1942; Grigg 1970; Hubbell and Sayre 1964; Hung and Shen 1971; Yang and Sayre 1971) suggest that both may be considered as being gamma distributions. A third distribution is considered which describes the stability of the soil and slope to flow conditions. This renewal distribution takes the form of an exponential tail-off of overall probabilities of movement. It is reset either if flow conditions increase, or by means of a random variable which represents the interaction of other soil processes. This is defined by a Poisson process. The governing equations of the sediment transport component of the model are described in detail in Wainwright and Thornes (1991). These convert the probability distributions for matrix movement into mass transfer volumes, which are then redistributed using a finite-difference technique. Both water and sediment are routed using the kinematic wave approximation (cf.

Dingman 1984) modified for two-dimensional flow, and solved by the tri-diagonal matrix algorithm (Reece 1976). Artifacts are distributed individually, using the three probability distributions described above together with a normally-distributed function for the across-flow movement, giving an angle of dispersion (Hung and Shen 1971).

The other principal interaction considered in the model is that with vegetation. Flow interaction is modelled by extension of Cowan's method (Chow 1959), which modifies Manning's $n$ as a function of vegetation cover. Vegetation also alters the infiltration rate. The effects on erosion have been derived from the literature (Elwell and Stocking 1976; Thornes 1988), and take the form of an exponential relationship. Vegetation growth is modelled using a form of the logistic growth curve, allowing for decreased soil fertility due to erosion.

### Examples of model results

Results from the model belong to four main categories – hydrology, slope form, soil texture and artifacts. Examples of each of these will be considered in turn.

#### *(a) Hydrology*

The earlier section on the hydrology of archaeological sites showed the importance of the distributions of hydrological parameters to the pattern of erosion on those sites. Here, the practical importance of this is demonstrated.

The infiltration rate used in this example is defined using the Philip curve ($f_t = a + bt^{-1/2}$, where $f_t$ is the infiltration rate, t is time and a and b are the experimentally-derived parameters). Variability in $a$ and $b$ as well as the soil storage parameter are studied here by assigning spatially-variable values. This is done by calculating the maximum deviation of outflow discharge between the spatially-variable and averaged (i.e. non-distributed) parameters. Random values were assigned using different levels of standard deviation about the mean for each parameter in turn under different rainfall conditions.

Widely different responses are demonstrated by the results. Discharge is largely insensitive to the $a$ parameter (Figure 20.8a). For all rainfall conditions and standard deviations, deviation from the average parameter discharge is consistently around the $10^{-13}$ $m^3 s^{-1}$ level. This would have little effect on the resultant erosion.

With low standard deviations of the $b$ parameter ($<10^{-5}$), the results are the same as those for the $a$ parameter (Figure 20.8b). Above this, there is a jump to much higher deviations, again independent of the rainfall intensity. Therefore, the amounts of erosion are much more sensitive to $b$ parameter variations.

Figure 20.8c shows a power–law relationship between the standard deviation of the storage parameter and the maximum discharge deviation. This clearly demonstrates the storage parameter to be the most sensitive parameter in the infiltration model. These differences would be greater in the real cases, as the values of the three parameters tested are not independent. Beyond the $b$ parameter threshold, and within a similar range for the storage parameter, the localized water flow patterns and corresponding erosion are extremely sensitive to spatial variability. This increases with distance from the slope divide. From the results of the field infiltration experiments, the parameter ranges actually found on sites fall within this sensitive range. It is thus possible to conclude that a distributed model is indeed necessary for the simulation of the erosion of archaeological sites.

#### *(b) Slope form*

The erosion simulation has been applied to real sites. An example of this is shown in Figure 20.9 for the site of Le Laouret. The result shown is for a single intense storm (40 mm in one hour: local records show the average rainfall event is approximately 11 mm), starting from a completely unvegetated slope. This is to simulate the conditions at site abandonment. Most erosion is concentrated around the edge of the limestone terrace upon which the site is located. Sedimentation has occurred around the bases of the huts, but most importantly at the foot of the terrace. The talus deposit formed reaches up to half a metre in thickness. Artifacts found within this deposit may thus be dated as synchronous to those on the terrace.

#### *(c) Soil texture*

As sediment is transported according to grain size, it is possible to observe the evolution of the soil texture through time. An example of this is presented in Figure 20.10. These results are shown as the enrichment factor of each grain size class, that is, the ratio of present to original sediment quantities in the erodible part of the soil profile. Under the conditions of the site and model configuration, the predominant process is that of the sealing of the soil surface. This occurs very rapidly towards the top of the slope (Figure 20.10a), but is still an ongoing process at the slope foot after 2000 years (Figure 20.10d). This is due to the progressive transport of the fine and medium sand fractions down and off the slope. Armouring forms a secondary process. This is again spatially and temporally distributed, being initially more important on the upper parts of the slope. There is a concentration on the lower slope in the medium term (250–750 years), with an equilibrium attained for the whole slope after about 1000 years. The pebble fraction, which is common at Le Laouret due to the nature of the bedrock weathering, tends to be rapidly buried or transported.

#### *(d) Artifacts*

Simulations of artifact movement at Le Laouret are shown in Figure 20.11. This demonstrates that most movement occurs in the period immediately following site abandonment. This is due to the combined effects

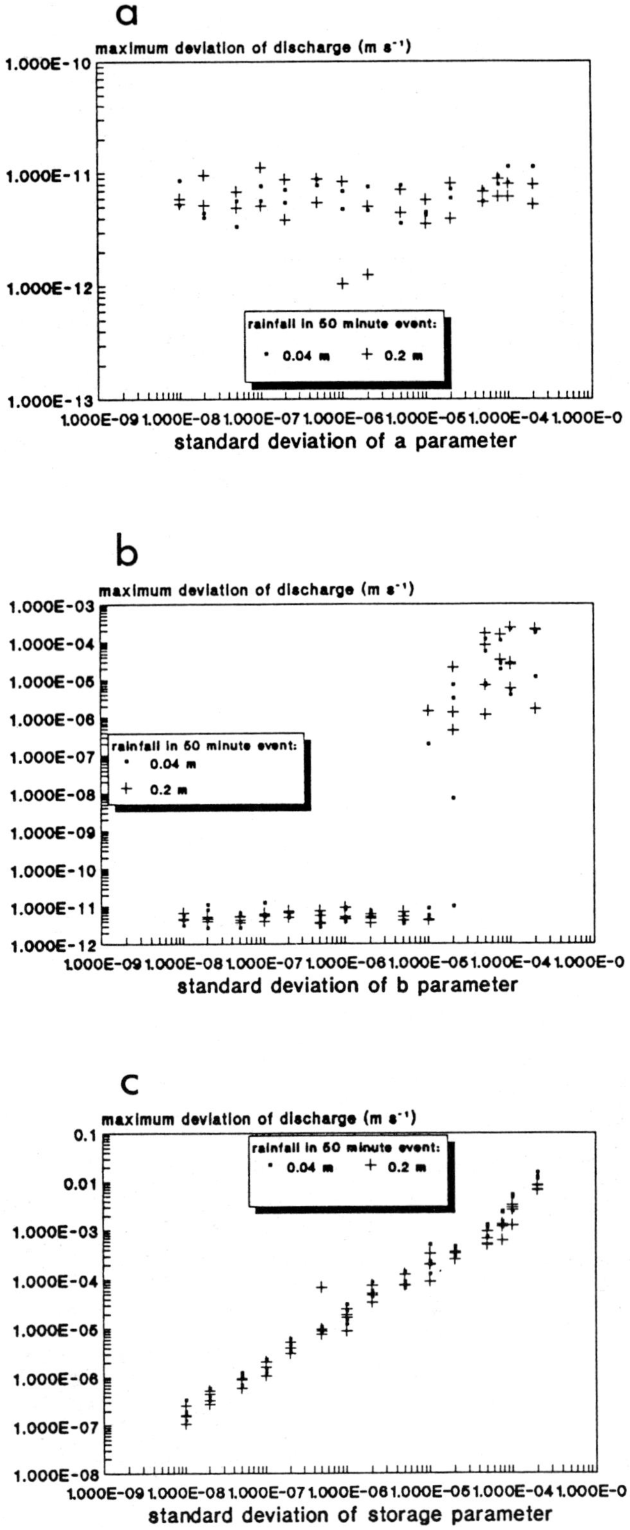


Figure 20.8 Sensitivity of outflow discharge to spatial distribution of hydrological parameters. Infiltration uses Philip equation: $f_t = a + bt^{-1/2}$. a. Sensitivity to the *a* parameter; b. Sensitivity to the *b* prameter; c. Sensitivity to the soil storage parameter.

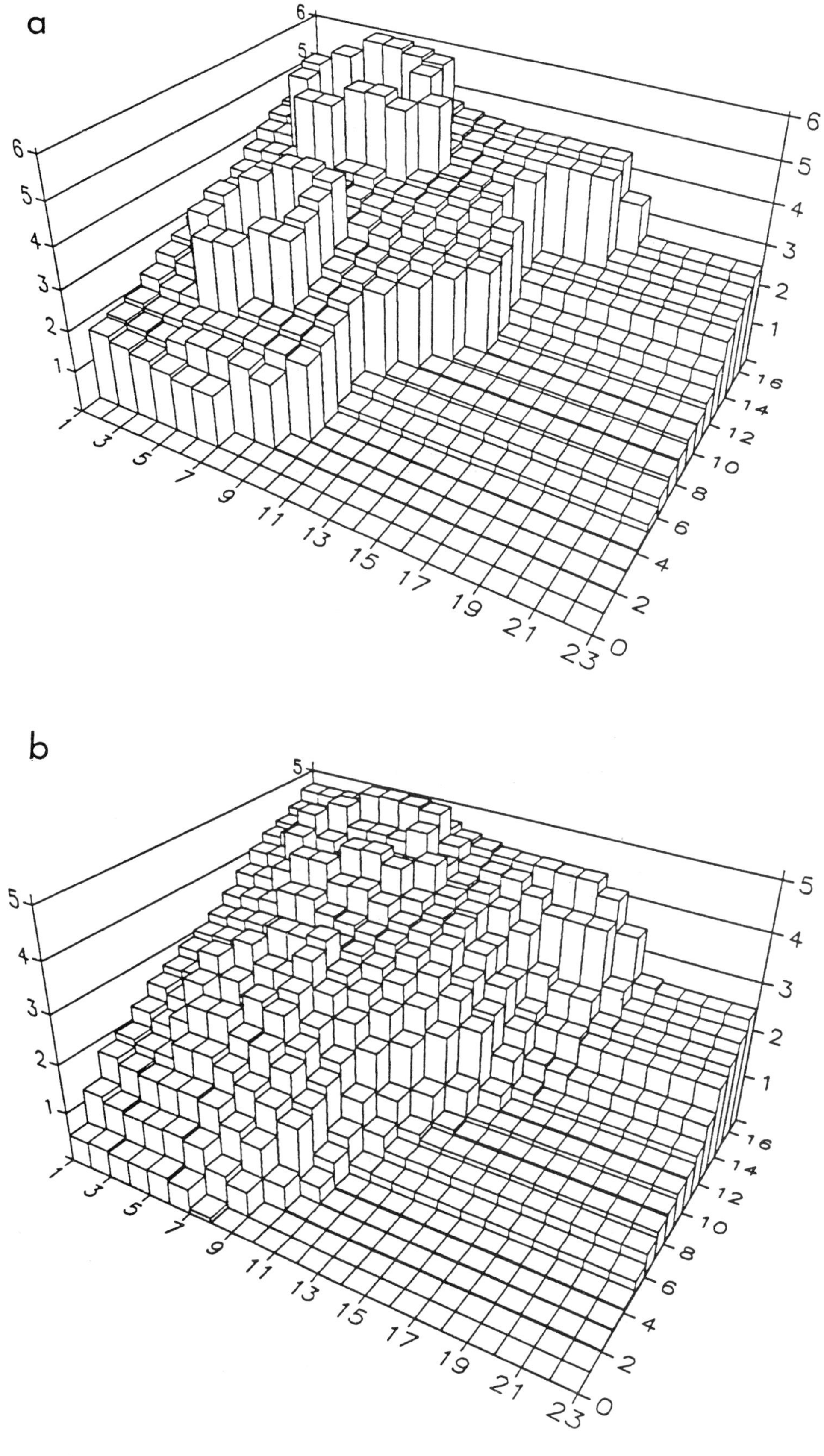


Figure 20.9 Example of erosion simulation on site of Le Laouret. Above: initial slope form; below: slope after single storm of 40 mm in one hour.

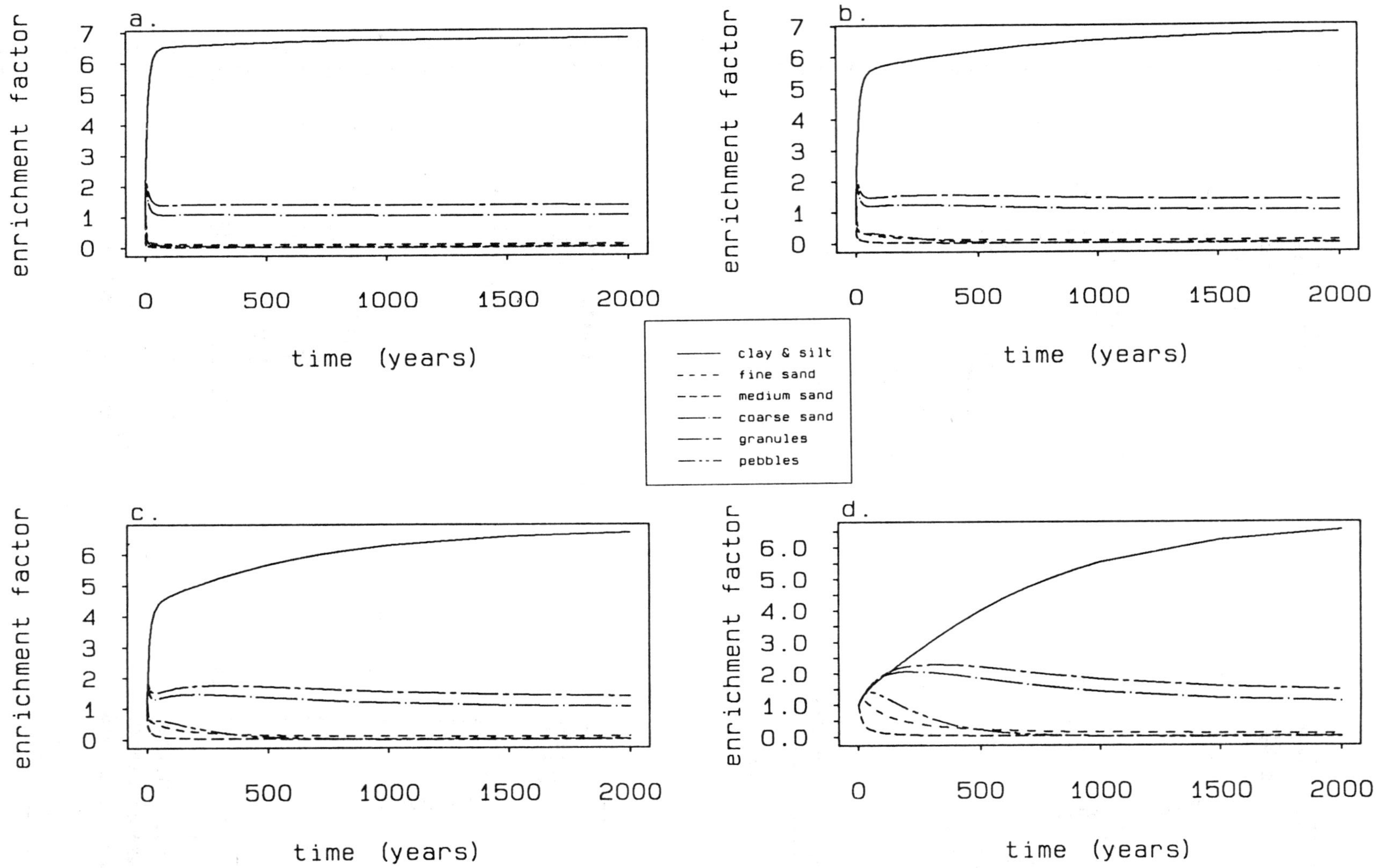


Figure 20.10 Modelled evolution of soil texture through time at different model cells, the locations of which are shown by the horizontal axes on Figure 20.9:- (a) at cell 5/12, (b) at cell 10/12, (c) at cell 10/7, (d) at cell 15/2.

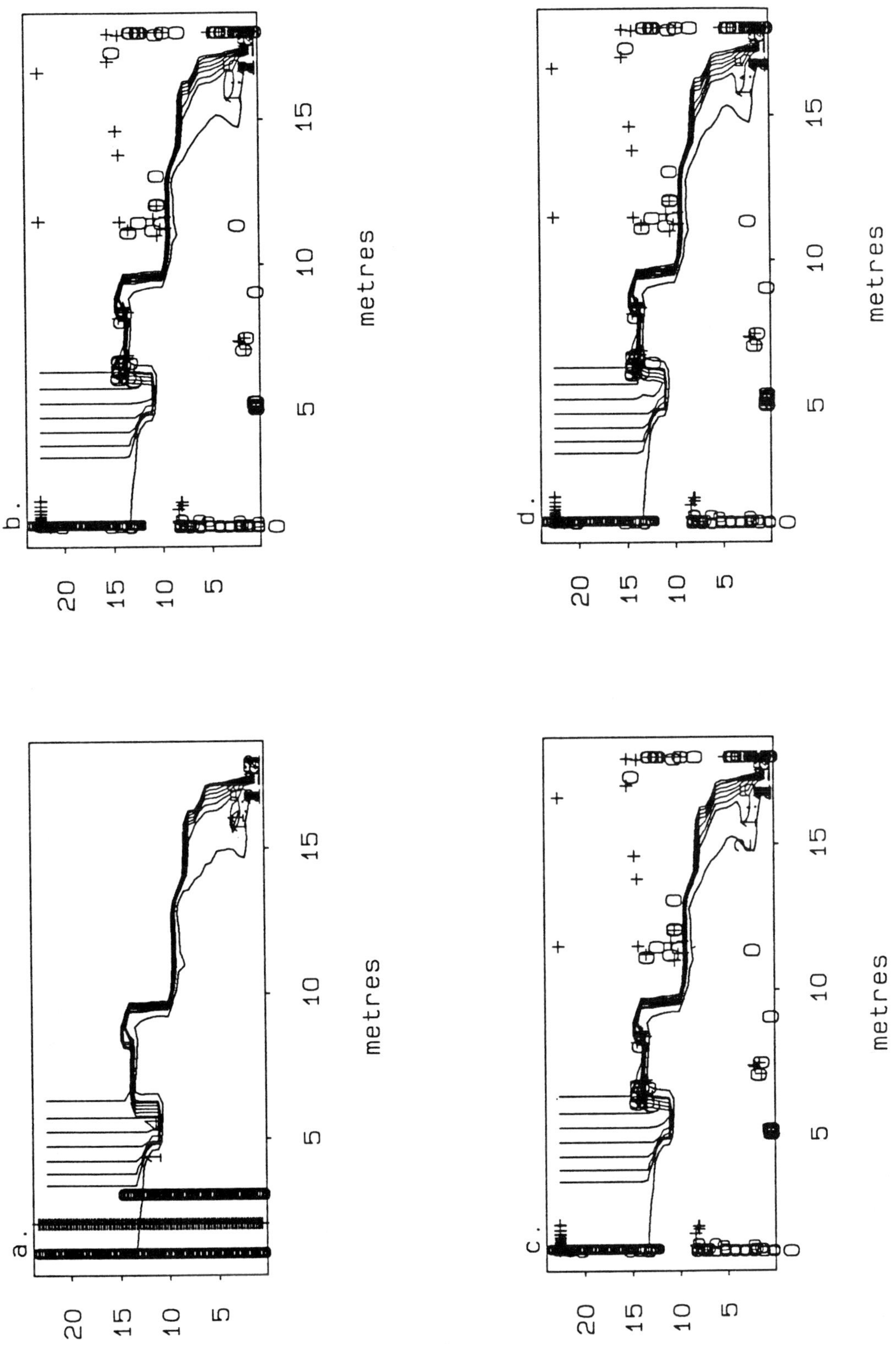


Figure 20.11 Example of simulated artifact movement on Le Laouret. a. initial positions; b. following one 40 mm storm event; c. two 40 mm events; d. three events. Particle sizes: · <2 cm; - 2 - 4 cm; o 4 - 6 cm; + 6 - 8 cm; O 8 - 10 cm.

of vegetation growth, artifact burial, and the renewal process component of the transport model. Concentrations of artifacts are to be found in the talus deposits (see above, section *b*), as well as other areas of the occupied terrace. Care must therefore be taken in interpreting the spatial distribution of artifacts in this zone.

### Conclusions and prospects

Erosion can be seen to be an important factor in the formation of archaeological sites. The hydrology of the site forms a crucial part in understanding this. A first approach based on theory and quantitative field analysis has been presented here. The implications have further been tested using numerical simulation.

For the transport of artifacts, overland flow has been shown to be the most important form, whether this be as sheet flow, or concentrated in rills. Splash transport can be shown to be ineffective with particles in the relevant size range.

In all this, the modelling approach can be seen to be useful both in integrating a range of studies on the erosion problem, and in allowing complex analysis. Both temporal and spatial patterns can be observed.

The techniques presented here have important applications in three major areas of archaeology. First, an understanding of site formation may be made. Areas may be defined within a site where materials are likely to be from derived contexts, so that the interpretations of the site may be adjusted correspondingly. An approach may also be made to the relative dating of deposits where there is either no fine-scale stratigraphic evidence, or the stratigraphy is discontinuous. Secondly, by modelling contemporary conditions, applications may be made in cultural resource management. Whereas this has already been attempted qualitatively (e.g. Mathewson and Gonzalez 1988), quantitative data may now be produced. This is vital, and allows easy assimilation of a scenario-based approach, either for a range of changing climatic or environmental conditions, or in the light of different proposed changes of land-use. Finally, by changing the scale of observation, the site can be seen in the context of its landscape. The human impact on this landscape may be modelled, giving important insights on prehistoric economy and society. Further work in these areas is ongoing.

### Acknowledgements

Jean Gascó and Jean Guilaine kindly allowed the infiltration experiments to be carried out on their sites. Jim Eastwood and Stuart Pope provided technical assistance with the Long Ashton experiments.

### References

Abrahams, A. D., Luk, S-H. and Parsons, A. J. 1988 Threshold relations for the transport of sediment by overland flow on desert hillslopes. *Earth Surface Processes and Landforms* 13, 407–19.

Al-Durrah, M. M. and Bradford, J. M. 1982 The mechanism of raindrop splash on soil surfaces. *Soil Science Society of America Journal* 46, 1086–90.

Bouwer, H. 1986 Intake Rate: Cylinder Infiltrometer. In, Klute, A. (ed.) *Methods of Soil Analysis, Part I. Physical and Mineralogical Methods – Agronomy Monograph no. 9 (2nd edition).* American Society of Agronomy – Soil Science Society of America, Madison, 825–44.

Brandt, C. J. 1990 Simulation of the size distribution and erodivity of raindrops and throughfall drops. *Earth Surface Processes and Landforms* 15, 687–98.

Chow, V. T. 1959 *Open-Channel Hydraulics.* McGraw-Hill, London.

Dingman, S. L. 1984 *Fluvial Hydrology.* Freeman, New York.

Einstein, H. A. 1942 Formulae for the transportation of bedload. *Transaction of the American Society of Civil Engineers* 107, 561–77.

Elwell, H. A. and Stocking, M. A. 1976 Vegetal cover to estimate soil erosion hazard in Rhodesia. *Geoderma* 15, 61–70.

Frostick, L. and Reid, I. 1982 Taphonomic significance of subaerial transport of vertebrate fossils on steep and semi-arid slopes. *Lethaia* 16, 157–64.

Gilman, A. and Thornes, J. B. 1985 *Land-use and Prehistory in South-East Spain.* The London Research Series in Geography 8. George Allen and Unwin, London.

Grigg, N. S. 1970 Motion of single particles in alluvial channels. *Proceedings of the American Society of Civil Engineers* 96, (HY12), 2501–18.

Hanson, C. B. 1980 Fluvial taphonomic processes: models and experiments. In: Behrensmeyer A. K. and Hill, A. P. (eds.) *Fossils in the Making.* University of Chicago Press, Chicago, 156–81.

Hills, R. C. 1970 The determination of the infiltration capacity of field soils using the cylinder infiltrometer. *British Geomorphological Research Group Technical Bulletin* 3.

Hodder, I. 1982 *Symbols in Action.* Cambridge University Press, Cambridge.

Huang, C., Bradford, J. M. and Cushman, J. H. 1982 A numerical study of raindrop impact phenomena: the rigid case. *Soil Science Society of America Journal* 47, 855–61.

Hubbell, D. W. and Sayre, W. W. 1964 Sand transport studies with radioactive tracers. *Proceedings of the American Society of Civil Engineers* 90 (HY3), 39–68.

Hung, C. H. and Shen, H. W. 1971 Research in stochastic models for bed-load transport. In: Shen, H. W. (ed.) *River Mechanics. Volume II: B.* H. W. Shen, Fort Collins, Colorado, 1–47.

Kirkby, A. V. and Kirkby, M. J. 1974 Surface wash at the semi-arid break slope. *Zeitschrift für Geomorphologie Supplementbd. NF* 21, 151–76.

Kirkby, A. V. and Kirkby, M. J. 1976 Geomorphic

processes and the surface survey of archaeological sites in semi-arid areas. In: Davidson, D. A. and Shackley, M. L. (eds.) *Geoarchaeology. Earth Science and the Past.* Duckworth, London, 229–53.

Marshall, J. S. and Palmer, W. McK. 1948 Relation of raindrop size to intensity. *Journal of Meteorology* 5, 165–6.

Mathewson, C. C. and Gonzalez, T. 1988 Protection and preservation of archaeological sites through burial. In: Marinatos, P. G. and Koukis, G. C. (eds.) *The Engineering Geology of Ancient Works, Monuments and Historical Sites. Preservation and Protection.* A.A. Balkema, Rotterdam, 519–26.

McIntosh, R. J. 1974 Archaeology and mud wall decay in a West African village. *World Archaeology* 6, 154–71.

McIntosh, R. J. 1977 The excavation of mud structures: an experiment from West Africa. *World Archaeology* 9, 185–99.

Moeyersons, J. and Ploey J. de 1976 Quantitative data on splash erosion, simulated on unvegetated slopes. *Zeitschrift für Geomorphologie Supplementbd. NF* 25, 120–31.

Moore, H. L. 1982 The interpretation of spatial patterning in settlement residues. In: Hodder, I. (ed.) *Symbolic and Structural Archaeology.* Cambridge University Press, Cambridge, 74–9.

Nichols, J. D., Brown, P. L. and Grant, W. J. (eds.) 1984 *Erosion and Productivity of Soils Containing Rock Fragments.* Soil Science Society of America Special Publication 13, Madison.

Poesen, J. 1987 Transport of rock fragments by rill flow. In: Bryan, R. B. (ed.) *Rill Erosion. Catena Supplement* 8. Catena, Cremlingen, 35–54.

Reece, G. 1976 *Microcomputer Modelling by Finite Differences.* Macmillan, Basingstoke.

Reid, I. and Frostick, L. 1985 Arid zone slopes and their archaeological materials. In: Pitty, A. (ed.) *Themes in Geomorphology.* Croom Helm, Beckenham, 141–57.

Roldán, J. and Woolhiser, D. A. 1982 Stochastic daily precipitation models. 1. A comparison of occurrence processes. *Water Resources Research* 18, 1451–9.

Schiffer, M. B. 1985 Is there a 'Pompeii premise' in archaeology? *Journal of Anthropological Research* 41, 18–41.

Schiffer, M. B., Baron, A. M., Cortes, P. P. and Sepuveda, J. T. 1987 Deterioration of adobe structures: a case study from San Pedro de Atacama, Northern Chile. In: Nash, D. T. and Petraglia, M. D. (eds.) *Natural Formation Processes and the Archaeological Record.* British Archaeological Reports, International Series 352, Oxford, 10–29.

Scoging, H. and Thornes, J. B. 1979 Infiltration characteristics in a semiarid environment. In: *The Hydrology of areas of low precipitation. Proceedings of the Canberra Symposium, December 1979.* IAHS-AISH Publication no. 128, 156–67.

Stevenson, M. G. 1985 The formation of artifact assemblages at a workshop/habitation site: models from Peace Point in Northern Alberta. *American Antiquity* 50, 63–81.

Thornes, J. B. 1988 Erosional equilibria under grazing. In: Bintliff, J., Davidson, D. and Grant, E. (eds.) *Conceptual Issues in Environmental Archaeology.* Edinburgh University Press, Edinburgh, 193–210.

Torri, D. and Poesen, J. 1988 Incipient motion conditions for single rock fragments in simulated rill flow. *Earth Surface Processes and Landforms* 13, 225–37.

Trimble, S. W. 1988 The impact of organisms on overall erosion rates within catchments in temperate regions. In: Viles, H. (ed.) *Biogeomorphology.* Basil Blackwell, Oxford, 83–142.

Vernet, J.-L. 1988 Les charbons de bois et le milieu végétal préhistorique. In: *Le Chasséen en Languedoc Oriental. Hommage à Jean Arnal. Actes des Journées d'Etudes,* Montpellier, 25, 26 et 27 Octobre 1985, Université de Montpellier, Montpellier, 367–71.

Wainwright, J. 1991 *Erosion of Semi-Arid Archaeological Sites: A Study in Natural Formation Processes.* Unpublished Ph.D. thesis, University of Bristol.

Wainwright, J. and Thornes, J. B. 1991 Computer and hardware modelling of archaeological sediment transport on hillslopes. In: Lockyear, K. and Rahtz, S. (eds.) *Computer Applications in Archaeology, 1990.* British Archaeological Reports, International Series 565, Oxford, 183–94.

Yang, C. T. and Sayre, W. W. 1971 Stochastic model for sand dispersion. *Proceedings of the American Society of Civil Engineers* 97 (HY2), 265–88.

Yellen, J. 1977 *Archaeological Approaches to the Present.* Academic Press, New York.

# INDEX

Compiled by Jennifer Foster